UNIVERSITY COLLEGE FRASER VALLEY

DISCARDED

QD 461 B974 1982

Burkert, Ulrich, 1949 -

Molecular mechanics

DATE DUE

FEB 2 4 2000

FEB - 5 2008

Molecular Mechanics

Ulrich Burkert
Norman L. Allinger

ACS Monograph **177**

AMERICAN CHEMICAL SOCIETY
WASHINGTON, D.C. 1982

Library of Congress Cataloging in Publication Data

Burkert, Ulrich, 1949–1982.
  Molecular mechanics.
  (ACS monograph, ISSN 0065-7719; 177)

  Includes bibliographies and index.

  1. Molecular structure. I. Allinger, Norman L.
II. Title. III. Series.

QD461.B974   1982   541.2'2   82-11442
ISBN 0-8412-0584-1
ISBN 0-8412-0885-9 (paperback)

Copyright © 1982
American Chemical Society

All Rights Reserved. The appearance of the code at the bottom of the first page of each chapter in this volume indicates the copyright owner's consent that reprographic copies of the chapter may be made for personal or internal use or for the personal or internal use of specific clients. This consent is given on the condition, however, that the copier pay the stated per copy fee through the Copyright Clearance Center, Inc., 21 Congress Street, Salem, MA 01970, for copying beyond that permitted by Sections 107 or 108 of the U.S. Copyright Law. This consent does not extend to copying or transmission by any means—graphic or electronic—for any other purpose, such as for general distribution, for advertising or promotional purposes, for creating a new collective work, for resale, or for information storage and retrieval systems. The copying fee for each chapter is indicated in the code at the bottom of the first page of the chapter.

The citation of trade names and/or names of manufacturers in this publication is not to be construed as an endorsement or as approval by ACS of the commercial products or services referenced herein; nor should the mere reference herein to any drawing, specification, chemical process, or other data be regarded as a license or as a conveyance of any right or permission, to the holder, reader, or any other person or corporation, to manufacture, reproduce, use, or sell any patented invention or copyrighted work that may in any way be related thereto. Registered names, trademarks, etc., used in this publication, even without specific indication thereof, are not to be considered unprotected by law.

PRINTED IN THE UNITED STATES OF AMERICA
Second printing 1984

ACS Monographs

Marjorie C. Caserio, *Series Editor*

Advisory Board

    Peter B. Dervan
    Ellis K. Fields
    Harold Hart
    John C. Hemminger
    Bruce R. Kowalski
    John Ivan Legg
    Judith P. Klinman

FOREWORD

ACS MONOGRAPH SERIES was started by arrangement with the interallied Conference of Pure and Applied Chemistry, which met in London and Brussels in July 1919, when the American Chemical Society undertook the production and publication of Scientific and Technologic Monographs on chemical subjects. At the same time it was agreed that the National Research Council, in cooperation with the American Chemical Society and the American Physical Society, should undertake the production and publication of Critical Tables of Chemical and Physical Constants. The American Chemical Society and the National Research Council mutually agreed to care for these two fields of chemical progress.

The Council of the American Chemical Society, acting through its Committee on National Policy, appointed editors and associates to select authors of competent authority in their respective fields and to consider critically the manuscripts submitted. Since 1944 the Scientific and Technologic Monographs have been combined in the Series. The first Monograph appeared in 1921, and up to 1972, 168 treatises have enriched the Series.

These Monographs are intended to serve two principal purposes: first to make available to chemists a thorough treatment of a selected area in form usable by persons working in more or less unrelated fields to the end that they may correlate their own work with a larger area of physical science; secondly, to stimulate further research in the specific field treated. To implement this purpose the authors of Monographs give extended references to the literature.

ABOUT THE AUTHORS

ULRICH BURKERT was born on February 22, 1949, in Stuttgart, Germany. He began his studies of chemistry at the University of Stuttgart, and received his diploma in 1972. He received his Ph.D. in 1974, having done his dissertation under the direction of R. R. Schmidt, on the chemistry of thiopyran anions. He spent a postdoctoral year with N. L. Allinger at the University of Georgia, and joined the group of J. C. Jochims as a faculty member at the University of Konstanz. He completed his habilitation in organic chemistry at the University of Konstanz. His main area of interest was conformational analysis with the aid of molecular mechanics calculations and NMR spectroscopy. He held the position of Privatdozent at the time of his death as the result of an automobile accident, March 14, 1982.

NORMAN L. ALLINGER, a native of California, received his B.S. at Berkeley and Ph.D. at UCLA working with D. J. Cram. After a postdoctoral year at Harvard with P. D. Bartlett, he joined the faculty at Wayne State University in Detroit in 1956. In 1968 he moved to the University of Georgia, and has been Research Professor there since that time. He is coauthor of the book "Conformational Analysis" (1965), coeditor of *Topics In Stereochemistry* (1967 to date), and editor of the *Journal of Computational Chemistry*. His main interest in recent years has been in the solution of problems in organic chemistry by computational methods, with emphasis on molecular mechanics.

CONTENTS

Preface, **ix**

1. Molecular Geometry and Energy, **1**

    Introduction, the Born–Oppenheimer Surface, **1**
    Molecular Structure—Experimental Methods, **6**
    Quantum Mechanical Calculations and Molecular Mechanics, **10**

2. Force Fields, **17**

    Force Fields in Vibrational Spectroscopy and Molecular Mechanics, **17**
    Potential Functions of Molecular Mechanics Force Fields, **22**
    Parameterization, **36**
    Force Fields for Molecules Containing Delocalized Pi Electrons, **52**

3. Methods for the Computation of Molecular Geometry, **59**

    Outline of the Computations. The Representation of Molecular Structure, **59**
    Energy Minimization, **64**

    > First Derivative Techniques, **66**
    > Second Derivative Techniques. Newton–Raphson Minimization, **67**

    Conformational Interconversion Pathways, **72**

4. Calculations on Hydrocarbons—Geometries and Relative Energies, **79**

    Acyclic Saturated Compounds, **79**
    Common, Medium, and Large Rings, **89**
    Fused Rings and Other Bicyclic Hydrocarbons, **108**
    Small Rings, **116**
    Alkenes and Cycloalkenes, **121**
    Alkynes, **143**
    Molecules with Cyclic Conjugated Pi-Electronic Systems, **144**

    > The Phenyl Group, **145**
    > General Treatment of Conjugated Pi-Electronic Systems, **150**

    Hydrocarbons Containing Deuterium, **156**

5. Steric Energy, Heats of Formation, and Strain, **169**

   Statistical Thermodynamics. Calculations of Vibrational Frequencies, **169**
   Heats of Formation, **173**
   Strain Energy, **184**
   Resonance Energies. Heats of Formation of Conjugated Hydrocarbons, **189**

6. Heteroatoms, **195**

   Electrostatic Interactions, Solvation, **195**
   Silicon, **202**
   Halogens, **205**
   Oxygen, **209**
   Nitrogen, **228**
   Sulfur, **235**
   Other Heteroatoms, **242**

7. Large Molecules, **253**

   Introduction, Steroids, **253**
   Carbohydrates, **257**
   Nucleosides, Nucleotides, and Nucleic Acids, **265**
   Peptides and Proteins, **274**

8. Stereochemistry and Rates of Chemical Reactions, **285**

   The S_N2 Reaction, **286**
   Reactions Involving Carbenium Ion, **288**
   Free Radical Reactions, **291**
   Addition–Elimination Reactions, **292**
   Electrocyclic Reactions, **300**

9. Applications to the Solid State, **307**

   Influence of Crystal Packing on Molecular Structure, **308**
   Crystal Packing Calculations, **309**
   Molecular Mechanics as a Tool for the Solution of X-ray Crystallographic Problems, **311**

   Appendix: Computer Programs Available from the Quantum Chemistry Program Exchange, 317

   Index, 321

PREFACE

COMPUTATIONAL CHEMISTRY has made enormous advances since the first digital computers became available some thirty years ago. The progress has been documented impressively in many books on the application of quantum mechanical calculations in chemistry. Surprisingly, no comparable survey has yet appeared of an alternative approach to the calculation of molecular geometry and energy, the empirical method that has become known as the "molecular mechanics" or "force field" method. This lack is surprising because of numerous and increasing applications of the method and the ready availability of related computer programs. This volume fills that void. This book is written primarily for organic chemists, but inorganic chemists and biochemists will also find much of interest. The level of the book is appropriate for graduate students and research chemists interested in the topic.

Our aim has been twofold: we want to give a more complete description of the basic principles and techniques of molecular mechanics calculations than has been possible in review articles, and we want to outline the present areas and usefulness limits of such calculations. Although the first half of the book will probably be read more intensely by people doing (or intending to do) their own calculations, we regard it as important for anyone who wants to use the results of such calculations to know something of the pitfalls in force fields and methods for geometry optimization that are described there. We hasten to add that one can make intelligent use of available programs without mastering all of the mathematical detail presented.

The second half of the book not only gives a compilation of reported molecular mechanics calculations, but will also be useful for those who want to find stereochemical data from experimental sources, which in general are cited. The number of papers in the literature that deal with molecular mechanics, and the current rate of publication of such papers, are both so great that it was impossible to cite them all. We must apologize for leaving out of this book discussion of many interesting research papers. We have tried to include examples of all possible types, and we have tried to show when the method works, and, especially, when it may not work, or may not work well, and why.

After a general introduction to the subject (Chapter 1), Chapter 2 is devoted to force fields, their theoretical background, their potential functions, and the sources of the necessary parameters. This chapter also compares several of the currently available force fields for hydrocarbons from the authors' point of view. Chapter 3 turns to the computer programs and the methods used for geometry optimization. Geometries were optimized in molecular mechanics calculations very much earlier than in quantum mechanical calculations, and for much larger molecules. Experience gained in molecular mechanics calculations may also be of interest to those writing related programs within the framework of quantum mechanical calculations. A field that has attracted much interest recently is the mapping of interconversion pathways between conformations. An important extension of this work, not very well exploited as yet, is to a study of reaction coordinates. Because reaction coordinates are fundamental in physical organic chemistry, future developments here are eagerly anticipated.

The molecular mechanics method has been applied most widely in hydrocarbon stereochemistry, and Chapter 4 is concerned with structural and conformational studies of hydrocarbons. A classical area to which the method has been widely applied is that of strain energies and heats of formation. This topic is discussed, and the present abilities of force fields are described, in Chapter 5. Also, some results are included on the heats of formation of molecules containing heteroatoms. The structures of the latter molecules are dealt with in detail in Chapter 6. This group of chapters completes the discussion of "small molecules" (up to about 10–12 carbons).

Chapter 7 then deals with "large molecules," where the objectives sought are generally at a lower level of precision. This chapter includes sections on steroids, carbohydrates, and peptides. Calculations on peptides are treated very briefly, and only a few technical aspects and principles of the problem are touched on, because this area is a whole field in itself. For applications here, the reader is referred to numerous excellent recent review articles. Homopolymers have been excluded from discussion here because they, too, constitute a field in itself.

A field where molecular mechanics has been applied more and more in recent years is the prediction of the rates and stereochemistries of selected chemical reactions. These topics are covered in Chapter 8.

Finally, we have collected work related to the calculation of crystal packing effects, starting with the deformations exerted by the crystal packing on individual molecules, and finishing with a priori calculations of crystal packing and molecular structure in the crystal. This rapidly growing field is outside of the area of interest of most organic chemists, but because x-ray crystallography is now used routinely for structure elucidation, there is a growing interest in being able to relate the structural properties of the molecule in the crystal to those of the isolated molecule.

In the opinion of the authors, molecular mechanics is one of the most powerful tools presently available for studies of molecular geometries, energies, and properties relating to them. The potential usefulness of the method to the average chemist is not yet appreciated as widely as it should be; perhaps this volume will help to improve that situation.

We would like to acknowledge the many helpful comments and discussions from several reviewers. L. S. Bartell, F.A.L. Anet, and D. H. Wertz read the manuscript and commented on it in detail. H. J. Fritz and H. J. Lindner each read and commented on one chapter of the manuscript. The authors did not accept completely and without qualification every suggestion from each reviewer, and if inaccuracies or obscurities remain in the manuscript, these should be charged to the authors, and not to the reviewers. The authors are indebted to Jo Antcliff for preparation of the manuscript.

ULRICH BURKERT

NORMAN L. ALLINGER

The University of Georgia
Athens, GA 30602

July, 1981

ULRICH BURKERT died from injuries sustained in an automobile accident while this book was in production. We hope that this volume is a fitting memorial to him and to his work.

1

Molecular Geometry and Energy

Introduction, the Born–Oppenheimer Surface

The expression "molecular mechanics" is currently used to define a widely used calculational method designed to give accurate a priori structures and energies for molecules. The method is a natural outgrowth from older ideas of bonds between atoms in molecules, and van der Waals forces between nonbonded atoms. It employs the fundamental formulations of vibrational spectroscopy, and some of the basic ideas behind this procedure can be traced back to D. H. Andrews (1930) (*1*). The basic idea is that bonds have "natural" lengths and angles, and molecules will adjust their geometries so as to take up these values in simple cases. In addition, steric interactions are included using van der Waals potential functions. In more strained systems, the molecules will deform in predictable ways with "strain" energies that can be accurately calculated.

While the basic ideas of molecular mechanics go back to 1930, serious attempts to use the method were not forthcoming until 1946. In that year three important papers appeared. T. L. Hill proposed that van der Waals interactions, together with stretching and bending deformations, should be used to minimize steric energies, and that this would lead to information regarding structure and energy in congested systems (*2*). Dostrovsky, Hughes, and Ingold (*3, 4*) simultaneously and independently utilized the same basic scheme in an effort to better understand the rates at which various halides underwent the S_N2 reaction. The complexity of the problem was so great, and the necessary available information of such limited accuracy, that the results were not very convincing at the time, but they certainly did foreshadow events to come. The third and most important paper, by Westheimer and Mayer (*5–8*) and also independent of the others, was successful in treating in a convincing manner a less complicated and thus more manageable problem than that attacked by the Ingold group. This problem concerned the relative rates of racemization of some optically active halo-substituted biphenyls. The methods and results were quite impressive. All of these papers together provided the basis for the subsequent development of the molecular mechanics method.

While Westheimer's calculations were important in showing that the method could be used to rationalize certain properties involving geometries and energies of molecules, the method was not widely useful at a practical level in the 1940s, because computers were not yet available. With the advent of computers during the 1950s and thereafter, interest in this approach to the determination and understanding of molecular structure rapidly increased, to such an extent that it can now be said that molecular mechanics is one of the standard methods of structural chemistry (9–20). The expression "Westheimer method" is synonymous with the molecular mechanics method, and the expression "(empirical) force field calculations" often is used to mean the same thing. Spectroscopists frequently use the term "force field" to mean a similar set of equations designed to reproduce or predict vibrational spectra. A really accurate force field would give both structure, spectra, and related properties. However, current spectroscopic force fields cannot ordinarily be used to determine structure, and current molecular mechanics force fields usually do not give good spectra.

The Born–Oppenheimer approximation, which is commonly used in quantum mechanics (discussed in most introductory books on quantum mechanics), states that the Schrödinger equation for the molecule can be separated into a part describing the motions of the electrons and a part describing the motions of the nuclei, and that these two sets of motions can be studied independently. It is a good approximation in general for studies involving molecules, and is usually used without comment in two different ways. First, with respect to electronic structure, it is common practice to establish the positions of the nuclei of the system by some method, and then to study the electronic structure using the fixed nuclear positions. In molecular mechanics or vibrational spectroscopy the opposite approach is taken. Namely, the motions of the nuclei are studied, and the electrons are not explicitly examined at all. They are simply assumed to find an optimum distribution about the nuclei.

The energy of the molecule in the ground electronic state is a function of the nuclear positions. The Born–Oppenheimer surface is the multidimensional "surface" that describes the energy of the molecule in terms of the nuclear positions. In molecular mechanics it is usually just called the potential energy surface. If we have a typical molecule, say butane, it can be represented by such a potential energy surface (21). If we consider only the dimension of the surface that corresponds to rotation about the central bond, there are three potential energy minima for a full rotation, and these are separated by maxima of moderate size. These three energy minima correspond to the three stable conformations, the *anti* and the two *gauche* forms (a *dl* pair). The *anti* form is of lower energy and is referred to as the global minimum, and the *gauche* forms as local minima. To proceed from one minimum to the other, the molecule must go over the saddle point that separates them. At the saddle point we have an eclipsed conformation, in which bonds have been stretched somewhat and bond angles are significantly deformed. Moving along the molecular coordinate that interconnects the *anti* and *gauche* conformations corresponds to torsion about the central bond in terms of internal coordinates. The energy of the system goes upward as the saddle point is approached, reaches a maximum, and then comes down as the other minimum is approached.

Certain terminology is useful in considering the structures involved along this potential surface (21). Any point that corresponds to an energy minimum can be referred to as a **conformer**. Points that can be defined (by symmetry considerations, for example) but that do not necessarily correspond to energy minima, can be referred to as **conformations**. Thus, we have three conformers for butane, corresponding to the three energy minima, but we have other conformations, such as the one in which the two methyl groups are eclipsed (which corresponds to a saddle point, with a unique energy and a geometry that can be exactly defined in terms of the torsion angle about the central bond).

With more complicated molecules, there will in general be a large number of energy minima of different depths. To a first approximation, the molecule is described by the structure corresponding to the deepest energy minimum. To the next approximation it is described by an equilibrium mixture of molecules at all of the minima in a Boltzmann distribution. A still more refined approximation allows for the fact that the molecules do not remain motionless at points at the bottom of these energy wells, but rather are vibrating over a portion of the surface around the energy minima. Finally, thermal motions carry some of the molecules across saddle points from one minimum to another, at a rate corresponding to the Gibbs free energy of activation. If this energy is lower than the zero vibrational energy level, the "conformation" corresponds to a large amplitude oscillation.

Molecular mechanics calculations employ an empirically derived set of equations for the Born–Oppenheimer surface whose mathematical form is familiar from classical mechanics (9–20). This set of potential functions, called the force field, contains adjustable parameters that are optimized to obtain the best fit of calculated and experimental properties of the molecules, such as geometries, conformational energies, heats of formation, or other properties. The assumption is always made in molecular mechanics that corresponding parameters and force constants may be transferred from one molecule to another. In other words, these quantities are evaluated for a set of simple compounds, and thereafter the values are fixed and can be used for other similar compounds. It is not possible to prove that this is a valid assumption. Indeed, the usual spectroscopic force fields (pp. 17–22) have parameters that are in general not transferable, except over limited sets of compounds. (This is because such force fields omit most or all of the important van der Waals interactions between nonbonded atoms.)

Simple molecular mechanics force fields include bond stretching, angle bending, torsion, and van der Waals interactions in their make-up,

$$V = \Sigma V_{\text{stretch}} + \Sigma V_{\text{bend}} + \Sigma V_{\text{torsion}} + \Sigma V_{\text{VDW}}$$

where the sums extend over all bonds, bond angles, torsion angles, and nonbonded interactions between all atoms not bound to each other or to a common atom (i.e., 1,4-interactions and higher). More elaborate force fields may also include either Urey–Bradley terms (1,3-nonbonded interactions, *see* pp. 17–22), or cross-interaction terms, electrostatic terms, and so on. For each of these there is in general a first approxima-

1,3-interaction   1,4-interaction

tion and, in many cases, further approximations have been developed. The sum of all these terms is called the **steric energy** of a molecule. In the simplest force fields, perhaps a dozen or so parameters are necessary to describe the alkanes. Such a force field gives a reasonable approximation to molecular structures and energy differences, but the values obtained from such a force field are clearly inferior to values measured experimentally. To better reproduce the available experimental data, further optimization of the parameters or the introduction of more of them by modifying the equations that make up the force field might be considered. In practice what has usually been done is to assume a certain set of equations, and to optimize the parameters involved. Systematic errors that remain will usually suggest the equations in which further improvement of the force field may be desirable, and the kind of modifications needed. Having modified the equations, it is again necessary to reoptimize the parameters. This process has been a continuing one for several years now, and it is generally agreed that certain equations are essential for a good force field. Others seem to be more borderline, and sometimes are used and sometimes not. *All of these equations* *are sufficiently simple that they can be solved very rapidly with modern computers and thus permit calculations on even very large molecules.*

Because of the intimate connection between structure and energy, molecular mechanics calculations always involve both. To find the structure, one necessarily has to examine the energy to find where energy minima occur. If only the structures and the relative energies of these structures (conformations and conformational energies) are evaluated, a great deal of information of chemical interest is amassed. Many force field calculations stop at this point. However, it is possible to carry the calculation to a much further degree of refinement. This has been done occasionally in the past, and will doubtlessly be done more in the future. If we know the equilibrium position of each atom as well as the potential surface in the vicinity of each atom, then we can calculate the vibrational levels within the energy minima (i.e., the vibrational frequencies and their energies). This gives us the frequencies to be expected in the vibrational spectra (infrared and Raman spectra, although selection rules must be employed to weed out forbidden transitions), and it also gives us, using a Boltzmann distribution, the populations of the various ground and excited states. We can thus find the zero-point energy, and the vibrational energy for each conformation. By mixing the conformations according to a Boltzmann distribution, we can find the heat content of the whole system. Further, it is possible to find the entropy of the substance from the populations, as it depends on the total number of occupied Boltzmann-weighted states (translation, rotation, vibration, internal rotation, plus some symmetry corrections). Finally, the thermodynamic functions, which depend on the way the populations change with temperature, can be found. While studies of this kind have not been pursued very far to date (*see* Chapter 5), they are potentially powerful tools for the investigation of, for example, ring closure reactions, and other processes of interest to chemists.

It should be clear from the foregoing that molecular mechanics lies within the broader subject of molecular physics, and one might be inclined to wonder if this is really going to apply to organic chemistry. Fifty years ago it would have been difficult to convince an organic chemist that there was anything in this area that really was of use to him. Now, however, there are few areas of organic or inorganic chemistry that have been unaffected by the advent of molecular mechanics (9–20). The authors thus feel that a reasonable awareness of the scope and limitations of this subject is useful background information for essentially all chemists of the present day. We point out, however, as with other subjects such as x-ray crystallography, it is not essential for each person to do his own molecular mechanics calculations any more than it is essential for him to do his own crystallography. But a knowledge of molecular mechanics (and also of crystallography) sufficient to indicate to the chemist when he might advantageously consult with a specialist in the area is highly desirable.

Molecular Structure—Experimental Methods

To a first approximation the organic chemist thinks of molecular structure in terms of a ball-and-stick model. This is often a useful approximation. But molecular mechanics has an accuracy far beyond that of such a model, and to solve most problems of current interest this accuracy must be more fully exploited. While the concept of a bond length, or of a bond angle, is simple enough in terms of a ball-and-stick model, it becomes less simple as greater accuracy in representing an actual molecule is demanded. With the real molecule the atoms are not at rest; they are undergoing vibrational motions. These vibrations in general are anharmonic, which means that the simplest kinds of ideas about the vibrational motion may not be accurate enough. Modern experimental techniques can often determine a bond length to within a few thousandths of an angstrom or better, while the vibrational motions of the nuclei carry them over distances of the order of several hundredths of an angstrom. So it becomes a nontrivial problem to interpret the exact meaning of highly accurate bond lengths, since some kind of averaging of the atomic positions over their vibrational motions is required.

Currently there are available three experimental techniques that are widely used for determining accurate molecular structures. These are x-ray diffraction, which is carried out on crystals, and electron diffraction and microwave spectroscopy, which are carried out on gases. Each of these techniques will give quantities nominally referred to as bond lengths and bond angles. But these techniques are actually measuring different physical quantities, and so it turns out, not surprisingly, that bond lengths and angles are slightly different, depending on the method of measurement.

Two points of difference between x-ray crystallography and electron diffraction on gases need to be pointed out here. In x-ray work, one determines the distance between the mean positions of atoms. In electron diffraction, one primarily determines the average distance between atoms. These quantities are not quite the same, as the following simple considerations will show. Suppose we have three atoms, A, B, and C, in a line. For simplicity, imagine that the two end atoms, A and C, are fixed, and B is vibrating between them along a vertical axis as shown in the diagram. The mean position of B is in the center as labeled, and it

Ⓐ—ab—Ⓑ—Ⓒ

moves to extreme positions on the ends of the arrows as shown. The distance between the mean positions of A and B is indicated by the line segment ab (an x-ray quantity). But, this is not the average distance between the atoms (an electron diffraction quantity). Rather, ab is actually the minimum distance between the atoms, and the average distance between them is significantly longer.

The second principle difference results from the fact that x-rays "see" (are scattered by) the electron clouds while electron diffraction "sees" nuclei. If the nucleus were centered within a spherical electron cloud the "atom" would be seen in the same place by both techniques, but this situation is only approximately true in practice. For atoms other than hydrogen the difference is not significant for our purposes. But the bonding electrons are the only electrons around the hydrogen, and they are not spherically distributed about the nucleus, but are pulled into the bonding region. Therefore, bond lengths involving hydrogens are usually much shorter when measured by x-ray diffraction than by other methods (22). Recently neutron diffraction measurements have become available for a few compounds (23). Neutrons are also scattered by the nuclei, and for bonds involving hydrogen, neutron diffraction bond lengths are quite similar to microwave or electron diffraction values. In molecular mechanics any of these values for bond lengths may be used for hydrogens, except x-ray values.

The importance of vibrational effects in structure determinations is further illustrated by the phenomenon that in electron diffraction is referred to as **shrinkage** (22). This term can be most easily understood in terms of a simple vibrating linear triatomic molecule such as CO_2. The bond lengths have the same value, call it d. The bond angle OCO is 180°. What is the distance between the oxygens? Experimentally it is found not to be $2d$, but a little less. The molecule shrinks relative to what simple geometry appears to require. In this case, the angle-bending vibrational motion of the molecule causes the oxygen to move along the arcs indicated. The distance $2d$ corresponds to the maximum separation

O—d—C—d—O

of the oxygens; the mean separation is slightly less. The phenomenon is a general one, and electron diffraction work that does not properly take shrinkage into account can yield inaccurate results.

The vibrational motion of a diatomic molecule, or of any two atoms bound together, is often described with the aid of a Morse curve. This curve is close to a parabola near the minimum (Hooke's Law), but at shorter distances the energy rises more steeply, and at longer distances more slowly, as shown in the Figure 1.1. This means that as the temper-

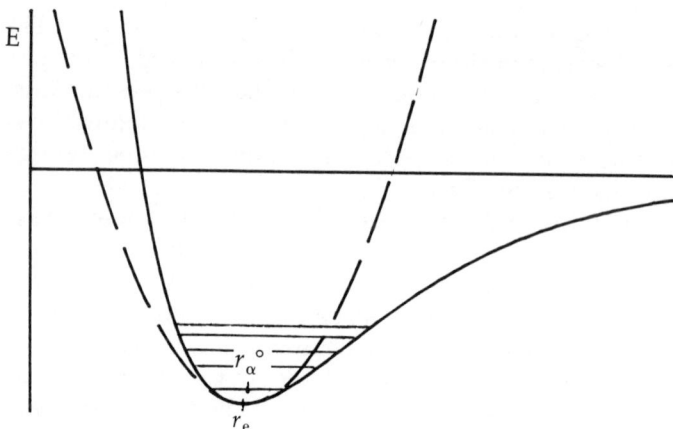

Figure 1.1. The energy–distance relationship between two atoms bonded together. Solid curve, Morse function; dashed curve, harmonic approximation.

ature is raised, and more molecules go into excited vibrational levels, bond lengths tend to be longer.

The definition of a bond length is not as simple as we would like. From electron diffraction (22, 24, 25) one commonly obtains interatomic distances that are labeled as r_a, r_g, r_α, and r_α°. From microwave spectroscopy the quantities obtained (22, 24) are r_0, r_s, and r_z. From both of these methods together the quantities r_{av}, and r_e are obtained. Of these, the latter (r_e) is the easiest to understand, although it is difficult to obtain experimentally. This is simply the internuclear distance corresponding to the rigid model where each of the nuclei is at the bottom of its potential well. This is the quantity obtained directly from quantum mechanical (ab initio) calculations. The quantity obtained directly from the electron diffraction radial distribution function is r_a. This is converted into r_g, which is typically about 0.002 Å longer than r_a, by averaging over all of the molecular vibrations. The quantity r_α is the distance between the mean positions of atoms at a particular temperature, as deduced from r_g by a further vibrational correction. The quantity r_α° is the value of r_α extrapolated to 0 K. It is r_α (or r_g, which is nearly identical to r_α) that is commonly used to compare with x-ray data.

The definitions of the spectroscopic quantities are in general more complicated, and these quantities differ from the diffraction values in that the latter are thermally averaged over occupied states, whereas the spectroscopic values apply to a particular state under examination (usually the ground state). Microwave spectroscopy gives directly the quantities r_0 and r_s, which are significantly different from the diffraction values. For a carbon–carbon single bond, r_0 is about 0.006 Å smaller than r_α and r_g on the average. But there is no simple general correction that allows conversion of r_0 and r_s values into r_α and r_g values.

When vibrations are taken into account in spectroscopic structures, r_z values are obtained, and these are now becoming available. Both r_α° and r_z represent the distance between the average positions of a pair of atoms in the ground vibrational level, and these values, derived from diffraction and spectroscopy, respectively, should agree. Unfortunately, most spectroscopic studies give only r_0 or r_s structures.

Molecular mechanics is based on experimental data. Although it could be parameterized to yield any type of structure obtained by experimental methods, present force fields have been designed to reproduce room-temperature vibrationally averaged structures, which are r_α structures obtained from x-ray or electron diffraction measurements. The small difference between r_α and r_g structures is usually neglected, as it is within the error limits of what can be expected from molecular mechanics calculations. Microwave r_0 and r_s structures are usually not used for parameterization, except when other kinds of data are lacking. In that case they can be regarded as close to what is desired, if not quite right.

The practical problem in molecular mechanics is not how to utilize all different kinds of structural data, but rather, for the most part, how to utilize whatever data (if any) happen to be available. This is not a serious problem in practice, but it places a limit (of a few thousandths of an angstrom) on how "good" the calculations can be expected to be.

One might ask at this point why we should be interested in calculating the structures and energies of molecules, since these can be determined by experimental methods? The answer to the question is at least twofold. First, there is the matter of understanding. A calculational method that gives good results contains within it a potential for understanding that does not come from a collection of experimental results. Second, there are practical matters to consider. To determine the structure of an average molecule, say by crystallography or by one of the other available methods, may involve a few weeks of work. This is assuming that a suitable crystal is available. Clearly, if it is necessary to go through a lengthy synthesis to obtain such a crystal, or if the compound is too unstable to survive irradiation by x-rays, then the difficulties of determining the structure can be much greater, even insurmountable. Furthermore, to experimentally determine the energy of the compound (heat of formation) and the energy differences between conformations obviously involves a good deal of effort and requires a sizable amount of pure material. Hence, the practical advantage of a molecular mechanics calculation is evident. Such calculations usually require a few minutes of computer time and perhaps an hour or two of the chemist's time, and one may then have most or all of this information that could be determined only by laborious, perhaps impossible, experiments. And a sample of the compound is not needed.

So then why do experiments? Indeed, in some cases the same information can be found by molecular mechanics calculations with much less work. The problem currently is one of reliability. The molecular mechanics method is empirical. It is developed by fitting equations and parameters to experimental data. As long as these equations and parameters are then applied to other situations so that one is interpolating among known data, one can be confident that the results will be of an accuracy comparable with that determined by the original data used to develop the force field. On the other hand, if one has to extrapolate some of the functions beyond the areas where they have been tested, then the reliability of the calculations is in doubt. In practice, small extrapolations are commonly used, and, therefore, the calculated values in such cases may not be as accurate and reliable as one would like.

So the situation at present is that molecular mechanics can be used to advantage in certain cases, probably in thousands, or tens of thousands of cases. But the method should not be used indiscriminately. An understanding of the current limitations of the method is necessary if mistakes are to be minimized.

Quantum Mechanical Calculations and Molecular Mechanics

For calculating molecular properties, quantum mechanics seems to be the obvious tool to use. Calculations that do not use the Schrödinger equation are acceptable only to the extent that they reproduce the results of high level quantum mechanical calculations.

In the type of calculation that we will here refer to as quantum mechanical or ab initio, the Schrödinger equation for the electronic system is actually formulated and solved by standard approximations. In the general case, this must be done for various assumed positions of the nuclei, and then the total energy minimum as a function of the internuclear positions must be found. In quantum mechanical methods empirical data of the sort used in molecular mechanics are, in principle, not needed. Returning to the Born–Oppenheimer approximation, the molecular mechanics calculation (which often is referred to as classical, but in fact only uses potential functions from classical mechanics) can be put on a quantum mechanical basis with respect to the nuclei, wherein the electrons are not explicitly treated, but are regarded as simply the cause of the potential field in which the nuclei find themselves. This potential is evaluated empirically. From that point on, the calculations of molecular geometries and energies as well as (harmonic) vibrational frequencies may be either quantum mechanical or classical, the results are the same (26).

To use the quantum mechanical approach to molecular structure the simplifications commonly used in quantum mechanics must be employed; these ordinarily include the self-consistent field (Hartree–Fock) approximation. While this approximation can be improved upon for small molecules, the calculations have rarely been refined past this point. To this point, the calculations are usually referred to as ab initio. Such calculations can be of limited accuracy or of very high accuracy with regards to molecular structure, depending on the accuracy of the atomic wave functions employed. A minimal basis set of Slater orbitals, or the so-called STO-3G basis set (27) (where the orbitals are written as Gaussian functions, of which three are used to approximate a Slater orbital) gives reasonably good results, which are probably one order of magnitude less accurate than the corresponding molecular mechanics results. Thus, bond lengths may be determined in this way to within a few hundredths of an angstrom (as opposed to a few thousandths in molecular mechanics), and relative energies to perhaps 10 or 20 kJ (as opposed to within a few kJ). If the calculations are refined by extending the basis set as far as including d-orbitals on first-row atoms, and p-orbitals on hydrogen, then geometries and relative energies are obtained that are competitive with molecular mechanics, where the definitions of properties like bond lengths really limit the accuracy more than the calculations themselves do. Because of the expense of carrying out these ab initio calculations with extended basis sets, very little such work has been done so far. We can give some estimates of cost here, but emphasize that they are a function of the particular computer used, and the efficiency of the programs, and these efficiencies have been steadily improving with time. But, the following are current reasonable estimates.

A STO-3G calculation on one geometry of a molecule such as ethylamine (three first-row atoms) requires about five minutes of computer (cpu) time, at a cost of perhaps thirty dollars. To optimize the structure by such a calculational method would increase the time required considerably. The cost would be a strong function of the accuracy to which the initial starting geometry was known, but if we say that perhaps ten calculations would be needed, the cost would go up to three hundred dollars. If we increase the size of the molecule, say from ethylamine to propylamine, the number of heavy atoms increases from three to four, and the number of orbitals involved increases in about the same proportion. The computational time goes up by approximately the fourth power of the number of orbitals, so the cost of a single calculation would be about one hundred dollars, and say one thousand dollars for the geometry optimization calculation.

But these calculations are only for the STO-3G level, and if we wish to use an extended basis set, we will have to increase the computer time and total cost by perhaps a factor of five or more. For propylamine, then, the cost is now of the order of five thousand dollars for a single conformation. There are five ground state conformations of interest, plus several rotational transition states, and we are in the thirty thousand dollar range for a good geometry study on this simple molecule. The corresponding cost with molecular mechanics is perhaps twenty-five dollars, so the cost differential is sizable. What is worse, as far as the ab initio calculation is concerned, is that the cost will go up as about the fourth power of the number of atoms, while for molecular mechanics, the cost goes up at a rate proportional to between the second and third powers.

Obviously, detailed ab initio calculations of structure are carried out only in special instances. Such calculations are not economically feasible at present, and it does not seem likely that they will be in the forseeable future, at least for molecules of the size organic chemists usually deal with.

To keep the discussion balanced, we should point out that an ab initio calculation can be carried out on any molecule or fragment, and no experimental information concerning the system need be known. With molecular mechanics, on the other hand, there are a great many parameters that go into the calculation, which for any given molecule need to be known from previous studies on other molecules of that general class. So the limitation on molecular mechanics calculations is quite severe, in the sense that the molecule studied must belong to a previously examined class. No such limitation exists in the case of an ab initio calculation, making the latter suitable for studies on really novel kinds of molecules.

Next, we might consider simplifications of the ab initio calculation that will give good results at a lower cost. Such modified methods (28, 29, 30) are usually referred to as semi-empirical methods, and were introduced by Pariser, Parr, and Pople as long ago as 1953, before the general availability of computers (28). The objective of such methods is always the same, although the details of the calculations may vary. We want to omit somehow calculation of the electronic integrals of the type

$$\int_{-\infty}^{\infty}\int_{-\infty}^{\infty}\int_{-\infty}^{\infty}\int_{-\infty}^{\infty}\int_{-\infty}^{\infty}\int_{-\infty}^{\infty} \psi_1(1)\ \psi_2(1)\ \frac{1}{r_{12}}\ \psi_3(2)\ \psi_4(2) dx_1 dy_1 dz_1 dx_2 dy_2 dz_2$$

insofar as possible.[†] While such integrals can be evaluated readily enough with a computer if the Ψs are Gaussian in form, they occur in such large numbers in the Schrödinger equation describing molecules

---

[†] The Ψs in the general case represent four different orbitals; (1) and (2) represent electrons 1 and 2, respectively; and r_{12} is the distance between them, integrated over the coordinates of the two electrons.

of average size that they lead to long computing times and the large costs previously mentioned. These integrals can be grouped into different families, and by omitting some of these various families, the computer time required can be cut down considerably. The hope is that if the families to be omitted are properly chosen, the calculation time can be greatly reduced without affecting the outcome of the results very much. The reason such methods are called *semi-empirical* is because attempts are made to compensate for the omission of these integrals by adjusting some of the numbers that go into the calculation so as to better match ab initio or experimental data. Unfortunately, the approximations introduced this way decrease the precision and the reliability that can be expected in the results.

These semi-empirical procedures were first applied to pi systems, because such calculations could be managed without computers, or with the early small computers of the 1950s. The CNDO type calculation (complete neglect of differential overlap) was the first successful widely used method of this kind applied to ordinary molecules (not just to pi systems). It has been modified considerably over the years in an attempt to improve the accuracy to which the calculations could reproduce desired results. The current most widely used program of this general type (*31*) is the MNDO program of Dewar (modified neglect of diatomic overlap). This program and the similar MINDO/3 (*32*) and PRDDO (*33, 34*) programs are highly successful in reducing the computation times required for ab initio calculations, but not very successful in terms of the accuracy of the results. In general the results are probably less good than the results of STO-3G calculations by factors of two to five. The computer times required for single geometry calculations are quite modest, but the energy minimizing routines that are available so far are relatively slow. The estimated average running times for small molecules are perhaps two orders of magnitude less than for an ab initio calculation, but one order of magnitude greater than for a molecular mechanics calculation. For some kinds of studies, for example, looking at reaction surfaces, if the molecules are of ordinary size, ab initio calculations are impossibly expensive. On the other hand, molecular mechanics is designed to work for stable molecules, and while it can be applied to transition states of chemical reactions and to other transient species, such applications tend to be difficult, because the necessary empirical data are not readily available. Thus, for many purposes, MNDO is the best method currently available. It seems likely that the semi-empirical method can be substantially improved (*35*), and the solution of problems that are virtually unassailable by either molecular mechanics or by ab initio calculations will in the future probably be both more accurate and less expensive.

Literature Cited

1. Andrews, D. H. *Phys. Rev.* **1930**, *36*, 544.
2. Hill, T. L. *J. Chem. Phys.* **1946**, *14*, 465.
3. Dostrovsky, I.; Hughes, E. D.; Ingold, C. K. *J. Chem. Soc.* **1946**, 173.
4. de la Mare, P.; Fowden, L.; Hughes, E. D.; Ingold, C. K.; Mackie, J. J. *Chem. Soc.* **1955**, 3200.
5. Westheimer, F. H.; Mayer, J. E. *J. Chem. Phys.* **1946**, *14*, 733.
6. Westheimer, F. H. *J. Chem. Phys.* **1947**, *15*, 252.
7. Rieger, M.; Westheimer, F. H. *J. Am. Chem. Soc.* **1950**, *72*, 19.
8. Westheimer, F. H. In "Steric Effects in Organic Chemistry"; Newman, M. S., Ed.; John Wiley & Sons: New York, 1956.
9. Allinger, N. L. *Adv. Phys. Org. Chem.* **1976**, *13*, 1.
10. Ermer, O. *Struct. Bonding (Berlin)* **1976**, *27*, 161.
11. Altona, C. L.; Faber, D. H. *Top. Curr. Chem.* **1974**, *45*, 1.
12. Engler, E. M.; Andose, J. D.; Schleyer, P. v. R. *J. Am. Chem. Soc.* **1973**, *95*, 8005.
13. Bartell, L. S. *J. Am. Chem. Soc.* **1977**, *99*, 3279.
14. White, D. N. J.; Bovill, M. J. *J. Chem. Soc., Perkin Trans.* 2 **1977**, 1610.
15. Dunitz, J. D.; Bürgi, H. B. In "Int. Rev. Sci. Phys. Chemistry"; Buckingham, A. D.; Robertson, J. M., Eds.; Butterworths: London, 1975; Ser. 2, Vol. 11; p. 81.
16. Niketic, S. R.; Rasmussen, K. "The Consistent Force Field"; Springer: New York, 1977.
17. Warshel, A. In "Semiempirical Methods of Electronic Structure Calculation"; Segal, G. A., Ed.; Modern Theoretical Chemistry Vol. 7; Plenum: New York, 1977; p. 133.
18. White, D. N. J. *Mol. Struct. Diffr. Methods* **1978**, *6*, 38.
19. Mislow, K.; Dougherty, D. A.; Hounshell, W. D. *Bull. Soc. Chim. Belg.* **1978**, *87*, 555.
20. Williams, J. E.; Stang, P. J.; von R. Schleyer, P. *Ann. Rev. Phys. Chem.* **1968**, *19*, 531.
21. Eliel, E. L.; Allinger, N. L.; Angyal, S. J.; Morrison, G. A. "Conformational Analysis"; Wiley-Interscience: New York, 1965; p. 9.
22. Robiette, A. G. *Mol. Struct. Diffr. Methods* **1973**, *1*, 160.
23. Speakman, J. C. *Mol. Struct. Diffr. Methods* **1973**, *1*, 201.
24. Beagley, B. *Mol. Struct. Diffr. Methods* **1975**, *3*, 52.
25. Seip, H. M. *Mol. Struct. Diffr. Methods* **1973**, *1*, 7.
26. Wilson, E. B.; Decius, J. C.; Cross, P. C. "Molecular Vibrations"; McGraw-Hill: New York, 1955.
27. Hehre, W. J.; Pople, J. A. *J. Am. Chem. Soc.* **1970**, *92*, 2191.
28. Parr, R. G. "Quantum Theory of Molecular Electronic Structure"; W. A. Benjamin: New York, 1963.
29. Pople, J. A.; Beveridge, D. L. "Approximate Molecular Orbital Theory"; McGraw-Hill: New York, 1970.
30. Dewar, M. J. S. "The Molecular Orbital Theory of Organic Chemistry"; McGraw-Hill: New York, 1969.
31. Dewar, M. J. S.; Thiel, W. *J. Am. Chem. Soc.* **1977**, *99*, 4899, 4907.

32. Dewar, M. J. S. In "Further Perspectives in Organic Chemistry"; Elsevier: Amsterdam, 1978; p. 109.
33. Halgren, T. A.; Kleier, D. A.; Hall, J. H., Jr.; Brown, L. D.; Lipscomb, W. N. *J. Am. Chem. Soc.* **1978**, *100*, 6595.
34. Halgren, T. A.; Lipscomb, W. N. *J. Chem. Phys.* **1973**, *58*, 1569.
35. Thiel, W. *J. Am. Chem. Soc.* **1981**, *103*, 1413.

2

Force Fields

Force Fields in Vibrational Spectroscopy and Molecular Mechanics

Molecular mechanics is also often called the force field calculation method. Force fields, which were originally developed in a physically more rigorous form in vibrational analysis, are employed in molecular mechanics. To understand molecular mechanics force fields, it is convenient to start with the formalism of force fields appropriate for vibrational analysis.

When a molecule with n atoms, defined in terms of $3n$ coordinates, x_i, is deformed from its geometry of minimum potential energy, V_0, and coordinates, x_0, the potential energy may be written in a Taylor series expansion as:

$$V_{\text{pot}} = V_0 + \sum_{i=1}^{3n} \left(\frac{\partial V}{\partial x_i}\right)_0 \Delta x_i + \frac{1}{2} \sum_{i,j=1}^{3n} \left(\frac{\partial^2 V}{\partial x_i \partial x_j}\right)_0 \Delta x_i \Delta x_j$$
$$+ \frac{1}{6} \sum_{i,j,k=1}^{3n} \left(\frac{\partial^3 V}{\partial x_i \partial x_j \partial x_k}\right)_0 \Delta x_i \Delta x_j \Delta x_k + \text{higher terms} \quad (2.1)$$

In the vibrational analysis of a molecule having a geometry corresponding to an energy minimum, the first term, V_0, is taken as zero.[†] From the definition of a potential minimum it follows that at this geometry the first derivative term also vanishes. For sufficiently small displacements, as ordinarily treated in vibrational analysis, the terms higher than quadratic are neglected (harmonic approximation) (1–4). The potential energy depends only on the third term, to a first approximation, and if we replace the second derivatives, which are called the force constants, by their symbols f_{ij}, we are left with the simple relationship of a harmonic force field, Equation 2.2. This equation defines a

---

[†] This term is just a constant for any particular molecule, and is only of importance when we want to calculate the heat of formation (Chapter 5).

system of coupled harmonic oscillators completely, and, within the harmonic approximation, precisely. Usually the force constants are arranged in the form of a matrix, the terms with $i = j$ being diagonal, and the cross-terms with $i \neq j$ being off-diagonal. If only the diagonal terms are considered, which implies that the oscillators are not coupled, Hooke's law results. In a general treatment, a large number of cross-terms corresponding to a simultaneous displacement in two coordinates must be included.

$$V_{\text{pot}} = \frac{1}{2} \sum_{i,j=1}^{3n} f_{ij} \, \Delta x_i \, \Delta x_j \qquad (2.2)$$

At larger displacement amplitudes, outside the range of validity of the harmonic approximation, the higher terms in Equation 2.1 become increasingly important. This means that anharmonicity effects will be treated in a more satisfactory manner if at least the cubic terms are included in the force field. Another possibility would be to replace the harmonic potential by a Morse type of function. However, these refinements cause such enormous mathematical problems in vibrational spectroscopy that they have rarely been used there (5), whereas no major problems arise with such terms in molecular mechanics calculations that exclude vibrational motions, as we shall see.

The following outline discusses harmonic vibrations only. The relationship between the force constants and the observed vibrational frequencies follows from Newton's equations of motion and is properly expressed in the secular determinant, the derivation of which is given in standard textbooks on vibrational spectroscopy (1–4). In the determinant, the ms are the atomic masses. The determinant is set equal to zero and solved for the values of λ. The vibrational frequencies are equal to $\sqrt{\lambda}/2\pi$. If the force constants are given, it is a straightforward procedure to calculate the vibrational frequencies by evaluating the eigenvalues of the matrix corresponding to this determinant. Six of these eigenvalues (and frequencies) are equal to zero for a nonlinear molecule (five for a linear molecule), corresponding to the translational and rotational degrees of freedom.

$$\begin{vmatrix} f_{11} - m\lambda & f_{12} & f_{13} & \cdots & f_{1,3n} \\ f_{21} & f_{22} - m\lambda & f_{23} & \cdots & f_{2,3n} \\ \vdots & & & & \vdots \\ f_{3n,1} & \cdots & & & f_{3n,3n} - m\lambda \end{vmatrix}$$

Note that in the foregoing outline no restrictions have been imposed on the choice of coordinates. The $3n$ Cartesian coordinates may be used, but more frequently internal coordinates have been employed to describe the molecular geometry (6). In this case the secular determinant is of lower rank because the number of independent internal coordinates is only $3n - 6$ or $3n - 5$ instead of $3n$, and the solutions attributable to translation and rotation do not occur. This has an advantage in terms of the computing time required to solve the determinant. If the solution of a large determinant must be found by hand calculation, reducing the rank by 6 is of utmost importance, and early work on vibrational spectra was always formulated in internal coordinates for this reason. With a computer this change in rank makes no difference, and Cartesian coordinates offer a real advantage with cyclic molecules (*see* pp. 61–64).

The unique coordinates of special interest for vibrational spectroscopists are the normal coordinates, which, in the harmonic approximation, are connected to other kinds of coordinates by linear relationships. The normal coordinates have the advantage that, by treating the vibrations in terms of these, all off-diagonal force constants are equal to zero. They give the normal modes of vibration for the molecule, that is, the directions in which the atoms vibrate in phase with the $3n - 6$ (or $3n - 5$) eigenfrequencies. The eigenvectors of the secular determinant are the coefficients in the linear equations that relate the normal coordinates to the coordinates used in evaluating the f_{ij} terms of the secular determinant (1).

The central problem of vibrational analysis is the determination of a force field from the vibrational frequencies, and in this process the normal coordinates are obtained automatically. For this purpose vibrational analysis employs the Wilson GF matrix formalism. Molecular mechanics works the other way around, because the force field is here given mostly from other sources, and the vibrational frequencies are calculated by solving the secular equation (pp. 169–173).

The main obstacle to a direct determination of the molecular force field from the frequencies in vibrational analysis is the very large number of force constants. As long as the normal coordinates are not known (and they are known only after the force field is known), the force constant matrix contains at least $1/2 (3n - 6)(3n - 5)$ independent force constants (more if Cartesian coordinates are used). This, of course, far exceeds the number of vibrational frequencies ($3n - 6$ for a nonlinear molecule). Therefore, a complete determination of all of the force constants requires the analysis of the spectra of many isotopically substituted molecules, since this gives more frequencies without introducing any additional (unknown) force constants (to the extent that anharmonicity can be neglected).

However, many of the off-diagonal force constants are so small that they practically vanish, which corresponds to a small physical significance for these cross-terms. Therefore, spectroscopists have developed systematic simplifications of the complete force field to set as many of the off-diagonal terms as possible equal to zero (1–4).

The first approach along these lines was the **central force field**, which is specified in terms of the interatomic distances in the molecules only. Cross-terms corresponding to a simultaneous displacement of two interatomic distances are usually neglected, in which case a diagonal force field results. The physical model is that harmonic forces act between all pairs of atoms, not discriminating between bonded and nonbonded pairs of atoms. This approach appears physically reasonable for an ionic crystal but not for an organic molecule, and it has never been widely used.

The simple force field that best fits ordinary chemical ideas about the nature of the forces acting in a molecule is the **valence force field**. It can be formulated with internal coordinates, which usually are all the bond distances, and a set of independent bond angles and torsion angles. This means that forces act along and perpendicular to covalent bonds, restoring the equilibrium bond lengths, r, bond angles, Θ, and torsion angles, ω. With this choice of coordinates, the simplest, but of course very rough, approximation is to neglect all off-diagonal force constants. This results in a force field with Hooke's law harmonic potentials (Equation 2.3).

$$V = \frac{1}{2} \sum_i f_{r,i}(r_i - r_{0i})^2 + \frac{1}{2} \sum_k f_{\Theta,k}(\Theta_k - \Theta_{0k})^2 + \frac{1}{2} \sum_l f_{\omega,l}(\omega_l - \omega_{0l})^2 \quad (2.3)$$

The normal coordinates of the molecular vibrations that correspond to this force field give an approximate description of the vibrations the molecule undergoes. These coordinates are sums of the internal coordinates in various proportions. Using a group of related molecules, the normal coordinates can be broken down to yield the force constants of the internal coordinates.

We might then hope that such force constants are typical of a bond type, for instance, and are transferable from one molecule to another, but this is usually not the case, because of the gross assumptions inherent in the neglect of all off-diagonal force constants and the related interactions. Of the utmost importance for the development of molecular mechanics, which rests on the assumption that transferable potentials exist, was therefore the finding by Schachtschneider and Snyder that a valence force field with transferable force constants can be obtained when a few off-diagonal terms are *not* neglected (7). They found that cross-terms are usually largest when two internal coordinates end on the same atom or on nearest neighbor atoms. If such an extended valence force field is used, the number of required force constants still exceeds the number of vibrational frequencies of one molecule, but when the frequencies of similar molecules like homologous alkanes are collected, the information is sufficient to derive a transferable force field. By least squares optimization with 270 vibrational frequencies in many unstrained hydrocarbons, Schachtschneider and Snyder obtained a set of force constants that includes 3 stretch–stretch, 3 stretch–bend, and 20 bend–bend interaction terms, up to those of second nearest neighbor carbon atoms. The agreement between the calculated and observed frequencies (±0.25%) is impressive. It shows that indeed a large number of force constants can be omitted, and that the remaining force constants are transferable, within the limitation, however, of open-chain, strainless molecules (7). Transferable force fields were also derived, with increasing difficulty, for olefins (8, 9), for ethers (10), and for halides (11).

Of the several deficiencies in diagonal valence force fields necessitating the cross-terms, the most important one is the neglect of nonbonded steric interactions. On the other hand, steric effects are considered—although in a very limited way—in the central force field. Urey and Bradley (12) therefore introduced a mixed force field that contains, in addition to the valence force field potential, two-center nonbonded interaction terms between the atoms bound to a common atom (1,3-interactions). Additional internal coordinates are introduced this way, which are, however, redundant over the $3n - 6$ coordinates already used in the valence force field. For completeness, linear terms must be added, so the complete **Urey–Bradley force field** has the form of Equation 2.4, where f and f' represent harmonic force constants, and f'' represents other force constants with different numerical values, but between the same groups of atoms.

$$V_{UB} = \frac{1}{2}\sum_i f_{r,i}(r_i - r_{0i})^2 + \frac{1}{2}\sum_k f_{\Theta,k}(\Theta_k - \Theta_{0k})^2$$
$$+ \frac{1}{2}\sum_l f_{\omega,l}(\omega_l - \omega_{0l})^2 + \sum_{i,j=1}^{3n} f' r_{ij}^2$$
$$+ \sum_{i,j=1}^{3n} f'' r_{ij} + \sum_i f''_{r,i}(r_i - r_{0i})$$
$$+ \sum_k f''_{\Theta,k}(\Theta_k - \Theta_{0k}) + \sum_l f''_{\omega,l}(\omega_l - \omega_{0l}) \qquad (2.4)$$

Linear terms have the effect of shifting the parabola of the harmonic potential function, mainly along the abscissa. This corresponds to an offset of the equilibrium geometry, but not to any change in the force constant. In vibrational spectroscopy the equilibrium geometry is the actually observed internal coordinate. In molecular mechanics the standard geometry terms are adjustable parameters, and instead of using a linear term, the same effect can be obtained by adjusting the standard geometry terms, making linear terms redundant in molecular mechanics.

Inclusion of nonbonded interactions in vibrational force fields eliminates some of the force constants, as many cross-terms become unnecessary, but new parameters must be introduced for the nonbonded interactions and the linear terms. Schachtschneider and Snyder found that as a result, to calculate vibrational frequencies with comparable accuracy, approximately the same total number of parameters is necessary in a valence force field as in a Urey–Bradley force field (7).

The Urey–Bradley force fields of vibrational analysis contain 1,3-nonbonded interaction terms, whereas the usual spectroscopic valence force fields do not. In molecular mechanics these terms are also used to distinguish the types of force fields; these force fields also contain longer range, nonbonded interactions. While the Urey–Bradley force fields of molecular mechanics calculate all nonbonded interactions, the valence force fields exclude 1,3-nonbonded interactions. Since 1,3-nonbonded interactions are strongly repulsive, at one time it appeared that they might swamp all other nonbonded interactions in the molecule. The effects of 1,3-interactions can also be thought of as being partly included in the bending potential, and other effects such as the bond stretching that results from deformation of a bond angle to a very small value (as in cyclobutane) can be handled in molecular mechanics valence force fields by cross-terms (for example stretch–bend). But a Urey–Bradley force field certainly is a reasonable alternative to the cross-terms in a valence force field in molecular mechanics.

Potential Functions of Molecular Mechanics Force Fields

Bond Stretching and Angle Bending. In the molecular mechanics model the atoms of a molecule may be thought of as joined together by mutually independent springs, restoring "natural" values of bond lengths and angles. As in a diagonal valence force field, one can then assume a harmonic potential with Hooke's law functions, Equations 2.5 and 2.6, for bond stretching and bending, and every molecular mechanics force field contains these functions. At very large deformations one would expect deviations from the harmonic potential, and a Morse function would be the more general potential (Figure 2.1) (13).

$$V_r = \frac{1}{2} k_r (r - r_0)^2 \qquad (2.5)$$

$$V_\Theta = \frac{1}{2} k_\Theta (\Theta - \Theta_0)^2 \qquad (2.6)$$

Figure 2.1. Harmonic and anharmonic stretching and bending functions: ———, Morse type of function; ———, harmonic potential; ----, harmonic potential with cubic term added.

Morse functions are not generally used in molecular mechanics force fields, however, because they require excessive amounts of computer time. Simpler approximations can be designed to give results which are just as good. The theoretically most appealing approximation is to truncate the Taylor series expansion of equation 2.1 after a term higher than quadratic for bond stretching and bond angle bending (14, 15). A potential function including a cubic term such as Equation 2.7 has the desired properties in a certain range, and works well for bond

$$V_r = \frac{1}{2} k_r (r - r_0)^2 + k'(r - r_0)^3 \qquad (2.7)$$

lengths that are unusually long. [Note that at large deformations cubic functions invert. When energy minimization is done with a very poor starting geometry this may lead to disaster—with the molecule flying apart. Cubic terms should therefore be used only during late stages of energy minimization unless the computer program is able to take care of this problem automatically. An even powered function behaves symmetrically, and the problem of the molecule escaping from the potential well does not arise. Such functions have been used for bending in the Schleyer (14) and MM2 force fields (16)[‡].] Evidently it is not difficult to

---

[‡] Force fields that have been widely used are sometimes available as computer programs with names representing the initials of the authors, or some specific expression (e.g., MM2, meaning Molecular Mechanics, program 2). These designations will commonly be used without comment.

devise useful anharmonic stretching and bending potentials. Yet it is still not clear where the limits of the harmonic approximation are for angle bending. For angles opened to unusually wide values, such as in di- and tri-*t*-butylmethane, the deviation of the angle from normal is only about 15°. At present there is no compelling evidence that the quadratic function is inadequate for deformations of this size. More severe angle deformations occur only in the opposite direction, especially in three- and four-membered rings. Most early force fields tried to fit ordinary alkanes and cyclobutane rings with the same bending parameters, but this is difficult to do and wrong in principle (17). If we think of a five-membered, or larger, ring, in which we do not include van der Waals interactions between the carbons bound to a common carbon (no Urey–Bradley term), we can make up for this lack of van der Waals treatment by adjusting the bending parameters of a valence force field appropriately. Alternatively, we can include such van der Waals interactions in the Urey–Bradley part of the calculation. There is one UB interaction for each ring angle.

When we think of the cyclobutane ring, however, there are four carbon–carbon–carbon angles, but only two van der Waals interactions between carbons bonded to a common carbon. Hence, a valence force

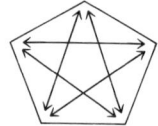

field cannot treat the cyclobutane system the same way it treats larger systems, but must somehow account for the fact that these van der Waals interactions, which are not explicitly included, are different. So the four-membered ring is a special case. In the valence force field, cyclobutane needs a separate set of bending parameters. In the Urey–Bradley force field, it is in principle possible to account for the cyclobutane case without any special effort; however, in practice it is not yet certain if this can be done. The Urey–Bradley type of interactions are going to be very large in cyclobutane, and the usual way of evaluating these interactions may be too crude to use here. Cyclopropane rings are even more different from larger rings in their bonding characteristics, and require still other bending parameters.

We recall that in vibrational analysis the molecular deformations are calculated from the equilibrium geometry of the particular molecule. The equilibrium values for each bond length and bond angle in a molecule are different (unless they are related by symmetry). The situation is quite different in molecular mechanics. Here a minimum number of transferable parameters is desirable, and bond lengths and angles are given standard transferable values. Although one can devise a sophisticated procedure and use, for example, different standard bond lengths between primary, secondary, and tertiary carbon atoms, this is not necessary. The use of only one bond length results in a good model. Similarly, a single force constant is used for all deformations of a given type. The equilibrium geometry is then reached automatically by allowing the molecule to relax to its minimum of energy. Take, for example, the central bond in hexamethylethane, which is stretched far beyond the standard C–C bond distance, but which is at its equilibrium distance due to van der Waals repulsions. A vibrational force field would assume a harmonic stretching potential about this equilibrium distance. In a molecular mechanics force field, a similar stretching potential with its minimum at the large value found for hexamethylethane results from a superposition of two contributions; the harmonic stretching function with its minimum at a small value, and the nonbonded repulsions, which prevent the distance from becoming too short and increase the apparent stretching force constant (Figure 2.2). If the effect of anhar-

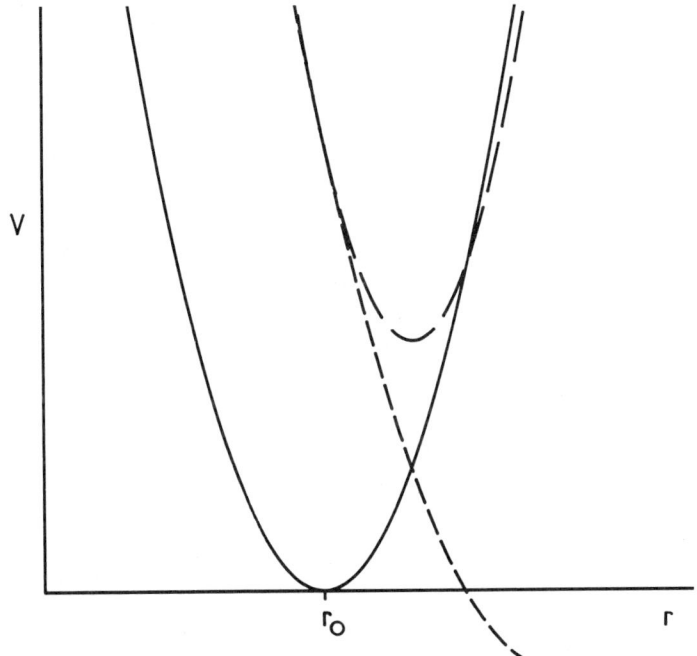

Figure 2.2. Harmonic stretching potential (V) of molecular mechanics for distorting a bond length from r_0 to r, not including van der Waals and other energy terms (———); nonbonded and other energy terms of molecular mechanics (- - -); harmonic vibrational stretching potential in strained hydrocarbon (– – –). Note that the force constant here (before inclusion of anharmonicity) is larger than the molecular mechanics force constant.

monicity were taken into account by adding a cubic or higher order stretching function, the left sides of the curves would be raised and the right sides lowered somewhat. Similar arguments also hold for angle bending. The assumption of transferable parameters is found to be a pretty good one in molecular mechanics force fields, certainly better than in vibrational spectroscopy, where it was shown to work only in limited areas (7).

Another basic difference between molecular mechanics force fields and those used in vibrational analysis is that the latter are based on $3n - 6$ independent coordinates (in a linear molecule, $3n - 5$), but the molecular mechanics fields are based on the much larger number of all possible internal coordinates. Here each and every bond length and angle, torsion angle, and nonbonded distance tries to assume its "optimum" value. At a tetrahedral atom, for example, five of the six bond angles are independent of each other, and a vibrational analysis force field will treat only these, whereas in molecular mechanics all six are used.

What this means is that molecular mechanics has adopted from vibrational analysis the potential functions for the single coordinates such as bond lengths and angles, but has adopted only part of the formalism. The molecular mechanics force field is not a force field in the sense of a normal coordinate analysis, but rather a collection of potential functions analogous to those of vibrational force fields. On the other hand, a vibrational analysis force field and normal coordinates may be derived from the molecular mechanics potential functions, just as can be done from other potential functions such as the Schrödinger equation, by forming the matrix of second derivatives of the potential energy. This basic difference between molecular mechanics force fields and conventional vibrational analysis force fields is essential in allowing the former to employ transferable parameters that are also applicable for large deformations. Of course this has the consequence that molecular mechanics force constants do not ordinarily have the same numerical values as those in vibrational analysis (pp. 38–39).

The analogies between molecular mechanics and vibrational analysis inspired the adoption of other elements of the latter into molecular mechanics force fields. At trivalent atoms that have their energy minimum at the planar geometry (olefin carbon, carbonyl carbon), a factorization of the bending potential familiar from vibrational analysis must be applied (18), because the deformations in the plane defined by the three atoms bound to the trigonal atom usually work against quite different restoring forces than do deformations perpendicular to this plane (out-of-plane deformations). An instructive example is cyclobutanone, which has an unusually large C—C=O bond angle (133°).

The associated angle strain might appear to be decreased easily by bending the oxygen atom out of the plane of the four-membered ring. This would in reality distort the pi bonding and lead to an enormous increase in the double bond pi energy, which would more than outweigh the relief of bond angle strain in the sigma framework. Bond angle deformations around double bonds must therefore be separated into those that do not distort the pi bond (in plane) and those that do (out-of-plane) (19–22), with different force constants for each.

Another example of the adaption of elements of normal coordinate analysis is the bond stretching observed to accompany bond angle reduction. This is especially notable for large deformations as in cyclobutane, and is considered an example of 1,3-interactions between the atoms forming the angle. In molecular mechanics valence force fields that do not include 1,3-repulsions, an additional term is necessary. The same problem was encountered by vibrational spectroscopists when setting up valence and Urey–Bradley force fields for hydrocarbons (7). In the valence force field, cross-terms (off-diagonal force constants) had to be included (7). In the same manner, a molecular mechanics force field that neglects 1,3-interactions reproduces the bond stretching upon angle deformation when a stretch–bend cross term is added (23, 24) (Equation 2.8). The inclusion of this term is not necessary in the Urey–Bradley type of force fields that specifically include 1,3-interactions (25).

$$V_{Str/b} = \frac{1}{2} k_{r\Theta}(r - r_0)(\Theta - \Theta_0) \qquad (2.8)$$

Nonbonded Interactions. For these, vibrational spectroscopy ordinarily uses potential functions that are very different from those of molecular mechanics. A harmonic potential is employed for 1,3-nonbonded interactions in most Urey–Bradley force fields, as in simple central force fields. Such a potential is clearly not suitable for atoms whose bonding separates them further than 1,3, since their interaction energies must tend to zero at large distances. Spectroscopic force fields simply ignore specific inclusion of the longer range interactions, and account for their effects indirectly by adjusting force constants for individual molecules. For transferability of force constants to cases where these interactions are severe, molecular mechanics has explicitly included these interactions from the earliest treatments. Assuming that intramolecular and intermolecular nonbonded interactions follow the same laws, the potential functions originally derived to describe the interaction potentials of rare gas atoms, usually known as van der Waals energies, have been employed. (Van der Waals' original model of a real gas assumed that the atoms are impenetrable hard spheres, while the nonbonded interaction potentials discussed below are soft-sphere potentials, improving the model of molecular mechanics over hard-sphere calculations.) A direct experimental access to nonbonded interactions is in scattering experiments with molecular beams of gases, or the deviations from ideal behavior of a real gas, measured in the second virial coefficients (26–30).

The theory of intermolecular interactions has been refined extensively over the past 50 years (26–30), and the potential functions used by molecular mechanics are among the simplest possible. The general shape of any van der Waals potential function results from two forces: a repulsive one at short distances, and an attractive one at larger distances—both of which asymptotically go to zero at still longer distances. The curve is characterized mainly by the minimum energy distance (related to the van der Waals radii), the depth of the potential well (related to the polarizabilities), and the steepness of the repulsive part (the hardness), as in Figure 2.3.

Noble gas atoms and nonpolar molecules with closed shells of electrons interact at large distances through induced electrical moments, the London dispersion forces. Second-order perturbation theory supplies us with the potential corresponding to this attraction (Equation 2.9). The first term, the instantaneous dipole/induced dipole interaction energy, prevails at large distances, and the attractive energy is usually described only by this r^{-6} term, the coefficient of which may be adjusted somewhat to allow for the neglected higher order terms. For molecules carrying a

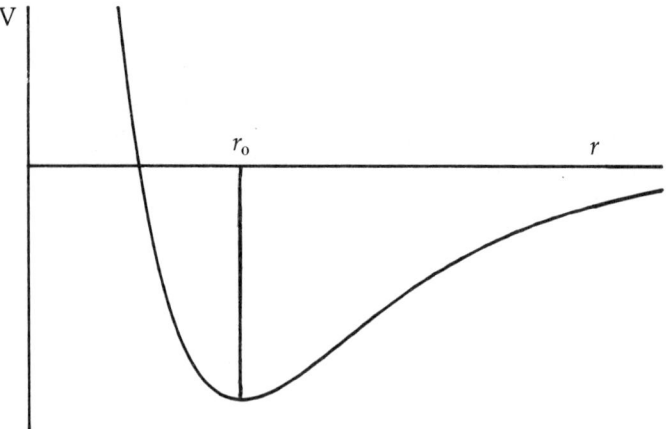

Figure 2.3. Van der Waals energy function.

permanent electric charge or dipole moment, the interaction terms of the permanent moments are definitely larger than those of induced moments. In classical electrostatic theory the interaction energy between diffuse charge clouds is calculated by another series expansion, containing charge/charge, charge/dipole, dipole/dipole and higher terms (*31*). For interactions between ions, the Coulomb potential (Equation 2.10) is the leading term. In noncharged polar molecules the dipole/dipole interaction energy is the leading term, which can, following classical electrostatics, be calculated by Equation 2.11, the so-called Jeans' formula, where D is the (effective) dielectric constant, χ is the angle between the two dipoles μ_i and μ_j, and the αs are the angles that the dipoles form with the vector connecting the two (Figure 2.4) (*32*).

$$V_{disp} = -c_6 r^{-6} - c_8 r^{-8} - c_{10} r^{-10} - \cdots \qquad (2.9)$$

$$V_{charge} = \frac{q_i q_j}{D r_{ij}} \qquad (2.10)$$

$$V_{dipol} = \frac{\mu_i \mu_j}{D r_{ij}^3} (\cos \chi - 3 \cos \alpha_i \cos \alpha_j) \qquad (2.11)$$

At sufficiently small distances we can be sure that a repulsive interaction results from Pauli exclusion, but the exact shape of the potential function used to calculate the repulsion energy is based on expediency.

The total interaction potential between a pair of atoms is the sum of the energies from the attractive and repulsive forces. The most general

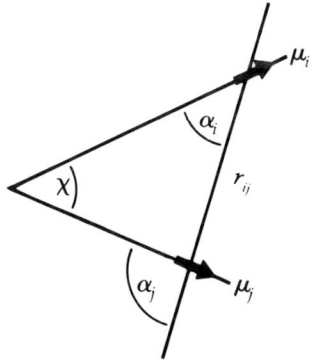

Figure 2.4. *Definitions of the geometric terms appearing in Equation 2.11.*

form of the interaction potential of neutral, nonpolar atoms or molecules, as originally deduced by Lennard–Jones in a study aimed at the determination of second virial coefficients of gases (33), has the form shown in Equation 2.12.

$$V_{\text{VDW}} = \frac{n\varepsilon}{n-m}\left[\frac{m}{n}\left(\frac{r_0}{r}\right)^n - \left(\frac{r_0}{r}\right)^m\right] \qquad (2.12)$$

The exponent of the attractive potential, m, is in all more recent studies set equal to 6 for the reasons outlined above, while there is no compelling reason to prefer any special value of n on theoretical grounds, as long as it is greater than 6. With some simple arithmetic it follows that the equation has an especially simple form when n is set equal to 12,

$$V_{\text{VDW}} = \varepsilon\left[\left(\frac{r_0}{r}\right)^{12} - 2\left(\frac{r_0}{r}\right)^6\right] \qquad (2.13)$$

Until recently these exponents were used most often. The factor of 2 between the attractive and repulsive parts of the potential can be justified only in a 6/12 potential, and can be regarded as an adjustable parameter. Epsilon (ε) defines the depth of the potential well, r_0 the minimum energy distance, and the exponent of the repulsive term the hardness of the function.

Another frequently used potential function (26) is that due to Buckingham, which has an exponential repulsive function. In the form of Equation 2.14 its minimum is at r_0 with a potential depth of ε. This function has a steepness similar to that of the Lennard–Jones 6/12 potential in the region of general interest (up to a repulsion of a few kJ) when the constant α is between 14 and 15.

$$V_{\text{VDW}} = \frac{\varepsilon}{1 - \frac{6}{\alpha}}\left[\frac{6}{\alpha}e^{\alpha(1-r/r_0)} - \left(\frac{r_0}{r}\right)^6\right] \qquad (2.14)$$

A specific, widely used example of the Buckingham potential (16), which is not quite the same as that given in Equation 2.14, is Equation 2.15.

$$V_{\text{VDW}} = 2.90 \times 10^5\, \varepsilon\, e^{-12.50 r_0/r} - 2.25\, \varepsilon \left(\frac{r_0}{r}\right)^6 \qquad (2.15)$$

Here the minimum energy is -1.1ε at a nonbonded distance of r_0. T. L. Hill showed that this equation gives a good description of the rare gases (and many simple molecules) (34), so that the problem of selecting the whole function seemed to be reduced to picking two parameters, the van der Waals radius, r_0, and ε. In molecular mechanics, however, extensive studies have shown that a better model can be obtained if all three of the parameters of Equation 2.14 are adjusted (or the equivalent, which is to adjust the superscript given as 12 in Equation 2.13). It was found that both the Lennard–Jones and Buckingham functions work well, and are in fact nearly indistinguishable in the region of general interest. The Lennard–Jones potential has certain advantages for the incorporation in computer programs because the calculation of an exponential function requires more time than a power of r, and also the Buckingham curve inverts at very small distances. (Thus with the latter, if we inadvertently get two atoms too close together they collapse to a common nucleus—nuclear fusion!)

The use of pair potentials for nonbonded interactions in general, and in molecular mechanics in particular, has been criticized. Margenau and Kestner (29) have pointed out that pair potentials can be too simple an approximation, because the induced dipole moments from two interacting atoms may be disturbed considerably by a third or additional atoms, introducing sizeable errors. This must be especially significant when highly polarizable atoms such as sulfur are close to polar groups, which might affect the nonbonded properties of the polarizable atom. An objection raised against the use of potential functions of intermolecular interactions for intramolecular forces is that through space, nonbonded interactions between atoms on the same side of the molecule will differ from those between atoms on different sides of the molecule, since the latter are shielded by the electron density of the body of the molecule (35). (This is the pair potential problem in a different form). No examples are yet known in molecular mechanics where either of these effects is significant, however.

Much more severe is the problem that atoms in a Lennard–Jones or Buckingham potential are treated as spherical, which is correct for isolated noble gas atoms, but clearly is an approximation for atoms in a molecule. Nonbonded interactions are determined, however, by the polarizabilities of the atoms as well as by their shapes (36). A measure of the atom's shape in the molecule may be obtained from the electron density, as studied by ab initio calculations or by high quality x-ray diffraction measurements (37).

The approximation of spherical atoms is better for larger atoms. Since the van der Waals interactions arise primarily from the electrons, the electrons of a large atom will better approximate a sphere centered on the nucleus. For the most part, atoms are taken to be spherical in molecular mechanics, and the center of the sphere is placed at the nucleus. For small atoms the approximation is not as good, and fluorine and oxygen, for example, are detectably nonspherical. In the case of hydrogen the deviation from a sphere centered at the nucleus is very large, because the atom has only one electron, which is used for bond formation. There are complex ways to deal with the problem, but the simplest one, which seems to work well in practice, is to treat the atom as spherical, but to offset the center of the sphere from the hydrogen nucleus towards the carbon. Williams found that a molecular mechanics force field in which the van der Waals center of the hydrogen is relocated inward by about 10% of the bond length can calculate the crystal packing of hydrocarbons, which is determined by van der Waals interactions, much better than when the van der Waals center is at the nucleus (38, 39). This procedure has been adopted in most recent force fields (15, 16, 25).

Other atoms for which an aspherical van der Waals potential was found useful are oxygen and nitrogen. The repulsions from oxygen, for example, should be stronger in the direction of the lone pairs. This can be simulated by introducing the lone pairs into the force field as pseudoatoms with their own van der Waals potentials (Figure 2.5). It is not quite yet clear, however, whether the aspherical electron density really requires such lone pairs, or whether they are merely convenient (*see* pp. 209–224) (40, 41).

Torsional Energy. Many attempts have been made to design molecular mechanics force fields on the basis of only stretching, bending, and van der Waals interactions by varying parameters and potential functions. It was impossible, however, to obtain even an approximately correct energy difference between staggered and eclipsed conformations in ethane when van der Waals parameters obtained from molecular beam scattering or from crystal packing studies were applied (42). If the van der Waals parameters were chosen to reproduce the torsional barrier, they were found to be unreasonable for the calculation of most other properties (43).

Completely different approaches like Scheraga's EPEN method (44–46) have been devised in which extensive use is made of electrostatic nonbonded interactions, but in a molecular mechanics scheme

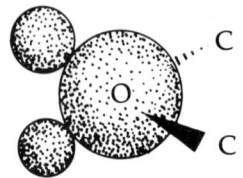

Figure 2.5. The van der Waals potential of an ether oxygen using lone pairs as pseudoatoms.

based on van der Waals interactions, an additional term, the torsional energy, is essential. For nearly twenty years a cosine function with energy minima at staggered conformations and maxima at the eclipsed geometries, Equation 2.16, was used in molecular mechanics calculations for this correction function for torsion about single bonds. This energy term has usually been thought of as resulting from a repulsion between the bonds not covered by van der Waals interactions. An alternative, but not entirely satisfactory, statement of this picture is that the torsional energy is a correction for the anisotropy of van der Waals repulsions, which are more repulsive at small angles to the bond than at large angles (47). Many other rationalizations have been given as the result of quantum mechanical studies (48–55). Accordingly, in one scheme the torsional energy was calculated only for torsion angles of less than 60° with the function Equation 2.16. But, for reasons which follow, all current force fields calculate the torsional energy over the full 360° rotation.

$$V_{tor} = k_\omega (1 - \cos 3\omega) \qquad (2.16)$$

Equation 2.16 is, of course, only a first approximation to the torsional energy necessary as a correction term for the force field. In vibrational spectroscopy it is common to write the complete torsional potential function for rotation about a bond (not just the part exceeding the van der Waals effects) as a Fourier series expansion, Equation 2.17, for which about six terms can sometimes be determined experimentally from overtones. For molecules with C_3 symmetry like ethane, all terms

$$V_{tor} = \sum_j \frac{1}{2} V_j (1 - \cos(j\omega)) \qquad (2.17)$$

with other than $3n$-fold periodicity must disappear for symmetry reasons. Even for ethane the data are barely sufficient for the determination of the sixth term. For this reason the torsional potential for ethane has usually been described by Equation 2.16. Probably because of the identification of bond repulsion with this equation, the threefold potential alone was transferred to molecular mechanics, generalizing to all kinds of alkane torsion angles, and not just the symmetrical rotors like methyl groups. Only in 1976 was it recognized that a better level of accuracy can be achieved with molecular mechanics force fields when the two first terms of Equation 2.17 are also used (56, 57). By giving certain V_1 and V_2 terms of CCCC torsion angles small values (of the order of 1 kJ/mol) a conspicuous improvement in the force field results (16, 25). One physical picture of these torsional terms is that the onefold term is a dipole–dipole interaction, the twofold term is from hyperconjugation in alkanes (or conjugation in unsaturated systems) and the threefold term is steric

(*58*). Ab initio calculations show this division of interactions clearly in molecules like 1,2-difluoroethane, where each term is large (*58*). Such calculations are not accurate enough to convincingly interpret (or really, even to see) the one- and twofold terms in alkanes. The basis for expecting such terms and attributing them to the same kinds of effects is, however, well founded.

Low periodicity (V_1 and V_2) torsional terms tend to be numerically much larger in systems carrying heteroatoms than in hydrocarbons. In the 1,2-difluoroethane case mentioned earlier it was found, both by molecular mechanics and by ab initio calculations, that the stability of the *gauche* arrangement due to hyperconjugative interactions (*58*) requires a V_2 term for the FCCF fragment, with a minimum at a torsional angle of 90° (*57, 58*). Resonance forms of the type $H^+HFC{=}CH_2F^-$ contribute to and stabilize the molecule only to the extent that the fluorines are out of plane. Other important examples are encountered when 1,4-interactions of lone pairs are involved. This is the case with the anomeric effect (pp. 219–220), which relates to the *gauche* preference of COCO torsion angles. This effect is not accounted for by nonbonded interactions (van der Waals and electrostatic), but is at least in part due

to a hyperconjugative interaction of the oxygen HOMO, a p_z orbital, and the σ^* orbital of the neighboring CO bond (*58–60*). Other cases where hyperconjugation is less important, but where the nonbonded interactions alone give erroneous results, are the torsional potentials in CCOC (*61*) and CCNC (*62*) fragments. Again, these can be well treated by additional low-order torsional potentials. In addition to these examples of molecules with heteroatoms where large V_1 and V_2 terms were necessary, small V_1 and V_2 terms are valuable for many kinds of torsion angles.

The torsional potential about the double bond in ethylene has two-fold symmetry, and is mainly determined by effects other than nonbonded interactions. A cosine function with twofold symmetry (Equation 2.18) was therefore employed so that the potential minima occur at the two eclipsed positions (*63, 22*). In one force field, a V_4 term was added to account for errors in the unusually soft V_2 potential that became apparent at large (>20°) torsion angles (*94*). When the symmetry in the Newman projection about the double bond is destroyed by out-of-plane bending, which is often the case in *trans*-cycloolefins and bridgehead olefins (64), the situation becomes more complicated. The torsional potential is mainly due to the disruption of the overlap of the pi-orbitals, and as long as their torsion angle (ω) is the same as that of the atoms attached to the double bond, the calculation will be correct (A). At small out-of-plane bendings (B) the torsional potential can be

A   B   C

approximated by calculating the (virtual) pi-orbital torsion angle (ω') as the bisectors of the torsion angles between the atoms attached to the double bond (22), or by individually considering the four torsional angles involved. Out-of-plane bending changes the hybridization of the two carbon atoms and is considered separately (19–22). The procedure works well, but some error is to be expected with extreme torsion/out-of-plane bend deformations as in adamantene (C) (64).

$$V_{tor} = \frac{V_2}{2}(1 - \cos 2\omega) \quad (2.18)$$

Olefins with a delocalized system of pi electrons must, of course, be handled differently, one of the main differences being the torsional potential about the partial double bonds (pp. 52–55).

Cross-terms involving torsional energy terms have also been applied in molecular mechanics, especially to reproduce the ring pucker and bond stretching in cyclobutane. The ordinary torsional interactions are insufficient to pucker this molecule, and one approach to the problem is to include a torsion-bend interaction term, Equation 2.19, which is effective only at very large angle bending deformations, but has little effect in compounds other than cyclobutanes (23). An alternative, which in fact works better, is simply to use a different (larger) threefold torsional constant for four-membered rings, along with other different constants as discussed in pp. 24–27. An additional cross-term, Equation 2.20, was found useful in another force field (65) for the calculation of vibrational frequencies.

$$V_{tor/bend} = k_{TB}(\Theta - \Theta_0)(\omega - \omega_0) \tag{2.19}$$

$$V_{TTB} = k_{TTB}(\Theta - \Theta_0)(\Theta' - \Theta_0')(\omega - \omega_0) \tag{2.20}$$

Parameterization

The quality of a molecular mechanics force field, that is, the precision of its predictions and their reliability, is critically dependent on the potential functions *and* on the parameters used. Parameter optimization has to be done with the same care that is used in the selection of the potential functions. Because different force fields usually differ not only in their parameters, but also in the mathematical form of one or more of the potential functions, it is usually hazardous to try to transfer the parameters from one force field to the other.

The parameter optimization may take one of two extreme forms. In principle, least squares methods are standard in optimization procedures. The alternative, and the method originally used in force field calculations, was simply to optimize by inspection. That is, one looked at where the errors were, and then made adjustments in such a way as to minimize those errors. On the surface of it, this method appears less efficient than the least squares procedure, because of the large number of simultaneous optimization conditions. Some force fields have been constructed using structural data alone, while others used both structural and energy data. Other force fields have added to these quantities vibrational frequencies obtained spectroscopically, and still others have added thermodynamic functions. In principle we would like to use all of these kinds of data, but in practice the amount of data available, or potentially available, is so immense that rather limited portions of it have been used. The more data being used, the more difficult it is to optimize the parameters by trial and error methods, and the more useful the least squares methods become.

Lifson and coworkers made major use of linear least squares parameter optimization (66–71), and referred to their method as a **consistent force field method**. The apparent advantage of such a method is that it optimizes the parameters in a precise and mechanical way. There are several disadvantages. The amount of computer time required can be immense, unless the person doing the optimization helps things along. A fundamental problem, however, is that least squares methods optimize variables that are all measured in the same units. To optimize variables measured in different units, let us say, for example, bond lengths and bond angles, a judgment must be made as to how much error in one case is equal to how much error in the other case. A weighting scheme can be applied (72, 73), and the procedure then becomes the normal least squares method. But since we are dealing with perhaps a dozen or more different kinds of variables, quite a lot of decisions have to be made about equivalencies of different parameter types. Once these are made, and the optimization is carried out, then it becomes clear in the typical case that some of the equivalencies decided upon are not, in fact, the best. So the weighting scheme is revised, and the process is repeated. A further difficulty is that the classical method of least squares optimization would require that the data are linear functions of the parameters to be optimized. However, the actual relationship is much more complex than even a single potential function indicates. To linearize the relationship it can be developed in a Taylor series, which is truncated after the linear term, a reasonable approximation when the initial guess of the parameters is close to the optimum values (Equation 2.21, where the y_i are data, and the k_j are parameters). The series truncation is the reason why the procedure must be applied iteratively.

$$y_i \text{ (improved)} = y_i \text{ (initial)} + \sum_{j=1}^{k} \frac{\partial y_i}{\partial k_j} \delta k_j \qquad (2.21)$$

The arithmetic details of the least squares procedure have been described elsewhere (66–73), and will not be duplicated here. The most time-consuming step of the procedure is the determination of the first derivatives of the data with regard to the parameters. In the original method of Lifson and Warshel (66) the derivatives were evaluated partly numerically and partly analytically (67). To make the procedure efficient, the derivatives that concern the most time-consuming data (geometries and vibrations) have been evaluated approximately (68). The geometry is reoptimized (pp. 64–72) with the improved parameters before the next cycle of parameter optimization (69). It is also useful to apply this procedure interactively, that is, with human interference, to avoid unreasonable parameter changes. Occasionally parameters go to zero, which allows removal of certain potential functions from the force field. This was encountered by Ermer and Lifson in their olefin parameterization, where they were able to eliminate charge interactions in hydrocarbons (22), and by Hagler et al., who found that a hydrogen-bond interaction was better simulated in their force field when all nonbonded interactions between the hydrogen-bonded atoms were omitted, rather than when a hydrogen bond potential function was employed (74).

In our experience, the method that has been used in practice with the most success is neither of the two extremes mentioned at the outset, but a combination of the two. That is, optimization proceeds by trial and error adjustments, using least squares methods to supplement these adjustments, particularly in cases where the number of data available becomes large and tends to be unmanageable (72, 73).

Lifson has used the expression **consistent force field** (CFF) to denote that structures, energies, and vibrational frequencies have been examined in its construction, and that least squares optimization has been used. The application of an optimization method in parameter refinement is a basic part of the definition of the CFF (66), and it has been proposed that the parameter adjustments observed during optimization allow a systematic improvement of our understanding of the nature of intramolecular interactions. The hope of gaining insight into the nature of intramolecular forces from force field calculations is not as fervent as it was in the late 1960s, when the first CFF was published. Instead, it is now realized that, because of the strong correlation between parameters, different force fields that give the same results with regard to the observables (like geometry and energy) differ considerably in the partitioning of the interactions into the individual interaction terms. Therefore, we must not look at a force field calculation and ask "what interactions are really occurring in the molecule"; rather the question must be "what interactions are really occurring in *our model* of the molecule." Of course the hope is that the answers to the last question will, in fact, converge upon the answers to the first question as force fields improve.

Bond Stretching and Angle Bending Parameters. In practically all of the early molecular mechanics force fields, the force constants used for the stretch and bend deformations were unmodified from vibrational spectroscopy, where Schachtschneider and Snyder had shown that a transferable set can be found that permits the calculation of vibrational frequencies of unstrained hydrocarbons (7). Later it was found that these force constants were not the optimum parameters for a molecular mechanics calculation, which is understandable. The molecular mechanics calculation includes nonbonded interactions in a way quite different (depending on the details of the force field) from that used by Schachtschneider and Snyder, and hence, the force constants common to both fields have different numerical values in general. As it turns out, the stretching force constants are usually similar or identical in the two cases. While bending force constants from vibrational analysis were employed in some force fields, smaller values have been used by Lifson and Warshel (66), Ermer (67), and Allinger (16, 72). While structures can be fit about equally well with the spectroscopic values, energies are more sensitive to these constants, and usually require that they be significantly reduced (20%–50%). (For tabulations of the stretching and bending parameters in current force fields, *see* Tables 2.1 and 2.2.)

The geometry parameters (standard or "natural" bond lengths, r_0, and bond angles, Θ_0) are optimized to fit the geometries of simple model

Table 2.1.

Parameters used for Bond Length Deformations in Current Force Fields

$$E_r = \frac{1}{2} k_r [(r - r_0)^2 - k_r'(r - r_0)^3]$$

	C–C		C–H		
	k_s	r_0	k_s	r_0	k_s'
MM2 (Allinger, 16)	4.40	1.523	4.60	1.113	2.0
MUB–2 (Bartell, 25)	2.34	1.166[a]	3.85	1.0203[a]	2.0
		1.534[b]		1.1068[b]	
EAS (Schleyer, 14)	4.40	1.520	4.60	1.100	0.0
CFF–3 (Ermer/Lifson, 22)	4.48	1.526	4.54[c]	1.105	0.0
			4.73[c]		
Boyd (95)	4.40	1.530	4.70[c]	1.090	0.0
			4.55[c]		

Note: Data given as mdyn/Å and Å. To convert to kJ/Å, multiply the force constants by 602.
[a] Used for quadratic term.
[b] Used for cubic term.
[c] Different force constants are used depending on the environment of the CH bond.

compounds. There has been some concern about the meaning of these parameters. The different force fields in the literature employ quite different values for these "strainless" bond lengths and angles, and in the opinion of the present authors, these quantities cannot be thought of as a universal, natural property of a molecular fragment. The numerical values of the parameters depend to a major degree on the van der Waals parameters. Take, for example, ethane; the C–C bond length will depend on the repulsion between the vicinal hydrogen atoms, and with a hard repulsive potential, the C–C reference bond length must be shorter than in the case of a soft hydrogen. These reference bond lengths have no known physical meaning at present, and are therefore valid only within a particular model or force field. It must be noted, however, that over the last twenty-eight years there has been a considerable convergence in the functions and parameters used in different force fields. The continuing applications of additional experimental data to determining and improving these functions appears to be leading inexorably to "best" values for the various kinds of parameters, and it will be tempting to try to attach a physical reality to these limiting values (75).

Nonbonded and Torsional Energy Parameters The parameters used for nonbonded interactions have undoubtedly been the most difficult part of the parameterization. Direct experimental data are available from the

Table 2.2.
Parameters used for Bond Angle Deformations in Current Force Fields

$$E_\Theta = \frac{1}{2} \cdot 3.046 \cdot 10^{-4} \cdot k_\Theta (\Theta - \Theta_o)^2 + k_\Theta'(\Theta - \Theta_o)^n$$

	CCC		CCH		HCH			
	k_Θ	Θ_o	k_Θ	Θ_o	k_Θ	Θ_o	k'	n
MM2 (Allinger, 16)	0.450	109.5	0.360	109.4	0.320	109.4	$7 \cdot 10^{-8}$	4
MUB–2 (Bartell, 25)	0.629	109.47	0.322	109.47	0.350	109.47	0.0	—
EAS (Schleyer, 14)	0.570	109.5	0.400	109.2	0.330	109.1	−0.55	3
		110.1		109.0		109.2		
		110.4		109.5				
CFF–3 (Ermer/Lifson, 22)	0.647	110.50	0.521	109.18	0.549	109.60	0.0	—
		109.47				106.40		
Boyd (95)	0.80	111.0	0.608	109.5	0.508	107.9	0.0	—

Note: Data given as mdyn/deg and deg. To convert to kJ/deg, multiply the force constants by 602. When more than one value is given for Θ_o, the actual value depends on the kinds of atoms attached to the central carbon (primary, secondary, tertiary).

van der Waals potentials of rare gas atoms, as determined from molecular beam experiments, or from second virial coefficients, to mention the two most frequently employed methods (26–30). The van der Waals radii must increase as one goes from right to left across the periodic table, which allows a rough estimate of the parameters for other atoms. The parameters finally employed in the molecular mechanics force field must, however, be derived by a fitting of observed quantities. There are at least two parameters to evaluate, ε, which measures the depth, and r_0, the van der Waals radius. Actually there are two more parameters to evaluate, the steepness of the repulsive potential, and the factor used to correlate the attractive and repulsive potentials (2 in the Lennard–Jones equation, but also an adjustable parameter).

There is a difficulty with the expression *van der Waals radius*. The same expression is used in two different ways to mean two completely different things. This has led to substantial confusion in the literature. In molecular mechanics, two atoms will show a minimum in their van der Waals function at that distance said to be the sum of the two van der Waals radii. The "distance of closest approach" between two atoms is referred to by crystallographers as the sum of their van der Waals radii. This meaning is quite different from that used in molecular mechanics. The difference between the two meanings can be seen in the following way. If we consider the hexane crystal as an example (76), molecules lie side by side in the crystal, as shown in the end on view below.

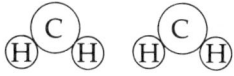

Imagine the pair of nearest nonbonded hydrogens (one from each molecule) being located at their distance of minimum energy from each other. By the molecular mechanics definition, they are at their van der Waals distance. But this is not the distance of closest approach. Because the energy of interaction of the nearest hydrogens is at a minimum, they are not exerting a force in any direction on each other. But the more distant atoms (the carbons and the more distant hydrogens) are all at distances exceeding their van der Waals distance, and, hence, they are all attracting each other. Thus, there is a net attractive force, and the molecules will approach each other. The crystal will consequently contract until the repulsion between the nearest pairs of hydrogens just balances the attraction between the rest of the molecules. In practice, it is found that the closest approach radius of hydrogens is about 1.2 Å, but this corresponds to a van der Waals radius of at least 1.5 Å. This appears not to have been widely recognized until about 1966, and earlier force fields that assigned hydrogen a van der Waals radius of 1.2 Å would yield very unsatisfactory crystal data, among other things.

In Figures 2.6 and 2.7 are shown some van der Waals curves described in the earlier literature. There was considerable disagreement among various authors, who deduced these curves in different ways. More recently, the set of curves in use has become much less diverse (Figures 2.8–2.10). These have been obtained empirically by molecular mechanics calculations. Crystal data can be used to substantiate van der Waals curves. In general, one has crystal spacing data, and also the heat of sublimation of the crystal. These data are not sufficient to evaluate fully the van der Waals parameters needed, but are criteria that any proposed parameter set must meet. n-Hexane was used first for this purpose (76), and other molecules have been similarly used (38, 39, 74, 77).

An important question concerns the optimum value of the steepness of the repulsive part of the van der Waals potential. The 6–12 law (or its equivalent exponential form), which fits well the data on the rare

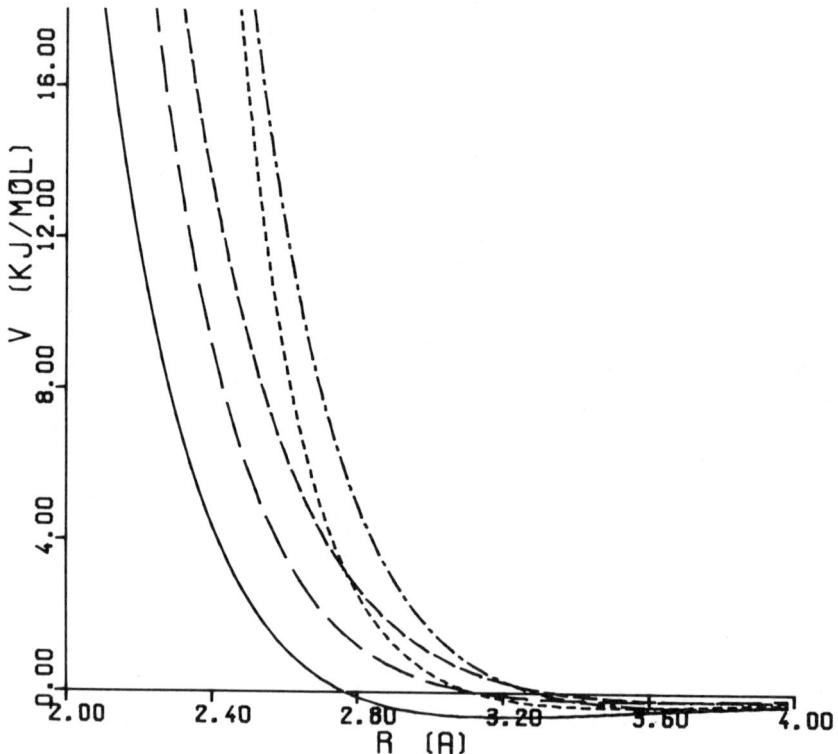

Figure 2.6 Van der Waals energy plots for C/C interactions from force fields of different generations: ———, Hendrickson 1961; ———, Allinger (MM1) 1973; ———, Allinger (MM2) 1977; ----, Bartell (MUB-1) 1967; -·-, Bartell (MUB-2) 1977.

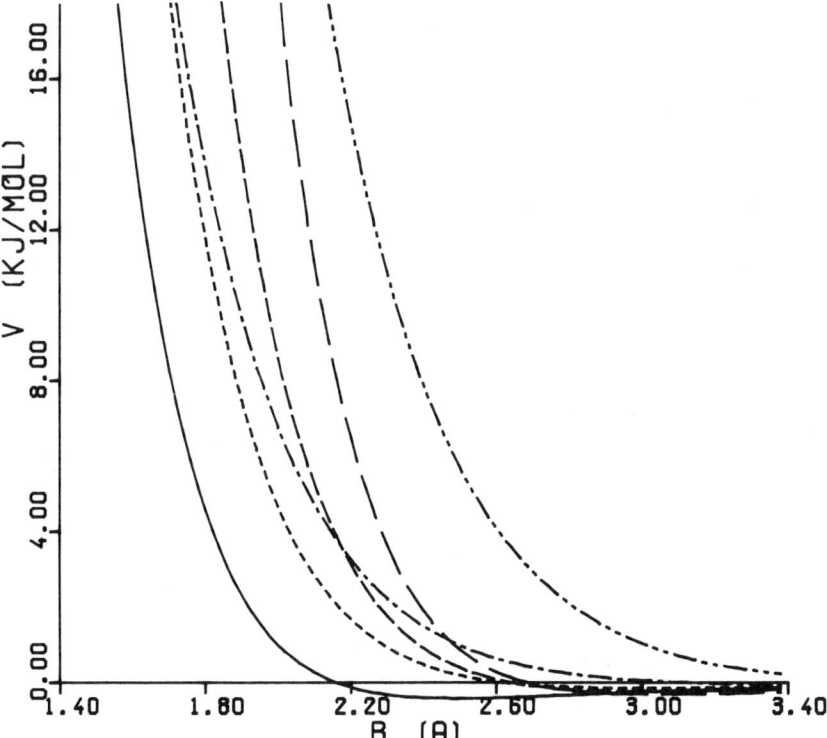

Figure 2.7. Van der Waals energy plots for H/H interactions from force fields of different generations: ———, Hendrickson 1961; — —, Allinger (MM1) 1973; ———, Allinger (MM2) 1977; ----, Bartell (MUB-1) 1967; –·–, Bartell (MUB-2) 1977; –··–, Mason/Kreevoy 1955.

gases, also proved to be reasonable for describing nonbonded interactions in a large number of hydrocarbon force fields. However, many years of effort by people in different laboratories have shown that this is not, in fact, the best available function. The repulsive twelfth power term, although appropriate for rare gases, is too steep or "hard" to describe hydrocarbon interactions. Something nearer to the ninth or tenth power is clearly better (70).

Finally, numerical values for the parameters for the different atoms must be established. This has been done in two different ways. The first method, utilized by Fitzwater and Bartell (25), was to accept theoretically (quantum mechanically) calculated functions. The early calculations were pretty crude, but the functions obtained were reasonably good. More recently, very careful ab initio calculations have been carried out on the hydrogen molecule (78), and have furnished more refined van der Waals information for when the hydrogen is bound to carbon. The assumption is made that the part of the hydrogen away from the bond has the same properties in either case. This function gives excellent results for hydrocarbons as far as structure is concerned.

The second approach used was empirical. Having decided on the rest of the force field, the van der Waals function to be used was chosen, and then the parameters were evaluated by consideration of the available data. An inverse twelfth power function (or exponential equivalent) was assumed in most early force fields, but was later revised downward as mentioned above. These functions were also found to give results in better agreement with ab initio calculations on the interactions between hydrogen molecules. These latter calculations indicated that there is a substantial difference in the interaction energy, depending on the orientation between the interacting molecules. As long as hydrogen is going to be treated as spherical, one has to accept that some kind of an average over the orientations is necessary. A number of van der Waals functions are in the current literature. These functions differ somewhat, depending on the data used to develop them. But as mentioned earlier, all van

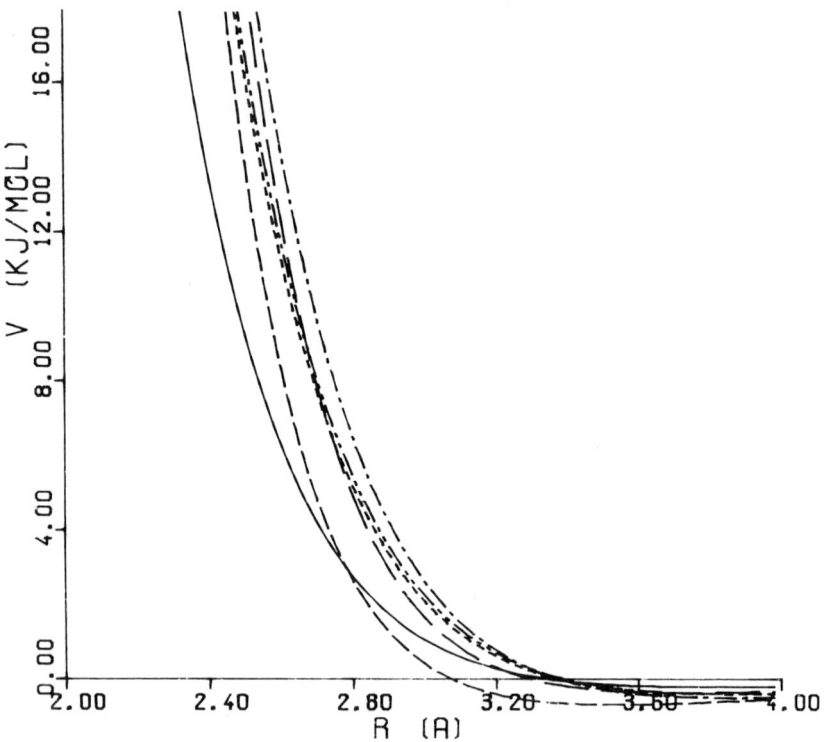

Figure 2.8. Van der Waals energy plots for C/C interactions from current force fields:
———, *Allinger (MM2) 1977;* ——, *Bartell (MUB-2) 1977;* ---, *Ermer/Lifson 1972;*
----, *Engler/Andose/Schleyer 1973;* -·-, *Bovill/White 1977;* -··-, *Boyd 1969.*

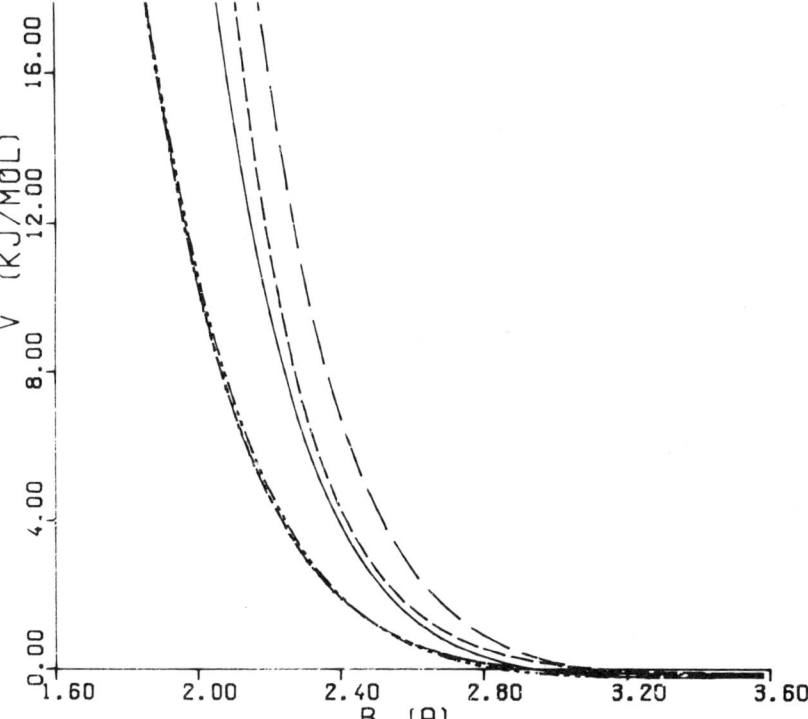

Figure 2.9. Van der Waals energy plots for C/H interactions from current force fields: ———, Allinger (MM2) 1977; ———, Bartell (MUB-2) 1977; ———, Ermer/Lifson 1972; ----, Engler/Andose/Schleyer 1973; —·—, Bovill/White 1977; —··—, Boyd 1969. The curve at the left is made up of overlapping curves from the Schleyer, White, and Boyd force fields.

der Waals functions in current use in force fields are relatively similar, and whether or not the remaining differences have any significance is not now known.

As the last step, the torsional energy parameters must be fixed, and the V_3 parameters are optimized quite easily by fitting the rotational barriers about single bonds in simple molecules. So the HCCH term is obtained from the barrier in ethane, the CCCH term from propane, and the CCCC term from n-butane, to complement the part of the barrier due to the nonbonded potential. V_2 terms about double bonds are also easy to fit to reproduce the rotational barrier in ethylene of 272 kJ/mol, but V_1 and V_2 terms about single bonds are arrived at by a less transparent rationale, with a large number of molecules being employed for the parameterization.

The problems in the parameterization of force fields are best illustrated by the butane problem. It was noted some time ago that the

46 MOLECULAR MECHANICS

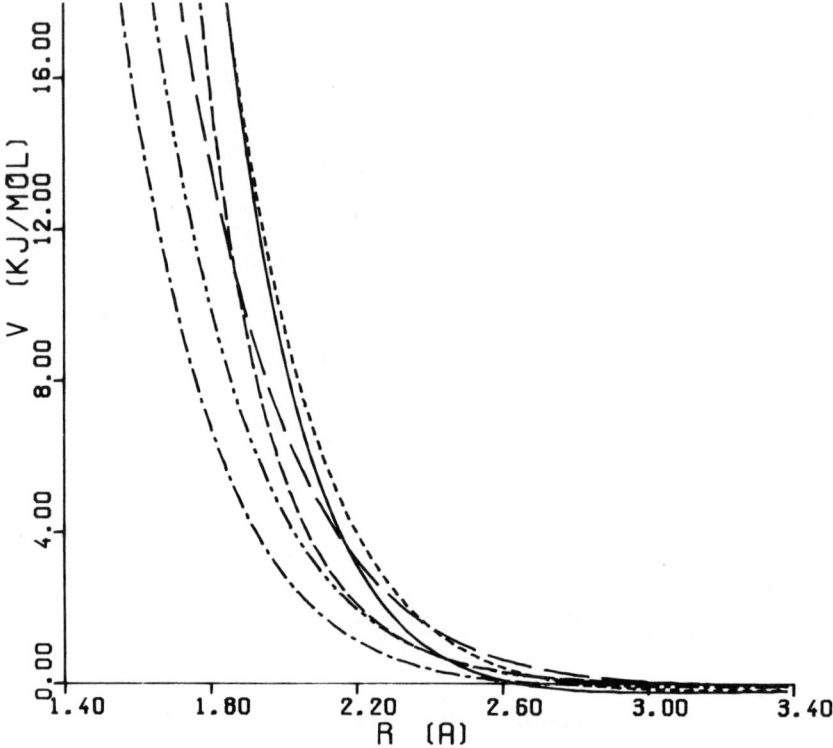

Figure 2.10. Van der Waals energy plots for H/H interactions from current force fields: ———, Allinger (MM2) 1977; — —, Bartell (MUB-2) 1977; – – –, Ermer/Lifson 1972; - - - -, Engler/Andose/Schleyer 1973; – · –, Bovill/White 1977; – · · –, Boyd 1969. Note that only in the force fields of Allinger and Bartell is the van der Waals center of hydrogen atoms relocated along the C–H bond.

calculation of the energy of *gauche*-butane relative to the *anti* conformation presented some significant difficulties. The early experimental values were not very accurate, and awareness of the problem came about somewhat gradually. An accurate value of 4.1 ± 0.2 kJ/mol (gas phase) was finally reported as recently as 1974 (*79*). A related quantity is the conformational energy of a methyl group on a cyclohexane ring. Here accurate numbers have been long known (*80*). Depending on the particular molecule, the value is about 7.1 kJ/mol in solution, and 8.0 kJ/mol in the gas phase. Simple considerations indicate that this value should be about twice the *gauche*-butane value, and this is now seen to be correct.

In his early force field (MUB-1, also sometimes referred to as JTB) (*81*), Bartell calculated a *gauche*-butane energy of 2.5 kJ/mol, which is in reasonable agreement with experiment. However, this number yielded 4.2 kJ/mol for the methylcyclohexane conformational energy, a number that was clearly much too small. Since Bartell was more interested in structures than in energies, he was not overly concerned with this point.

The Allinger group, on the other hand, was anxious to fit the methylcyclohexane value, since this number is really the cornerstone of conformational analysis. Within the constraints then employed in force fields, the only apparent way to do this was to make the hydrogens larger and/or harder. The energy of the *gauche*-butane can thus be increased to any desired amount, and in particular, the MM1 force field gave the axial methylcyclohexane energy as 6.6 kJ/mol (*15*). It was noted at the time that the structures of certain strained molecules that involve a good deal of hydrogen–hydrogen repulsion were not calculated in very good agreement with experiment. However, the differences were not so great that one could not question the experimental data. A key case was cyclodecane. The MM1 force field calculations yielded carbon–carbon–carbon bond angles that were a little larger than those found experimentally in crystals (but this could be blamed on crystal packing forces, since the errors were only about one degree). However, when a neutron diffraction study of a cyclodecane derivative became available, and the hydrogens in the interior of the ring could be located, it was clear that the MM1 force field calculated them to be too far apart, by about 0.18 Å. Hydrogens are rarely located with much precision, but this number is significantly beyond experimental error.

Schleyer, in developing his force field (EAS) (*14*), was aware that the Bartell force field gave better structures and used a small hydrogen, while the Allinger force field gave better energies and used a large hydrogen. Schleyer found a way around this dilemma by introducing another variable into the force field. This was the van der Waals interaction between carbon and hydrogen. Previously all authors had assumed that the interaction between carbon and hydrogen would be a mean of the interaction between two hydrogens and the interaction of two carbons, and theoretical considerations had indicated this was reasonable. Schleyer found that by making his hydrogen–carbon interaction very small, he could keep a small hydrogen, and still get a good energy for the *gauche*-butane interaction. The reason behind this can be seen in the following way. If we look at the Newman projection for butane in the *gauche* and *anti* conformations:

Me/Me	0	1
Me/H	4	2
H/H	2	3

We can see that there are interactions in these conformations as indicated. One of the basic ideas of conformational analysis as formulated by Barton was that the methyl groups exert a repulsion between each other, thereby destabilizing the *gauche* form (*82*). With the MM1 force field, it turns out that there is indeed a repulsion between the methyls, but there is an even bigger repulsion between the *gauche* hydrogens. The Bartell MUB-1 force field (*81*) avoids the difficulty by simply accepting an energy for the *gauche* conformation that is much too low. There are several ways to get the correct energy; the earliest was simply to use a large hydrogen, and then observe this "*gauche* hydrogen" effect.

Since the interactions between the hydrogens and methyls are not observable individually, but only the difference as it occurs in butane, Schleyer (*14*) noted that he could reduce the methyl–hydrogen interaction, thereby stabilizing the *anti* conformation (in which there are more Me/H interactions), and shift the equilibrium towards the *anti* form. By suitable adjustment of this methyl–hydrogen interaction, any desired value for the energy difference between the conformations could be achieved, and, therefore, small hydrogens could be used.

This scheme originally had only intuition going against it. Schleyer (*14*) found that the carbon–hydrogen interaction had to be very small indeed (comparable with the hydrogen–hydrogen interaction), and nowhere near the mean of the interactions as discussed above. While intuition suggests this is not reasonable, the force field model being developed here is empirical, and if a value works, it can be used. But in the long run, it doesn't work as well as one would wish. Looking at the heats of formation of a wide variety of hydrocarbons, the results with the Schleyer force field overall (standard deviation 3.47 kJ/mol) were not as good as those with the MM1 force field (2.55 kJ/mol) (Chapter 5). The obvious interpretation of these numbers is that the Schleyer force field does not really remove the error in the energy, but spreads it around so that it cannot be pinpointed as conveniently. In the MM1 force field the *gauche*-butane energy error is relatively small, whereas in the Bartell force field it is large and localized in the butane interaction. In the Schleyer force field, the error is still there, but cannot be said to reside in any one particular kind of interaction.

This situation of three force fields, all about equally good, but none as good as we would want, had some complicated consequences. If the individual energy terms were considered instead of the overall results, which were rather similar for the three force fields mentioned, say in *gauche*-butane, the following was observed. The MM1 force field attributed much of the *gauche*-butane energy to the hydrogen–hydrogen repulsion, whereas the Schleyer force field attributed it to other things. The Bartell MUB-1 force field attributed it to other things too, but we must remember that the Bartell force field gave a poor result for the *gauche*-butane energy. So, which of these force fields, if any, corresponds to physical reality?

Before we answer that question, let us recall the following. If the Schrödinger equation is solved with sufficient accuracy for butane in the *gauche* and *anti* conformations, the energy difference calculated agrees with experiment. What can we learn about the physical situation from looking at the details of such an ab initio calculation? The answer is, unless we have a preconceived idea we are trying to support, probably nothing. What we find is that in the calculation for each conformation there are literally millions of numbers, which sum to give the result observed. The final result should be accurate and physically correct and meaningful, but the individual (millions of terms) energies are not interpretable in any simple way. Nature only tells us that if we solve the Schrödinger equation, we will get the right answer. It does not tell us that we will find out all kinds of other things that we might like to know.

With a force field calculation, the situation is completely different. In this case, we are developing a mechanical model to represent our molecule. Here we can do the calculations exactly, and get answers that portray precisely the behavior of our model. The problem, however, is that our model may or may not represent accurately the molecule we are studying. Of course, the chemist wants his model to represent the molecule so that he can talk about bond energies, bond lengths, and so on as he always has. But, the nature of the empirical model is such that we are really talking about the model, and not about the molecule. There is no guarantee that we can ever learn anything about the molecule by this kind of study, but of course, we are going to assume that we can.

The next question usually asked is "which force field is the best?" The general answer to this question is, "it depends on what you are trying to calculate." People who devise force fields always devise them to solve certain kinds of problems. If you want to solve a particular kind of problem, the force field devised to solve it should work well. But if you want to solve a somewhat different kind of problem, then some other force field, which was designed to solve that particular kind of problem, will probably work better. So an attempt has been made from the beginning to construct very general force fields that would solve all kinds of problems. But in practice, different force fields have emphasized different kinds of data.

The consistent force field (CFF-3) (22, 66, 67, 69) was designed to calculate geometries, energies, and vibrational frequencies (for details see pp. 36–38), and in the last respect, CFF-3 is superior to other force fields. As will be discussed in pp. 169–173, the CFF-3 evaluates explicitly the vibrational contribution to the enthalpy. This is also true of the Wertz force field (72, 73), while the other force fields assume additivity of the vibrational energies. In principle the more detailed approach should improve the energy results significantly, but this has not been convincingly shown. The Wertz field gave better heats of formation than contemporary fields, but the importance of low-order torsional potentials was subsequently recognized, and heats of formation were further improved so that those from MM2 are comparable with those obtained by Wertz.

The philosophy of the consistent force field is certainly correct in principle. In practice, it has been developed by slighting a massive and important body of data, heats of formation. It can therefore be expected to give better results than other force fields with respect to vibrational frequencies, but poorer results with heats of formation. Boyd's force field was also "consistent" in that it included a full selection of different kinds of data (95). However, it was never really developed by including enough data.

With respect to the particular question of structure and energy, the early Schleyer, Bartell, and Allinger force fields, together with several others, were in reasonable agreement in their predictions. Each of them did a few things more poorly than would be liked, but on the whole they were quite good. The particular items that were poor varied from one force field to another.

More recently, what seems to be the "real" solution to the butane problem was uncovered independently and simultaneously by two research groups. This solution to the problem introduces low-order torsional terms (onefold and twofold barriers) into a molecule such as butane (*16, 56, 57*). When the MM1 force field was modified by first including this twofold torsional term, and then reducing the size of the hydrogen so as to properly fit the *gauche*-butane, and other structural data, it was found that a good force field for hydrocarbons was obtained (*16, 57*). Bartell approached the problem from the point of view of a force field that already gave good structures, and lacked only a low-order torsional function to give good energies (*56*). His force field now gives quite good structures, and also a good *gauche*-butane energy. The latest version of Bartell's force field (MUB-2, with low-order torsional terms added) (*25*) and of the Allinger force field (MM2) (*16*) are rather similar, at least when compared with previous force fields. These force fields give results that we believe are superior to all earlier force fields, as far as structure and energy are concerned.

To decide, then, if the *gauche* hydrogen effect (*15*) is important, we can only conclude as follows. The MM1 force field (*15*), in which this was an important effect, gave good results. That is, a model developed to reproduce the properties of molecules based on this interaction works fairly well. But, the MM2 force field reduces markedly the significance of this interaction, and attributes much of the energy to low-order torsional terms, rather than to van der Waals interactions. This is a much better model in the sense that the results more accurately portray what is observed experimentally in terms of structure and energy. There is still no guarantee that this is the best model (accepting that we can never know what is the correct model). But it now seems that we are approaching asymptotically a best molecular mechanics model. The approach is somewhat erratic, but it is leading to a convergence of existing force fields.

The importance of correctly fitting butane in establishing some of the basic parameters for molecular mechanics can hardly be overemphasized. Foregoing sections have discussed in detail the *gauche/anti* energy difference found experimentally (*79, 83*), and how it has been reproduced by various force fields. The remainder of the conformational energy surface is also worthy of comment. Specifically, we might inquire as to the heights of the two barriers (corresponding to CCCC torsion angles of 0° and approximately 120°). The second of these is relatively low, similar to that found in propane (13.8 kJ/mol). The height of the 120° barrier (above the *anti* form) has been experimentally measured by ultrasonic relaxation (*84*), and by extrapolation of spectroscopic data from the torsional levels (*83*), and it has been calculated by ab initio methods (*85, 86*). All of these methods are in approximate agreement, assigning a value of 13.4–15.2 kJ/mol to this barrier.

The height of the 0° barrier above the *anti* form is much less well known. It has not been measured by any direct experiment. Extrapolation of vibrational data from the torsional levels gives a value of 18.9

kJ/mol (83). The extrapolations here are subject to considerable uncertainty. The potential function developed contains significant V_5 and V_6 terms, but lacks a V_2 term. While it fits the observed data well, it is not apparent how the molecule can have such characteristics. Rather, it would seem that the V_5 term in particular must occur in the Fourier series to make up for other errors, for example, those introduced by the rigid rotor approximation used in analysis of the spectroscopic data.

Several ab initio calculations of this barrier have been carried out. The full potential surface was studied in a minimum basis set calculation by Peterson and Csizmadia (86). All of the expected features regarding maxima and minima were found, but the rigid rotation approximation was used, and grid points were spaced at 30° intervals in torsion angles. Hence, the results can be taken as only semiquantitative. The *gauche*–*gauche* barrier height was 33.5 kJ/mol. While the rigid rotation is a fair approximation for the smaller barrier, because there is very little bending and stretching in that case, the approximation is much poorer for the higher barrier. Here the relaxation is quite important, and the rigid rotation value is clearly too high. A more realistic number was obtained for the barrier height based on molecular mechanics geometries for the energy minima and maxima. A minimum basis set Hartree–Fock calculation gave the value 25.1 kJ/mol (85). This value did not change much with increasing basis set size (26.6 kJ/mol at the 6–31G level) and 27.1 kJ/mol after polarization functions were added (96). It was predicted (85) that electron correlation would be important here, because of the large methyl–methyl repulsion in the eclipsed conformation. Indeed, a configuration interaction calculation including 10^4 singly and doubly excited configurations (one-third of the total) yielded an energy difference of 19.2 kJ/mol, in good agreement with the spectroscopic value and the MM2 value (19.8 kJ/mol). These ab initio results have important implications regarding energies obtained in this way in congested molecules.

Force Fields for Molecules Containing Delocalized Pi Electrons

Conjugated systems pose special problems not discussed in earlier sections. The nature of these can be seen by examining two typical molecules, benzene and naphthalene. As previously described, suitable bond lengths and other parameters can be chosen so that the structure of benzene is correctly reproduced. If these parameters are then applied to the naphthalene molecule, the calculated bond lengths are essentially all the same, and the same as in benzene. Experimentally, it is found that certain bond lengths in naphthalene are about two-thirds of the way between a single and a double bond, while other bond lengths are one-third of the way, as is predictable from the Kékulé forms where each bond is either double in two and single in one, or the reverse. Thus, the benzene parameters do not reasonably reproduce the naphthalene structure. The reason for the failure of the benzene parameters is clear enough. In naphthalene, the bond lengths are determined largely by the bond orders. In benzene, the bond orders are all equal and were not included in the parameterization. But to correctly reproduce the naphthalene structure, and the structures of conjugated molecules in general, the effects of bond order variation in different molecules must be considered. This problem was recognized at an early date, and two somewhat different solutions to it have been proposed. Both have been found to be satisfactory and will be outlined here. Both depend upon the usual molecular mechanics procedures as well as the inclusion of a pi system molecular orbital treatment in the calculations.

The most straightforward approach to the problem, in principle, was that used by Warshel and Karplus (87, 88). The structure in question, for example, naphthalene, starts with an initial approximate structure, and a self-consistent field (SCF) calculation on the pi system (Pariser–Parr–Pople or PPP calculation) is carried out, and separately a molecular mechanics treatment of the sigma system is done. Because the bond length deformations in the conjugated olefin part of the molecule are considered to be outside the range of the harmonic approximation, a Morse function is used for stretch deformations there. The energies of the pi and sigma systems are obtained and are summed. The structural optimization is then carried out to minimize this sum. This requires that the SCF energy of the pi system be obtained as an analytic function of the molecular coordinates so that the first and second derivatives may be obtained.

To the level of approximation of the SCF method (essentially a semiempirical Hartree–Fock limit), the procedure gives good results for planar systems. To account for effects of nonplanarity of the pi system, the PPP calculation was modified by including nearest neighbor overlap. Since the virtual orbitals of the pi system as well as the occupied orbitals are obtained, excited pi states can also be studied by this scheme. For such calculations a limited configuration interaction treatment is included. The parameter set was devised to reproduce several different molecular properties, including atomization energies, electronic excitation and ionization energies as well as structures and vibrational frequencies. This approach has not been extensively applied to the calculations of structures or heats of formation.

The alternative approach is that of Allinger and Sprague (MMP1) (89). This method was devised to avoid the necessity of assuming that the pi–sigma separation condition is met for nonplanar systems. Beginning with an initial geometry, a SCF calculation is carried out on the pi system. If the pi system is nonplanar, it is flattened into a planar form for purposes of the SCF calculation, and from the coefficients of the wave functions obtained, the bond orders are calculated in the usual way. It is then assumed that there are linear relationships between the bond order of a bond and the following three quantities: the stretching force constant, the natural bond length, and the torsional barrier. The necessary force constants and bond lengths are calculated using these relationships, then used in the molecular mechanics force field. From this point on, the quantum mechanical calculation is no longer considered. The structural optimization then proceeds in the usual way, since at this point the problem has been reduced to the ordinary case.

From the above it is apparent that correct bond lengths should be obtained for molecules like naphthalene; this is found to be so. Perhaps of greater interest, however, are the results that are obtained when the system is deformed from planarity. In a planar system, the pi–sigma separation is a good approximation, which has been well studied and justified. As the system is deformed from planarity, this approximation breaks down. In the Allinger–Sprague approach, it is assumed that the pi–sigma separation is valid, and the force constants are obtained for the planar system. If the system is deformed from planarity, the force constants do not change, we simply move to a different place on the potential energy curve. Hence, nonplanar systems also should be well treated by this method, and this is found to be the case, at least for small or moderate distortions from planarity.

In the Allinger–Sprague (MMP1) approach, the use of a linear relationship between bond order and natural bond length, and between bond order and stretching force constant, is straightforward and leads to good results for planar systems. More difficult is the problem of the torsional potential function about a conjugated bond. In the original work a quantity called the conjugation energy, E_c, was defined for each bond, which was equal to a constant times the square of the bond order for that particular bond. The rationale was that the binding energy for this bond should be proportional to $2p_{ij}\beta_{ij}$ (where p_{ij} and β_{ij} are, respectively, the bond orders and resonance integrals between atoms i and j), the factor of two arising from the fact that the orbital is doubly occupied (two electrons). As the bond is twisted, both p_{ij} and β_{ij} will approach zero in general. Therefore, E_c is approximately proportional to p_{ij}^2. Rotational barriers for ethylene, stilbene, and butadiene were fit against this function, and a proportionality constant was obtained. This scheme worked reasonably well for a great variety of nonplanar systems. In more recent work (MMP2) this relationship has been modified slightly. (The program MMP2 will be available from the Quantum Chemistry Program Exchange. See Appendix.) In ethylene, for example, the pi bond order is always 1.0, regardless of the distance between the atoms. Conjugation energy, as defined above, therefore, is also independent of the distance between the atoms. What we really want is a conjugation energy that will fall off to zero as the atoms separate, and this is conveniently done by setting

$$E_c = kp_{ij}\ \beta_{ij}$$

as perturbation theory requires, where k is an adjustable parameter.

At the time of the original Allinger–Sprague work (1969–1970) the rotational function for butadiene was not known very accurately. The studies by Carreira (90) subsequently gave this function much more accurately, and the more recent function was reproduced in the MMP2 work.

The original Allinger–Sprague method (MMP1 program) utilized the variable electronegativity self consistent field (VESCF) (91, 92) calculation for the pi system. In this method the unsaturated carbon atoms are not given the atomic value for the carbon p-orbital ionization potential, but rather a value that depends on the substituents attached to the carbon, and also on the pi electron density at the carbon. While this modification is superior to the standard SCF scheme for the calculation of electronic spectra, the two methods are essentially the same as far as the calculation of structures. The MMP1 treatment gave good structures, but it proved to be difficult to calculate heats of formation from this point. The calculations were possible, but the parameter set developed (93) was not consistent with that developed for unconjugated molecules.

In the later improved version of the treatment (MMP2), the standard SCF scheme was used to calculate structures, and a consistent parameter set (for conjugated and unconjugated systems) that gave very good heats of formation was developed. If electronic spectra are desired, the structure is first determined (MMP2) and then a standard VESCF calculation (91, 92) is carried out on that structure. While the idea of one parameter set (87, 89) that is suitable for calculations of both ground and excited states is attractive, our current view is that the semiempirical MO parameterization can be done better for the ground state if the excited states do not need to be considered simultaneously, and that the two-stage approach is superior.

Literature Cited

1. Wilson, E. B., Jr.; Decius, J. C.; Cross, P. C. "Molecular Vibrations"; McGraw-Hill: New York, 1955.
2. Gans, P. "Vibrating Molecules"; Chapman & Hall: London, 1971.
3. Herzberg, G. "Infra-Red and Raman Spectra"; Van Nostrand-Reinhold: New York, 1945.
4. Fadini, A. "Molekülkraftkonstanten"; Steinkopff: Darmstadt, 1976.

5. Ibid., p. 472.
6. Wilson, E. B., Jr.; Decius, J. C.; Cross, P. C. "Molecular Vibrations"; McGraw-Hill: New York, 1955; p. 26.
7. Schachtschneider, J. H.; Snyder, R. G. *Spectrochim. Acta* **1963**, *19*, 117.
8. Snyder, R. G.; Schachtschneider, J. H. *Spectrochim. Acta.* **1965**, *21*, 169.
9. Califano, S. *Pure Appl. Chem.* **1969**, *18*, 353.
10. Snyder, R. G.; Zerbi, G. *Spectrochim. Acta Part A* **1967**, *23*, 391.
11. Snyder, R. G.; Schachtschneider, J. H. *J. Mol. Spectrosc.* **1969**, *30*, 290.
12. Urey, H. C.; Bradley, C. A., Jr. *Phys. Rev.* **1931**, *38*, 1969.
13. Hagler, A. T.; Stern, P. S.; Lifson, S.; Ariel, S. *J. Am. Chem. Soc.* **1979**, *101*, 813.
14. Engler, E. M.; Andose, J. D.; Schleyer, P. v. R. *J. Am. Chem. Soc.* **1973**, *95*, 8005.
15. Wertz, D. H.; Allinger, N. L. *Tetrahedron* **1974**, *30*, 1579.
16. Allinger, N. L. *J. Am. Chem. Soc.* **1977**, *99*, 8127.
17. Wiberg, K. B. *Comput. Chem.* **1977**, *1*, 221.
18. Wilson, E. B., Jr.; Decius, J. C.; Cross, P. C. "Molecular Vibrations"; McGraw-Hill: New York, 1955; p. 58.
19. Allinger, N. L.; Tribble, M. T.; Miller, M. A. *Tetrahedron* **1972**, *28*, 1173.
20. Allinger, N. L.; Sprague, J. T. *Tetrahedron* **1975**, *31*, 21.
21. Winkler, F. K.; Dunitz, J. D. *J. Mol. Biol.* **1971**, *59*, 169.
22. Ermer, O.; Lifson, S. *J. Am. Chem. Soc.* **1973**, *95*, 4121.
23. Allinger, N. L.; Tribble, M. T.; Miller, M. A.; Wertz, D. H. *J. Am. Chem. Soc.* **1971**, *93*, 1637.
24. Ermer, O. *Tetrahedron* **1974**, *30*, 3103.
25. Fitzwater, S.; Bartell, L. S. *J. Am. Chem. Soc.* **1976**, *98*, 5107.
26. Fitts, D. D. *Ann. Rev. Phys. Chem.* **1966**, *17*, 59.
27. Buckingham, A. D.; Utting, B. D. *Ann. Rev. Phys. Chem.* **1970**, *21*, 287.
28. Hirschfelder, J. O., Ed. *Adv. Chem. Phys.* **1967**, *12*.
29. Margenau, H.; Kestner, N. R. "Theory of Intermolecular Forces"; Pergamon: Oxford, 1969.
30. Hirschfelder, J. O.; Curtiss, C. F.; Bird, R. B. "The Molecular Theory of Gases and Liquids"; John Wiley & Sons: New York, 1954.
31. Böttcher, C. J. "Theory of Electric Polarization"; Elsevier: Amsterdam, 1952.
32. Lehn, J. M.; Ourisson, G. *Bull. Soc. Chim. Fr.* **1963**, 1113.
33. Lennard-Jones, J. E. *Proc. R. Soc. London, Ser. A* **1924**, *106*, 463.
34. Hill, T. L. *J. Chem. Phys.* **1948**, *16*, 399.
35. Williams, J. E.; Stang, P. J.; Schleyer, P. v. R. *Ann. Rev. Phys. Chem.* **1968**, *19*, 531.
36. Pauling, L. "The Nature of the Chemical Bond"; Cornell Univ. Press: Ithaca, 1960.
37. Coppens, P.; Row, T. N. G.; Leung, P.; Stevens, E. D.; Becker, P. J.; Yang, Y. W. *Acta Crystallogr. Sect. A* **1979**, *35*, 63.
38. Williams, D. E. *J. Chem. Phys.* **1965**, *43*, 4424.
39. Williams, D. E.; Starr, T. L. *Comput. Chem.* **1977**, *1*, 173.
40. Allinger, N. L.; Chung, D. Y. *J. Am. Chem. Soc.* **1976**, *98*, 6798.
41. Burkert, U. *Tetrahedron* **1977**, *33*, 2237.

42. Mason, E. A.; Kreevoy, M. M. *J. Am. Chem. Soc.* **1955**, *77*, 5808.
43. Van-Catledge, F. A., unpublished data.
44. Burgess, A. W.; Shipman, L. L.; Nemenoff, R. A.; Scheraga, H. A. *J. Am. Chem. Soc.* **1976**, *98*, 23.
45. Snir, J.; Nemenoff, R. A.; Scheraga, H. A. *J. Phys. Chem.* **1978**, *82*, 2497.
46. Nemenoff, R. A.; Snir, J.; Scheraga, H. A. *J. Phys. Chem.* **1978**, *82*, 2504.
47. Kitaigorodski, A. I. *Chem. Soc. Rev.* **1978**, *7*, 133.
48. Wilson, E. B., Jr. *Adv. Chem. Phys.* **1959**, *2*, 367.
49. Cignitti, N.; Allen, T. L. *J. Phys. Chem.* **1964**, *68*, 1292.
50. Lowe, J. P. *Prog. Phys. Org. Chem.* **1968**, *6*, 1.
51. Lowe, J. P. *Science* **1973**, *179*, 527.
52. Payne, P. W.; Allen, L. C. In "Applications of Electronic Structure Theory"; Schaefer, H. F., Ed.; Modern Theoretical Chemistry Vol. 4; Plenum: New York, 1977; p. 29.
53. Pitzer, R. M.; Lipscomb, W. N. *J. Chem. Phys.* **1963**, *39*, 1995.
54. Brunck, T. K.; Weinhold, F. *J. Am. Chem. Soc.* **1979**, *101*, 1700.
55. Gavezzotti, A.; Bartell, L. S. *J. Am. Chem. Soc.* **1979**, *101*, 5142.
56. Bartell, L. S. *J. Am. Chem. Soc.* **1977**, *99*, 3279.
57. Allinger, N. L.; Hindman, D.; Hönig, H. *J. Am. Chem. Soc.* **1977**, *99*, 3282.
58. Radom, L.; Hehre, W. J.; Pople, J. A. *J. Am. Chem. Soc.* **1972**, *94*, 2371.
59. David, S.; Eisenstein, O.; Hehre, W. J.; Salem, L.; Hoffmann, R. *J. Am. Chem. Soc.* **1973**, *95*, 3806.
60. Jeffrey, G. A.; Pople, J. A.; Radom, L. *Carbohydr. Res.* **1974**, *38*, 81.
61. Burkert, U. *Tetrahedron* **1979**, *35*, 1945.
62. Profeta, S., Jr., Ph.D. Thesis, Univ. of Georgia, 1978.
63. Allinger, N. L.; Sprague, J. T. *J. Am. Chem. Soc.* **1972**, *94*, 5734.
64. Burkert, U. *Chem. Ber.* **1977**, *110*, 773.
65. Warshel, A.; Lifson, S. *Chem. Phys. Lett.* **1969**, *4*, 255.
66. Lifson, S.; Warshel, A. *J. Chem. Phys.* **1968**, *49*, 5116.
67. Ermer, O. *Struct. Bonding (Berlin)* **1976**, *27*, 161.
68. Warshel, A. In "Semiempirical Methods of Electronic Structure Calculation"; Segal, G. A., Ed.; Modern Theoretical Chemistry Vol. 7; Plenum: New York, 1977; p. 133.
69. Niketic, S. R.; Rasmussen, K. "The Consistent Force Field"; Springer: New York, 1977.
70. Warshel, A.; Lifson, S. *J. Chem. Phys.* **1970**, *53*, 582.
71. Hagler, A. T.; Lifson, S. *Acta Crystallogr., Sect. B* **1974**, *30*, 619.
72. Wertz, D. H.; Allinger, N. L. *Tetrahedron* **1979**, *35*, 3.
73. Wertz, D. H., Ph.D. Thesis, Univ. of Georgia, 1974.
74. Hagler, A. T.; Huler, E.; Lifson, S. *J. Am. Chem. Soc.* **1974**, *96*, 5319.
75. Wiberg, K. B. *Comput. Chem.* **1977**, *1*, 221.
76. Allinger, N. L.; Miller, M. A.; Van-Catledge, F. A.; Hirsch, J. A. *J. Am. Chem. Soc.* **1967**, *89*, 4345.
77. Nemethy, G.; Scheraga, H. A. *J. Phys. Chem.* **1977**, *81*, 928.
78. Kochanski, E. *J. Chem. Phys.* **1973**, *58*, 5823.
79. Verma, A. L.; Murphy, U. F.; Bernstein, H. J. *J. Chem. Phys.* **1974**, *60*, 1540.
80. Eliel, E. L.; Allinger, N. L.; Angyal, S. J.; Morrison, G. A. "Conformational Analysis"; Wiley-Interscience: New York, 1965; p. 43.

81. Jacob, E. J.; Thomson, H. B.; Bartell, L. S. *J. Chem. Phys.* **1967,** *47,* 3736.
82. Eliel, E. L.; Allinger, N. L.; Angyal, S. J.; Morrison, G. A. "Conformational Analysis"; Wiley-Interscience: New York, 1965; p. 13.
83. Compton, D. A. C.; Montero, S.; Murphy, W. F. *J. Phys. Chem.* **1980,** *84,* 3587.
84. Piercy, J. E.; Rao, M. G. S. *J. Chem. Phys.* **1956,** *25,* 943.
85. Allinger, N. L.; Profeta, S., Jr. *J. Comput. Chem.* **1980,** *1,* 181.
86. Peterson, N. R.; Csizmadia, I. G. *J. Am. Chem. Soc.* **1978,** *100,* 6911.
87. Warshel, A.; Karplus, M. *J. Am. Chem. Soc.* **1972,** *94,* 5612.
88. Warshel, A. *Comput. Chem.* **1977,** *1,* 195.
89. Allinger, N. L.; Sprague, J. T. *J. Am. Chem. Soc.* **1973,** *95,* 3893.
90. Carreira, L. A. *J. Chem. Phys.* **1975,** *62,* 3851.
91. Allinger, N. L.; Tai, J. C. *J. Am. Chem. Soc.* **1965,** *87,* 2081.
92. Allinger, N. L.; Tai, J. C.; Stuart, T. W. *Theor. Chim. Acta* **1967,** *8,* 101.
93. Kao, J.; Allinger, N. L. *J. Am. Chem. Soc.* **1977,** *99,* 975.
94. Ermer, O.; "Aspekte von Kraftfeldrechnungen"; Wolfgang Bauer Verlag: Munich, 1981.
95. Chang, S.; McNally, D.; Shary-Tehrany, S.; Hickey, M. J.; Boyd, R. H. *J. Am. Chem. Soc.* **1970,** *92,* 3109.
96. Van-Catledge, F. A.; Allinger, N. L. *J. Am. Chem. Soc.* **1982,** in press.

Methods for the Computation of Molecular Geometry

Outline of the Computations. The Representation of Molecular Structure

Calculation of the energy of one molecular geometry using classical potential functions is so fast that, unlike the situation with quantum mechanical calculations, geometry optimization with respect to energy (energy minimization) is standard in molecular mechanics. The necessity for energy minimization is generally acknowledged. Energies calculated with standard geometries are usually only qualitatively correct. Geometry relaxation changes the energies of the different conformations by different amounts, with consequences for conformational equilibria, especially in strained molecules. The largest energy changes from relaxation are usually found at transition states for conformational interconversions. Rotational barriers, for example, are calculated to be far too large, often by several hundred percent, with the rigid rotor approximation.

Gradient techniques for the location of energy minima and transition states were applied in molecular mechanics shortly after digital computers first became available (*1*). These early calculations were, however, carried out with programs tailored to the particular molecules studied. The first general computer program for energy minimization was Wiberg's, based on the steepest descent method (*2*). This method was explored in depth in the 1960s (*3–6*), while at the same time methods utilizing Gauss–Newton and Newton–Raphson algorithms were developed (*7–9*). The details of these minimization techniques will be discussed in pp. 64–72, and applications for the calculation of transition state geometries and energies are dealt with in pp. 72–76.

Some important points concerning the validity of the geometry found by any energy minimization technique may be noted here. Energy minimization is an iterative geometry *optimization,* so unless there is only one potential well for a molecule, the minimum energy geometry obtained will depend on the starting geometry of the optimization, that is, in which potential well we begin with our rough input structure. No known general method finds the global energy minimum; what are found are local minima. In a thorough study, a number of reasonable conformations will serve as trial geometries, such as the chair, boat, and twist-boat conformations for cyclohexane, but in large molecules, the number of necessary trial conformations may be very large. Which ones are actually chosen will usually be determined from Dreiding models, or other similar considerations, but in this way important conformations may be overlooked. So unless a systematic study of all geometrically feasible structures is carried out, the minimum energy conformation found may depend on which conformations seemed appropriate in the eyes of the investigator.

Another important point to note here is that most energy minimization procedures do not, strictly speaking, locate energy minima, but rather they locate extrema, or points on the potential surface where the first derivatives of the potential energy are zero (so-called stationary points). Therefore, such procedures may lead to a saddlepoint on the potential surface rather than to a minimum. Thus, the classical boat conformation of cyclohexane (C_{2v}) will appear as an "energy minimum," as will a conformation that has all six carbon atoms in a plane (D_{6h}). With most methods, the structure obtained will be that corresponding to the stationary point nearest to the starting structure. Thus, the results of conformational analysis by molecular mechanics (and many other methods) will depend upon the trial structures.

The effect of the symmetry of the trial geometry should also be mentioned here. Some minimization techniques do not reduce the symmetry of the system, and if a plane or axis of symmetry is present in the trial geometry, on purpose or by chance, or if it is introduced during the course of the calculations, it may remain, because a symmetrical structure often corresponds to a saddlepoint between two less symmetrical (mirror-image) structures. Most reliable results are obtained from a large number of trial structures with low symmetry. If a minimum is in fact symmetrical, the minimization procedure from an unsymmetrical structure will find it, but under some circumstances, the reverse will not be true. We will return to this point when the different energy minimization techniques are discussed (pp. 64–72).

Occasionally it may save much time to carry out the energy minimization not from different starting geometries, but rather from a single geometry, and to employ methods for the calculation of conformational interconversions (pp. 72–76), for example, the "driving" of a torsion angle, to find other low energy conformations. This method was used by Anet and Yavari, for example, in the elucidation of the conformations of 1,4-cyclooctadiene. A boat–chair and a twist-boat conformation were found as the low energy forms, and actually do correspond to energy minima (10). Earlier starting geometries had been taken from Dreiding models (11), and the boat–boat instead of the twist-boat conformation resulted from the energy minimization; when this symmetrical geometry was slightly twisted, it returned to a symmetrical geometry, and so was thought to be an energy minimum. However, it is actually a saddlepoint (10).

The final structure obtained from geometry optimization is, of course, sensitive to the criteria used for deciding when to stop the iterative procedure. Usually the criterion has been a small energy threshold. When the energy decrease per iteration is less than this value, the energy minimization is terminated. Alternatively, a geometry change threshold has been used. If the slope of the potential surface is shallow, either of these criteria may cause the minimization to stop before the energy minimum has been reached. The obvious solution is then to lower the energy threshold. The practical problem is that a great deal of computer time may then be expended, with no significant change in geometry or energy. In practice, some compromise is chosen. It has also been proposed that the actual values of the derivatives of the energy with respect to the coordinates be examined to see how closely they have approached zero, that is, the use of a first-derivative threshold (12), but this method does not appear to offer any significant advantage.

One of the main differences between molecular mechanics computer programs and programs for quantum mechanical calculations is that, in the former, the handling of large amounts of structural information for bonds, bond angles, and torsion angles is involved. To make such programs convenient, the amount of input data should be minimal; also it is preferable to have great versatility available for modifying the force field. While the first condition is easily met by careful programming, the possibilities for varying the force field are usually limited. The various force fields that have been developed generally contain not only different parameters, but also slightly different potential functions, which makes it difficult to transfer parameters from one force field to another. While it is easy to read changed parameters into a calculation, changing a potential function in most programs is a sizable reprogramming task. Most available programs are consequently tailored to one force field. Programs have been reported (13, 14, 41) that contain a large number of available potential functions, from which selections may be made at the time the structural input data are read in, allowing fairly full flexibility in the calculation.

The flow diagram of a general molecular mechanics computer program is given in Figure 3.1. Several such programs are now available from the *Quantum Chemistry Program Exchange* (*see* Appendix). Figure 3.1 shows as the minimum input information the types of atoms and their trial coordinates that define the molecular structure. The coordinates may be either internal or Cartesian, and can be easily estimated

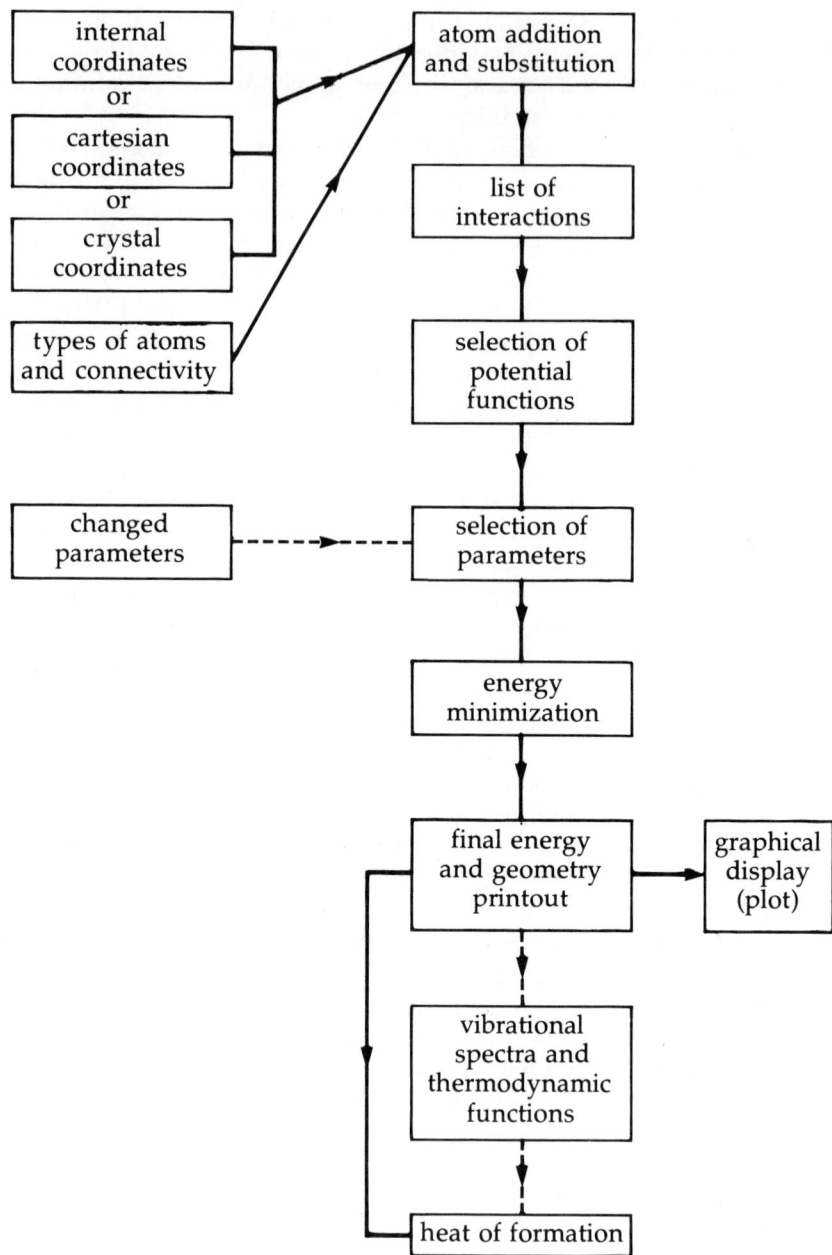

Figure 3.1. Flow diagram of a general molecular mechanics computer program. Input information is given on the left side.

from Dreiding models (for Cartesian coordinates a precision of ±0.3 Å is sufficient). Often the trial structures are available from x-ray fractional unit cell coordinates, which can be converted readily into Cartesian coordinates. It is easy to define the molecular geometry by $3n - 6$ independent internal coordinates for acyclic molecules; the bond lengths, all independent bond angles at each polyvalent atom (five out of six at tetravalent centers), and one torsion angle per rotational degree of freedom constitute one simple possibility. In rings, however, and especially in polycyclic molecules, it may become very troublesome to define the set of independent internal coordinates, because bond angles and torsion angles are correlated, which is not the case with Cartesian coordinates. Vector formulas for the interconversion of Cartesian and internal coordinates are given in Equations 3.1 and 3.2 (details can be found in any book on vector analysis).

Bond angles: $$\cos \Theta_{ijk} = -\frac{r_{jk} \cdot r_{ij}}{r_{jk} r_{ij}} \qquad (3.1)$$

Torsion angles: $$\cos \omega_{ijkl} = \frac{(r_{jk} \times r_{ij}) \cdot (r_{kl} \times r_{jk})}{r_{ij} r_{jk}^2 r_{kl} \sin \Theta_{ijk} \sin \Theta_{jkl}} \qquad (3.2)$$

For calculation of the energy, the list of all internal coordinates must be generated (all bond distances, bond angles, torsion angles, and nonbonded distances). To calculate internal from Cartesian coordinates, the connectivity must be known, and it is ordinarily given when internal coordinates are read in. In principle, the connectivity can be derived from the Cartesian coordinates by searching for reasonable bond distances (13, 15). It is safer, especially when using rough input coordinates, to read in the connectivity, for example, as a connected atom list or matrix (16).

An extremely simple input form is possible when the technical facilities of a writing tablet are available. Programs have been developed that convert a structural formula written on such a device into Cartesian coordinates (17). Using a molecular mechanics program can be reduced in the general case to drawing a chemical formula, with some notation of the stereochemistry.

Finally, not all atomic coordinates must be given as input, but only those of the molecular backbone. Substituent atoms in the trial structure (e.g., all hydrogen atoms) can be calculated automatically from standard geometric parameters, and this may also be done for frequently occurring substituent groups such as methyl or phenyl. By replacing successive hydrogens by methyl groups, even large and complicated alkyl groups can be generated easily.

Next, the computer program constructs a list of interaction terms. It searches through the molecular graph for connected atoms (bond distances), atoms attached to a common atom (bond angles), and atoms attached to adjacent atoms (torsion angles), as well as nonbonded interactions (separated into 1-3 and other nonbonded interactions), and picks the force field parameters for these interactions from a list stored in the program. At this step the force field parameters may be modified by input statements. Then the energy of the input geometry can be calculated, or the energy minimization process can begin.

The question as to whether or not energy minimization is to be performed on an independent set of internal coordinates or on Cartesian coordinates is, in most cases, decided in favor of the latter. The large number of internal coordinates that must be calculated for use in the potential functions is obtained more easily from Cartesian coordinates than from the internals, and calculation time as well as programming time is shorter with Cartesian coordinates. Currently, only one program for complete energy minimization (7, 14), which was originally designed to calculate properties of acyclic molecules, minimizes the energy using internal coordinates. Such a minimization is useful for the calculation of rotational barriers and of other transition states of noncyclic molecules (14), because the torsion angle defining the rotation can easily be fixed. More complicated procedures (pp. 72–76) may be required when Cartesian coordinates are used. If energy minimization is done on torsion angles using only rigid bond lengths and angles (and many such calculations have been done on relatively large molecules like peptides and others, discussed in Chapter 7), energy minimization is, of course, done on internal coordinates.

Energy Minimization

The methods that have been developed to find a molecule's equilibrium structure, at which the condition for the first derivatives of the potential energy, V, Equation 3.3, holds simultaneously for each coordinate x_i, may be divided into first derivative and second derivative minimizations. Simple search techniques (18) make use of only the slope of the potential surface (the first derivative), which may be determined numerically or analytically, whereas the more sophisticated minimization techniques utilize both the slope and the curvature of the potential surface (first and second derivatives).

$$\frac{\partial V}{\partial x_i} = 0 \tag{3.3}$$

The principles of these methods are easily understood in a one-dimensional minimization problem (Figure 3.2). The objective is to find x, the coordinate where the energy ($f(x)$) is a minimum, starting from x_0, an initial guess of the value of x. The approach of the first derivative method is to calculate $f(x_0)$ and, in numerical methods like the **steepest descent**, to compare it with a value $f(x_0 + \Delta x)$. The sign of the change in the value the function gives the direction of the minimum, and its absolute value indicates how large a step toward the minimum is to be taken. In the analytical approach, the first derivative $f'(x_0)$ gives the direction and absolute value of the correction step. The step size also depends on an undetermined scaling parameter, which in turn depends on the curvature of the function. Using condition (Equation 3.3), the minimization problem can also be formulated as an iterative determination of the value of x for which $f'(x) = 0$. Curvature information is utilized for this purpose in the second derivative minimization generally known as the **Newton–Raphson** (NR) minimization technique. In this method, which can be deduced by expanding $g(x) = f'(x)$ in a

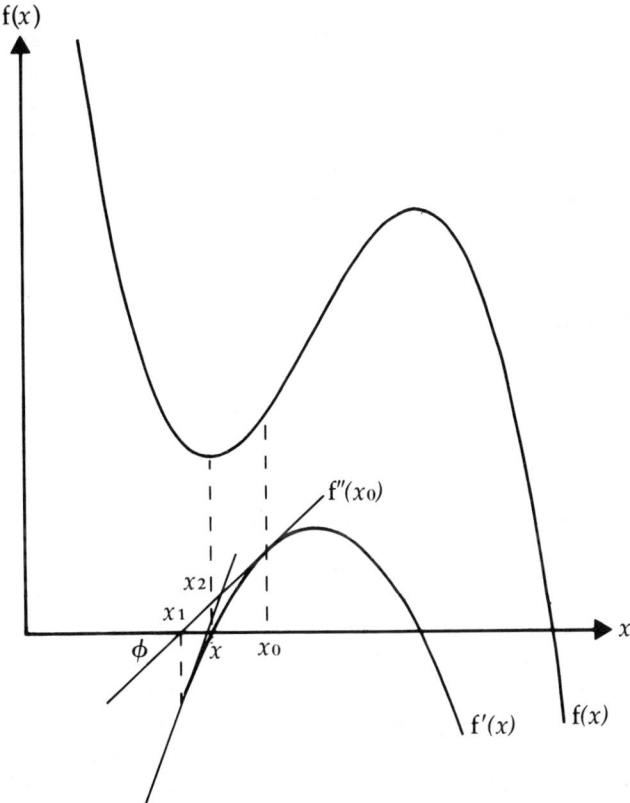

Figure 3.2. Minimization of a function f(x) by the Newton–Raphson procedure. Iteration starting at x_0 improves the solution of $f'(x) = 0$ successively to x_1 and to x_2, approaching the true solution x.

Taylor series and truncating after the first derivative term, an improved solution is found (Figure 3.2) as Equation 3.4. The geometric construction shows that

$$x_1 = x_0 - \frac{g(x_0)}{g'(x_0)} = x_0 - \frac{f'(x_0)}{f''(x_0)} \qquad (3.4)$$

$$\tan \varphi = f''(x_0) = \frac{f'(x_0)}{\Delta x} \quad \text{or} \quad \Delta x = \frac{f'(x_0)}{f''(x_0)}$$

The methods are applied in molecular mechanics for the minimization of the sum of potential energy functions with $3n - 6$ coordinates as variables.

Algorithms other than the steepest descent and Newton–Raphson methods have rarely been used for this purpose (19–22). Occasionally, the Simplex algorithm has been applied (20). The Davidon–Fletcher–Powell algorithm has become popular in quantum mechanical energy minimization procedures because it provides a fast, though only approximate, calculation of second derivatives. In molecular mechanics, it is easy to calculate second derivatives analytically for a Newton–Raphson procedure. The Davidon–Fletcher–Powell algorithm has not been competitive here because procedures using precisely calculated second derivatives converge much faster than those using very approximate ones, and it has been used only occasionally (21, 22).

First Derivative Techniques

To approach the energy minimum in $3n - 6$ dimensional space, the atoms must be moved in the direction of the energy surface descent as defined by the force field. Wiberg, in the first general computer program for energy minimization, used the steepest descent method with numerically calculated first derivatives of the energy (2). As in the one-dimensional case, the energy was calculated for the initial coordinates, and then again when an atom was moved in turn by small increments in the directions of the three coordinate axes. This is, in essence, a numerical determination of the first derivatives. The process was repeated for all atoms, which were finally moved simultaneously to new positions downhill on the energy surface by an amount dictated by the steepness of the surface (the energy change found before). The entire sequence was repeated until the predetermined minimum condition was fulfilled.

The steepest descent method converges rapidly and reliably when the derivatives are large (i.e., one is far from the minimum), but convergence is slow when the derivatives are small, which they are when close to the minimum or in a valley with a small slope. A modified steepest descent technique, called a pattern search, was applied by

Schleyer et al. (*18, 23*), and compares favorably in this regard. Here the computer program "remembers" its last step and uses this additional information to accelerate movement, especially along valleys in the energy surface. As before, the direction of steepest descent is sought for one atom, which is then moved to an improved position, but the size and direction of this movement is saved. After the program has cycled through all atoms, in the next cycle the atoms are moved by the sum of the new steepest descent step plus the last step, and so on, thus accumulating a moving pattern. A pattern is abandoned when the direction of steepest descent changes.

The main problem with steepest descent techniques is their scale dependence. There are two critical processes, the numerical determination of the first derivatives and the step of moving the atoms. If the search increments are too large, minimization—especially that of torsional strain—is hampered. In the latter case the energy increase due to bond stretching and bond angle bending may easily be larger than the energy gained from relieving torsional strain (*3*), and torsion angles were often not well minimized by this procedure. Reducing the increment used for the search [0.002 Å (*4*) vs. 0.01 Å (*2*)] largely avoids this obstacle, which is eliminated when the slope is determined analytically. However, the scale factor for the atom-moving step must be chosen empirically. If it is too small, the process is too slow; if it is too large, the resulting geometries are not reproducible, and the final energies are higher than when a finer grid is used.

Second Derivative Techniques. Newton–Raphson Minimization

Applying Equation 3.4 to the minimization of the energy of a molecule with regard to $3n - 6$ coordinates results in a system of $3n - 6$ linearly independent equations, Equation 3.5. Written in matrix notation, the equations are as Equation 3.6, where $\mathscr{F}$ is the complete matrix of second derivatives $\dfrac{\partial^2 V}{\partial x_i \partial x_j}$, the force constant matrix of pp. 17–22. The improved geometry x_1 is then given in Equation 3.7.

$$\frac{\partial V}{\partial x_i} + \frac{\partial^2 V}{\partial x_i \partial x_j} \Delta x_i = 0 \qquad i = 1, \ldots, 3n - 6 \tag{3.5}$$

$$\nabla V(x) + \mathscr{F}(x)\Delta x = 0 \tag{3.6}$$

$$x_1 = x_0 - \mathscr{F}^{-1}\nabla V(x) \tag{3.7}$$

The calculation of the derivatives was carried out numerically in some early programs (*8*), but this is done analytically in more recent programs (*13, 24–26, 41*), since the time required for convergence is much less than with numerical differentiation. Compared with the steepest descent techniques, the programming is very tedious, and altering a potential function in a NR procedure using analytical differentiation involves changes in the potential function itself as well as in its first and second derivative functions. In the steepest descent techniques with numerical differentiation, only the potential function itself appears, which is much easier to modify than the large number of functions of the derivatives. However, the advantages of the NR methods in terms of computer time (factors of 10–100) are so large that almost all present programs utilize this method and variants thereof.

A special problem of full-matrix NR minimization with Cartesian coordinates involves inversion of the F matrix. The problem arises because the $3n \times 3n$ matrix obtained when all coordinates are employed is sixfold-singular due to three translational and three rotational degrees of freedom (only $3n - 6$ eigenvalues correspond to nonzero vibration frequencies). Three ways around this obstacle have been used. In the simplest one, the reduced matrix method, translation and rotation of the molecule are prevented by fixing one atom at $x = y = z = 0$, a second along one axis ($x = y = 0$), and a third in a plane ($x = 0$). One can then delete six equations, which removes six rows and six columns of the F matrix, and the remaining reduced matrix is no longer singular. Oscillation problems with this approach have been found occasionally when the trial geometry was extremely poor (13, 42). Another problem with this approach is that one must be careful not to restrict the coordinates such that a bond length or angle is restricted.

A frequently employed technique that doesn't have these problems is that of the generalized inverse (28). In this method the $3n \times 3n$ F matrix is diagonalized, and the six zero eigenvalues and the corresponding eigenvectors are removed. To form the inverse, the eigenvectors and eigenvalues of the F matrix are used. Let $\mathcal{A}$ be the eigenvector matrix, and Γ be the eigenvalue matrix; after removing the six zero eigenvalues and the corresponding eigenvectors, the inverse F matrix is obtained as in Equation 3.8.

$$\mathcal{F}_{3n,3n}^{-1} = \mathcal{A}_{3n,3n-6}{}^t\Gamma_{3n-6,3n-6}^{-1}\mathcal{A}_{3n-6,3n} \tag{3.8}$$

A third approach to the singularity problem of the Cartesian force constant matrix is that of applying Eckart constraints (13, 42). The Eckart conditions fix the center of mass in space, and constrain infinitesimal rotations of the molecule. Although in this method a $(3n + 6) \times (3n + 6)$ matrix must eventually be inverted, energy minimization is reported to be faster here than with the other two inversion methods (13, 42).

For a deeper understanding of the NR process it is useful to compare it with the steepest descent method. In the latter method the correction terms are proportional to the first derivatives of the energy, which in matrix notation corresponds to Equation 3.9, and where c, the scaling constant, has the same value for all atoms. Comparing equations 3.7 and 3.9 we find that the steepest descent method is related to the NR process, but with a multiple of the unit matrix (the c matrix) in place of the F matrix. The correction terms are, of course, much better evaluated employing the true second derivatives instead of using, as the steepest descent techniques do, a matrix with no off-diagonal terms and diagonal terms that are equal to some constant, arbitrary value. This explains why NR minimizations converge in far fewer iterations than the steepest descent methods do.

$$x_1 = x_0 + c\nabla V(x_0) \tag{3.9}$$

A major improvement in the steepest descent method is possible even without taking off-diagonal second derivatives into account, if the moving step is obtained as the product of the gradient with the diagonal second derivative term $\frac{\partial^2 V}{\partial x_i^2}$ (25, 28, 29), rather than with a constant scaling factor as in Equation 3.9. This method is scale invariant. The time required for the evaluation of the moving steps is only slightly longer than for steepest descent methods, but convergence is faster by a factor from five to thirty and more.

The major drawbacks of the full-matrix NR method as described above are the large amounts of computer time necessary to evaluate the force constant matrix and its inversion, and the large amount of information that must be stored. Simplifications were therefore introduced by several groups (24, 25, 29, 30). Consider the force constant matrix written in terms of the $3n$ Cartesian coordinates. The full F matrix can be divided into 3×3 submatrices along the diagonal that belong to terms associated with the x, y, z coordinates of a single atom, and into 3×3 blocks off the diagonal with cross terms for interactions of two atoms (*see* Figure 3.3). The 3×3 matrices along the diagonal are all considerably different from zero because of the strong correlation of the energy changes during movement of one atom in the Cartesian directions, while many off-diagonal matrices are comparatively small. If these are completely neglected, a block-diagonal F matrix of 3×3 blocks is left. The computer time required to evaluate the matrix elements is smaller for the **block diagonal method** than for the full matrix NR. The number of matrix elements that must be evaluated in the first case goes up with n, and in the second with n^2, but there is no such simple relationship between the computer time required for the two processes. The dominant term for the time required for an energy minimization iteration is the time required for matrix inversion, and here the block-diagonal scheme is far superior. The time required per iteration is there-

70 MOLECULAR MECHANICS

$$\begin{array}{|ccc|ccc|ccc|}
\hline
\dfrac{\delta^2 V}{\delta x_1^2} & \dfrac{\delta^2 V}{\delta y_1 \delta x_1} & \dfrac{\delta^2 V}{\delta z_1 \delta x_1} & — & — & — & — & — & — \\
\dfrac{\delta^2 V}{\delta x_1 \delta y_1} & \dfrac{\delta^2 V}{\delta y_1^2} & \dfrac{\delta^2 V}{\delta z_1 \delta y_1} & — & — & — & — & — & — \\
\dfrac{\delta^2 V}{\delta x_1 \delta z_1} & \dfrac{\delta^2 V}{\delta y_1 \delta z_1} & \dfrac{\delta^2 V}{\delta z_1^2} & — & — & — & — & — & — \\
\hline
\dfrac{\delta^2 V}{\delta x_1 \delta x_2} & \dfrac{\delta^2 V}{\delta y_1 \delta x_2} & \dfrac{\delta^2 V}{\delta z_1 \delta x_2} & \dfrac{\delta^2 V}{\delta x_2^2} & \dfrac{\delta^2 V}{\delta y_2 \delta x_2} & \dfrac{\delta^2 V}{\delta z_2 \delta x_2} & — & — & — \\
\dfrac{\delta^2 V}{\delta x_1 \delta y_2} & \dfrac{\delta^2 V}{\delta y_1 \delta y_2} & \dfrac{\delta^2 V}{\delta z_1 \delta y_2} & \dfrac{\delta^2 V}{\delta x_2 \delta y_2} & \dfrac{\delta^2 V}{\delta y_2^2} & \dfrac{\delta^2 V}{\delta z_2 \delta y_2} & — & — & — \\
\dfrac{\delta^2 V}{\delta x_1 \delta z_2} & \dfrac{\delta^2 V}{\delta y_1 \delta z_2} & \dfrac{\delta^2 V}{\delta z_1 \delta z_2} & \dfrac{\delta^2 V}{\delta x_2 \delta z_2} & \dfrac{\delta^2 V}{\delta y_2 \delta z_2} & \dfrac{\delta^2 V}{\delta z_2^2} & — & — & — \\
\hline
— & — & — & — & — & — & & & \\
— & — & — & — & — & — & & \text{etc.} & \\
— & — & — & — & — & — & & & \\
\hline
\end{array}$$

Figure 3.3. Elements of the block-diagonal and full F matrix. The block-diagonal contains only 3 × 3 blocks where x, y and z belong to the same atom (top left and lower right). The full matrix also contains all other blocks.

fore very much smaller for a block-diagonal NR optimization. On the other hand, the geometry optimization per iteration cycle also is less with this method, and it becomes a question of the balance between time reduction for one iteration, against the number of additional iterations necessary, as to which method is faster overall. For large molecules of more than fifty atoms, the full-matrix method is not competitive with the block-diagonal scheme, because of both the computer storage requirements and the computation time to invert the matrix. However, it is useful to compare some general properties of the full-matrix and block-diagonal NR minimization procedures for small molecules.

The full-matrix NR minimization converges rapidly when the derivatives are small (i.e., near the minimum), but it converges slowly or even diverges if the derivatives are large (i.e., if one is far from the minimum) (32). The minimization characteristics of the block-diagonal method can approach this second-order convergence characteristic of the full matrix NR when the errors in the position of each atom in the trial structure are uncorrelated with the errors in the positions of other atoms. However, the errors are almost always correlated, and for an atom to get to its final position a cooperative movement of other atoms is required. When cooperative movements are required, the block-diagonal minimization has the characteristics of a first-order minimizer, like the steepest descent technique. That is to say it minimizes rapidly and reliably when far from the minimum, but convergence is slow when near the minimum or when moving along a valley having a small slope. This is especially noticeable when torsion angles must be adjusted. In simple cases such as the adjustment of the torsion angle of a methyl group, the block-diagonal scheme works well. But when a large number of correlated torsion angles must be optimized simultaneously (such as in pseudorotational movements of rings) the potential surface is such that the full-matrix method, which accounts for the correlation of different parts of the molecule with the off-diagonal force constants, can become more efficient. In most cases, however, the block-diagonal procedure was found to require less computer time than the full-matrix minimization. It has also been reported that in the case of a very flat energy surface, the block-diagonal scheme approaches the energy minimum rapidly to within an energy of about 0.04 kJ/mol, but torsion angles were found to change from such a position by up to 10° when a full-matrix NR minimization was employed afterwards (12). There is very little physical significance to an energy difference of 0.04 kJ/mol within the context of molecular mechanics, however, and consequently not much significance can be claimed for structural variation involving an energy difference of this size.

To speed the convergence of the block-diagonal method when dealing with cooperative movements of the atoms on a surface of small slope, the principles of pattern search can be applied (25). A characteristic situation where cooperative movements slow the convergence of the block-diagonal method is when the movements of the atoms are about the same from one iteration to the next. Convergence can be accelerated when the net movement of each atom over the previous few iterations (or some fraction of this movement) is added to the coordinates of the atom several times, until the energy increases. For a typical trial structure this method was found to speed the convergence of the block-diagonal method by a factor of two to three. If a large conformational change occurs in going from the trial structure to the minimum, this method can reduce the minimization time by a factor of ten to twenty. This accelerated convergence method has the advantage of programming ease and practicability for large and small molecules. However, while this method can reduce the likelihood that the block-diagonal method will stop prematurely when on a gentle slope, it cannot be expected to completely eliminate the problem.

The methods where one first uses the steepest descent method followed by a full-matrix NR minimization (12, 15), or where block-diagonal minimization is followed by the full-matrix NR minimization (12, 28, 29), has the advantage of using each minimization method where it performs best. The first-order minimization is used initially far from the minimum, while the second-order minimization method is used close to the minimum, so that each method converges rapidly.

With some of the more sophisticated methods of energy minimization, it is frequently observed that the geometry ends up on a saddlepoint or even a hilltop on the potential surface. This is advantageous when transition states are sought (see pp. 72–76), but is usually unwanted. Ending up on a saddlepoint has been reported to be a problem with full-matrix NR optimization, less so for block-diagonal methods, and to be essentially absent in the diagonal and steepest descent methods. An energy minimization procedure is especially prone to this difficulty if it first calculates all derivatives for all atoms at one geometry before moving all atoms at one time, which is the case for all full-matrix methods. Many programs are written, however, so that after the derivatives have been determined for one atom, it is moved to its new optimized geometry. Thus movement away from the saddlepoint is accomplished by cycling through the molecule. This works very well for a block-diagonal method, adding another advantage to this method. The same effect can be achieved, of course, if a steepest descent minimization is done before the full-matrix NR. An alternative method to avoid saddlepoints was proposed by Hilderbrandt (27), who minimized the energy by the full-matrix NR process, but on Wilson S-vectors. When the minimum is to be calculated in his method, the eigenvalues of the F matrix are restricted to their absolute values, and the restriction is removed as minimization proceeds.

Conformational Interconversion Pathways

Interconversion pathways can be studied experimentally in much less detail than can the minimum energy conformations. The heights of the energy barriers between the conformers can often be obtained, but only indirect information about the geometry at the transition state can be obtained from activation parameters. The lower parts of the potential surface can be explored by vibrational spectroscopy, but the exploration of the full potential surface is so far possible only by computational methods.

The information usually of most interest about a transition state is its geometry and energy. Since it is a saddlepoint on the potential surface that, like an energy minimum, satisfies Equation 3.3, we can in principle evaluate a transition state geometry as we do for an energy minimum. A trial geometry close to the transition state is assumed, and is subjected to the computational procedures described in pp. 64–72. From Figure 3.2, it is evident that the NR method will adjust the geometry to the saddlepoint (i.e., maximize the energy in a certain degree of freedom) if it begins closer to that point (to the right of the maximum in the first derivative function) than to the minimum. In $3n - 6$ dimensions this happens especially easily with symmetrical starting geometries, and more reliably with the full-matrix NR method. The later scheme has been reported to even optimize the geometry towards unsymmetrical transition states (15, 33). As mentioned in the previous section, optimization to a saddlepoint occurs rarely with the block-diagonal or diagonal NR method unless we begin on the saddlepoint, since the energy is minimized only locally. With the steepest descent method, it is extremely uncommon to end at such a point, even with a symmetrical starting geometry.

Thus, for molecular mechanics calculations on transition states, the full-matrix method is the most reliable way of ending in the right place. The tendency of the full-matrix method to go to saddlepoints as readily as it goes to energy minima is, in this case, an asset and not a difficulty.

While in simple cases we can decide by inspection whether the calculations have ended on a saddlepoint or at an energy minimum, for more complex situations this may not be so. If energy minimization methods other than the full-matrix method are used, it has been common in the past to simply distort the final structure. If the molecule does not return to the previous structure, but the energy continues to minimize from this distorted structure to another structure of lower energy, then obviously the original "minimum" energy structure was on a saddlepoint. But rare cases are known where the distorted structure returned to the starting point, which was in fact a saddlepoint (10, 11). Whether the structure corresponds to a minimum or to a saddlepoint can, however, be determined unambiguously from the second derivatives. If, in addition to the zero first derivatives, all of the eigenvalues of the F matrix (which give the vibrational frequencies) are positive, the molecule is at an energy minimum. If one eigenvalue is negative, it is at a saddlepoint, and if more than one is negative, it is at a hilltop (34). This check necessitates the evaluation and diagonalization of the full F matrix, and is therefore especially well suited to the full-matrix NR minimization, where the necessary F matrix is available from the last iteration. One must recall that just as all minimizations locate the nearest minimum—not necessarily the lowest minimum—so will the full NR method locate the nearest saddle point—not necessarily the lowest one. Calculation of all of the transition states appears to be the only presently known way to determine which one is the lowest in energy (35), and for more complicated molecules, this is clearly not feasible. The transition vector will sometimes be of use. This is the normal coordinate with the negative frequency obtained at the saddlepoint. It is an indicator of whether the correct transition state has been found, because it should point towards the two minima that it connects.

An alternative way to arrive at a transition state is to start at an energy minimum and move the molecule along the minimum energy pathway across the saddle. The minimum energy pathway follows the path of minimum Gaussian curvature (36). It has been suggested that, at least close to the energy minimum, this is the normal coordinate with the smallest eigenvalue, corresponding to the lowest vibrational frequency (27). It should be kept in mind, however, that while a normal coordinate depends also on the atomic masses, the minimum energy pathway should not. Hilderbrandt (27) has developed a scheme of moving the atoms along normal coordinates which he claims goes directly to the saddlepoint, which would seem a convenient method for simulating conformational interconversion. In his procedure, the atoms are initially moved along this coordinate from the minimum by a small amount. In the following minimization cycle, which also gives the new normal coordinate, the atoms are allowed to relax in all directions except the one having the largest scalar product with the displacement coordinate chosen before. Other transition states can be explored following other normal coordinates (27).

The most frequently used method of calculating transition state structures consists of restricting certain degrees of freedom, usually torsion angles. In the course of conformational interconversions, one or more torsion angles vary through a region of maximum strain. So, by iteratively giving certain torsion angles fixed values, but allowing all other internal degrees of freedom to relax, an interconversion can be forced across a saddlepoint. This procedure is easy to carry out if the energy is minimized on internal coordinates (7, 14), but is more complicated when working with Cartesian coordinates. Alternatively, in simple cases the molecule can be aligned with the coordinate axes, and movements along certain Cartesian coordinates are not carried out.

In a general method to be used in connection with Cartesian coordinates described by Wiberg and Boyd (37), a torsion angle can be "driven" through a range of values by modifying the force field during energy minimization through the addition to the usual force field of a very strong torsional potential (large potential constant V_{dr}) with its minimum at the desired torsion angle ω_{dr} (Equation 3.10).

$$V = \frac{1}{2} V_{dr} (1 + \cos 3(\omega - \omega_{dr})) \qquad (3.10)$$

Energy minimization using this force field now gives a geometry with the torsion angle at the desired value ω_{dr}, and with all other coordinates relaxed as usual. Then the strong V_{dr} potential is removed, and the energy is calculated (not minimized) with the original force field. This now gives the potential energy of the geometry corresponding to the fixed value of ω_{dr}. This method is easy to implement in all kinds of minimization procedures. It does not depend on full-matrix NR minimization, and has become the most popular means of mapping conformational interconversions.

For computer programs with full matrix NR minimization, it has been recommended (41) that Eckart constraints be applied to fix the internal coordinates as an alternative to torsion angle driving, but this method has not yet been widely tested.

The calculation of transition state geometry by fixing a torsion angle with either of the methods outlined in the preceding section still contains a conceptual deficiency. The minimum energy pathway is characterized by the requirement that the energy is minimized with regard to all coordinates *except* the reaction coordinate. But the reaction coordinate is hardly ever described exactly by a torsion angle; it is (like a normal coordinate, *see* pp. 17–22) made up from several torsion angles and/or other internal coordinates. In simple molecules like ethane or butane the problem is a minor one. Here a correct transition state is arrived at. As one torsion angle is driven in ethane down from 60° to 0°, the two other hydrogen torsion angles going to 0° are found to lag behind during the early phase of the rotation. Here the torsional and van der Waals forces tend to keep these torsion angles large, which can be achieved at the expense of some angle bending. As the rotation goes on, the slopes of the torsional and van der Waals potentials decrease, bending becomes the dominant force, and the hydrogens which are not driven catch up. The transition state that is arrived at has a C_3 symmetry axis, as has the energy minimum, but a C_3 axis is not retained during driving. The correct reaction coordinate would retain C_3 symmetry throughout, and would consist of a linear combination of the three torsion angles becoming eclipsed. To simulate this coordinate all three torsion angles would have to be driven simultaneously. In less symmetrical molecules it is more difficult to choose the correct combination of torsion angles. Thus, the lagging behind of the hydrogens of butane cannot be relieved in a straightforward manner without introducing a second constraint. The C_3 symmetry restraint in ethane comes from other sources, but in butane no symmetry is retained on the reaction coordinate. However, when only the CCCC torsion angle was driven, the hydrogens did catch up near the transition state, as in ethane (43).

In larger molecules it is possible that the correct transition state is not reached. In the transition state of 1,1,2,2-tetra-*t*-butylethane, two pairs of eclipsed *t*-butyl groups are found. When only one torsion angle between the *t*-butyl groups is driven to 0°, the other *t*-butyl groups lag behind, and apparently the molecule slides into a side valley of the potential surface, which prevents the *t*-butyl groups that are not driven from catching up (*38*). This is a consequence of the failure of the energy minimization routines to pass over ridges in the potential surface. What is found when driving only one torsion angle in this molecule is that the second pair of *t*-butyl groups never pass each other, but highly strained geometries are arrived at that have nothing to do with the interconversion (*38*). Only when both torsion angles are driven simultaneously can the interconversion be achieved (*38*); however, the transition state has two restricted degrees of freedom.

A related phenomenon is found in calculations of interconversion pathways between the conformations of rings. A large number of such studies has been reported by Anet (*39*). Interconversions to different families of ring conformations were found in his calculations, depending on which torsion angle was driven. Sometimes certain interconversions could not be achieved by driving only one torsion angle, but only when two torsion angles were driven (*40*) (see also Chapter 4) as in the tetra-*t*-butylethane case discussed previously. An undesirable side effect was observed during the interconversions of cyclooctane conformations, where different transition states were found, depending on the direction of the interconversion between the boat–chair and twist–boat conformations (*40*). This is contradictory to the principle of microscopic reversibility, and is caused by the inadequacy in the identification of the reaction coordinate with only one torsion angle; the precise structures of transition states obtained in this way may be only approximate.

When more than one torsion angle was driven in cyclohexane during the chair–boat interconversion, transition states with higher energies were found than when only one torsion angle was driven (*37*). This now appears to be a consequence of insufficiently searching the potential surface with different combinations of driven torsion angles (*see* Figure 4.8, p. 94). The true transition state of the cyclohexane chair–boat interconversion has been reported to pseudorotate freely (*see* pp. 92–93), whereas the driver routine leads to a defined transition state in which one torsion angle (the one that is driven) has passed through 0° already, while the rest of the ring shows the lagging behind discussed with ethane and butane (*43*).

Literature Cited

1. Hendrickson, J. B. *J. Am. Chem. Soc.* **1961**, *83*, 4537.
2. Wiberg, K. B. *J. Am. Chem. Soc.* **1965**, *87*, 1070.

3. Allinger, N. L.; Miller, M. A.; Van Catledge, F. A.; Hirsch, J. A. *J. Am. Chem. Soc.* **1967**, *89*, 4345.
4. Allinger, N. L.; Hirsch, J. A.; Miller, M. A.; Tyminski, I. J.; Van-Catledge, F. A. *J. Am. Chem. Soc.* **1968**, *90*, 1199.
5. Allinger, N. L.; Hirsch, J. A.; Miller, M. A.; Tyminski, I. J. *J. Am. Chem. Soc.* **1968**, *90*, 5773.
6. Ibid., **1969**, *91*, 337.
7. Jacob, E. J.; Thompson, H. B.; Bartell, L. S. *J. Chem. Phys.* **1967**, *47*, 3736.
8. Boyd, R. H. *J. Chem. Phys.* **1968**, *49*, 2574.
9. Lifson, S.; Warshel, A. *J. Chem. Phys.* **1968**, *49*, 5116.
10. Anet, F. A. L.; Yavari, I. *J. Am. Chem. Soc.* **1977**, *99*, 6986.
11. Allinger, N. L.; Viscosil, J. F., Jr.; Burkert, U.; Yuh, Y. *Tetrahedron* **1976**, *32*, 33.
12. White, D. N. J.; Ermer, O. *Chem. Phys. Lett.* **1975**, *31*, 111.
13. Faber, D. H.; Altona, C. *Comput. Chem.* **1977**, *1*, 203.
14. DeTar, D. F. *Comput. Chem.* **1977**, *1*, 141.
15. Ermer, O. *Struct. Bonding (Berlin)* **1976**, *27*, 161.
16. Balaban, A., Ed. "Chemical Applications of Graph Theory"; Academic: London, 1976.
17. Wipke, W. T., Whetstone, A. *Comput. Graphics* **1971**, *5*, 10.
18. Williams, J. E.; Stang, P. J.; Schleyer, P. v. R. *Ann. Rev. Phys. Chem.* **1968**, *19*, 531.
19. Skorczyk, R. *Acta Crystallogr. Sect. A* **1976**, *32*, 447.
20. Brunel, Y.; Faucher, H.; Gagnaire, D., Rassat, A. *Tetrahedron* **1975**, *31*, 1075.
21. Broyde, S.; Wartell, R. M.; Stellman, S. D.; Hingerty, B. *Biopolymers* **1978**, *17*, 1485.
22. Niketic, S. R.; Rasmussen, K. "The Consistent Force Field"; Springer: New York, 1977.
23. Engler, E. M.; Andose, J. D.; Schleyer, P. v. R. *J. Am. Chem. Soc.* **1973**, *95*, 8005.
24. Allinger, N. L. *Adv. Phys. Org. Chem.* **1976**, *13*, 1.
25. Wertz, D. H., Ph.D. Thesis, Univ. of Georgia, 1974.
26. Warshel, A.; Lifson, S. *J. Chem. Phys.* **1970**, *53*, 582.
27. Hilderbrandt, R. L. *Comput. Chem.* **1977**, *1*, 179.
28. White, D. N. J.; Sim, G. A. *Tetrahedron* **1973**, *29*, 3933.
29. White, D. N. J. *Comput. Chem.* **1977**, *1*, 225.
30. Allinger, N. L.; Tribble, M. T.; Miller, M. A.; Wertz, D. H. *J. Am. Chem. Soc.* **1971**, *93*, 1637.
31. Ollis, W. D.; Stoddart, J. F.; Sutherland, I. O. *Tetrahedron* **1974**, *30*, 1903.
32. Pennington, R. H. "Introductory Computer Methods and Numerical Analysis"; Macmillan: New York, 1965; Chap. 10.
33. Ermer, O. *Tetrahedron* **1975**, *31*, 1849.
34. Murrell, J. N.; Laidler, K. J. *Trans. Faraday Soc.* **1968**, *64*, 371.
35. McIver, J. W. *Acc. Chem. Res.* **1974**, *7*, 72.
36. Fischer, S. F.; Hofacker, G. L.; Seiler, R. *J. Chem. Phys.* **1969**, *51*, 3951.
37. Wiberg, K. B.; Boyd, R. H. *J. Am. Chem. Soc.* **1972**, *94*, 8426.
38. Osawa, E.; Shirahama, H., Matsumoto, T. *J. Am. Chem. Soc.* **1979**, *101*, 4824.

39. Anet, F. A. L.; Anet, R. In "Dynamic Nuclear Magnetic Resonance Spectroscopy"; Jackman, L. M.; Cotton, F. A., Eds.; Academic: New York, 1975; Chap. 4, p. 543.
40. Anet, F. A. L.; Krane, F. *Tetrahedron Lett.* **1973,** 5029.
41. van der Graaf, B.; Baas, J. M. A. *Rec. Trav. Chim. Pays-Bas* **1980,** 99, 327.
42. Thomas, M. W.; Emerson, D. *J. Mol. Struct.* **1973,** 16, 473.
43. Burkert, U.; Allinger, N. L. *J. Comput. Chem.* **1982,** 3, 40.

Calculations on Hydrocarbons–Geometries and Relative Energies

Acyclic Saturated Compounds

It has long been known that the geometries of even strainless, open chain hydrocarbons differ considerably from what a Dreiding model with ideal tetrahedral bond angles and standard bond lengths would indicate. One of the milestones in the development of the molecular mechanics method was its ability to explain and predict the nonideal geometries of saturated hydrocarbons at a time when the role of quantum mechanical effects such as hyperconjugation, inductive effects, and antibonding overlap (as an equivalent to the nonbonded repulsions of molecular mechanics) was not clear (*1–3*). The force fields that include 1,3-nonbonded or stretch–bend interactions can account for even more details of molecular geometry, such as the extension of bond lengths when the angle between them is very small—as in cyclobutane. These force fields are in part related to the valence shell electron pair repulsion (VSEPR) method (*4*). The latter successfully predicts semiquantitatively the geometries of a large number of compounds by considering the interactions between spheres of valence electrons (including spherical lone pairs) to explain changes in bond lengths and angles. (For a detailed comparison of the two methods see Ref. *3*). In most force fields that include 1,3-nonbonded interactions, only one standard value (tetrahedral) is used for all bond angles in saturated hydrocarbons. Valence force fields, on the other hand, allow for part of the 1,3-nonbonded interactions in the bending parameters, and the standard angles are usually different. For example, the standard bond angles for the CCC, CCH, and HCH groups are different. The effects of the remaining 1,3-interactions on geometry are sometimes included in valence force fields as cross terms. The final resulting geometries and relative energies are then quite similar for the modified Urey–Bradley and valence force fields, and we will not differentiate these two types of force fields in the remainder of this chapter.

Most structural data on small hydrocarbons come from electron diffraction (ED), with additional data from microwave spectroscopy. Most force fields have been parameterized to fit ED structures because of the wide availability of such data, and also because these agree fairly well with structures obtained from x-ray diffraction. The structures of the smallest members of the alkane family, methane to propane, have been employed in the parameterization of every hydrocarbon force field, so these data are reproduced by all force fields with high precision. In the ED structure determination for larger molecules, the number of internal coordinate parameters describing the molecular structure, which, therefore, are to be varied in the determination, is quite large. Approximations are often used, such as using the same bond length for all C–H bonds or assuming a certain symmetry, which cuts down the number of independent parameters. The geometries calculated by molecular mechanics are more refined in this regard. Bond length extensions of C–H bonds from 1.108 Å in methane to 1.113 Å for secondary C—H (as in cyclopentane) and 1.120–1.130 Å for tertiary C–H are faithfully reproduced, and the same is true for bond angle variations in CCC and CCH bond angles over a range of 15° (5–10). In Figures 4.1–4.4 are shown some calculated vs. experimental bond lengths and angles. On the whole the agreement is good, and especially so for the more recent force fields.

An important, often most important, way of reducing nonbonded strain in a molecule is through rotation about a single bond, but depending on the situation, angle bending may be of similar importance. For example, the CCC bond angle in n-butane opens from 111.8° in the *anti* form to 113.5° in the *gauche* form, while in the latter, the repulsion between the methyl groups increases the CCCC torsion angle from the ideal 60° value to 64.7° [calculated with MM2 force field (10)]. The interplay of bond angle bending and torsional adjustments is especially delicate in highly hindered compounds like tri-t-butylmethane, where a (chiral) twisted geometry with C_3 symmetry is more stable than the C_{3v} geometry (8, 11, 12, 325) according to both molecular mechanics calculations and electron diffraction. This compound is a fine example of where the combined application of molecular mechanics calculations and electron diffraction solved the structural problem (8). The number of independent diffraction parameters was so large here that they could not be uniquely determined without assistance from molecular mechanics calculations. The same combined method has also been used in many other cases including *trans*-cyclooctene (13), cyclodecane (14), *trans*-1,4-di-t-butylcyclohexane (15), *trans*-2-decalone (16), 2,2,6-trimethylcyclohexanone (17), 10-methyl-*trans*-2-decalone (18), and the *cis*- and *trans*-bicyclo[4.2.0]octanes (19). Examples of this approach of combining molecular mechanics with experimental techniques are also known in other areas where the obtainable direct experimental information is insufficient to determine the structure. The experimental methods used include x-ray diffraction, which is discussed in Chapter 9, and microwave spectroscopy (20–22). A combination of molecular mechanical and quantum mechanical calculations has also been fruitful. Geometries optimized by molecular mechanics have been used in ab initio calculations (23–29) to avoid the expense of geometry optimiza-

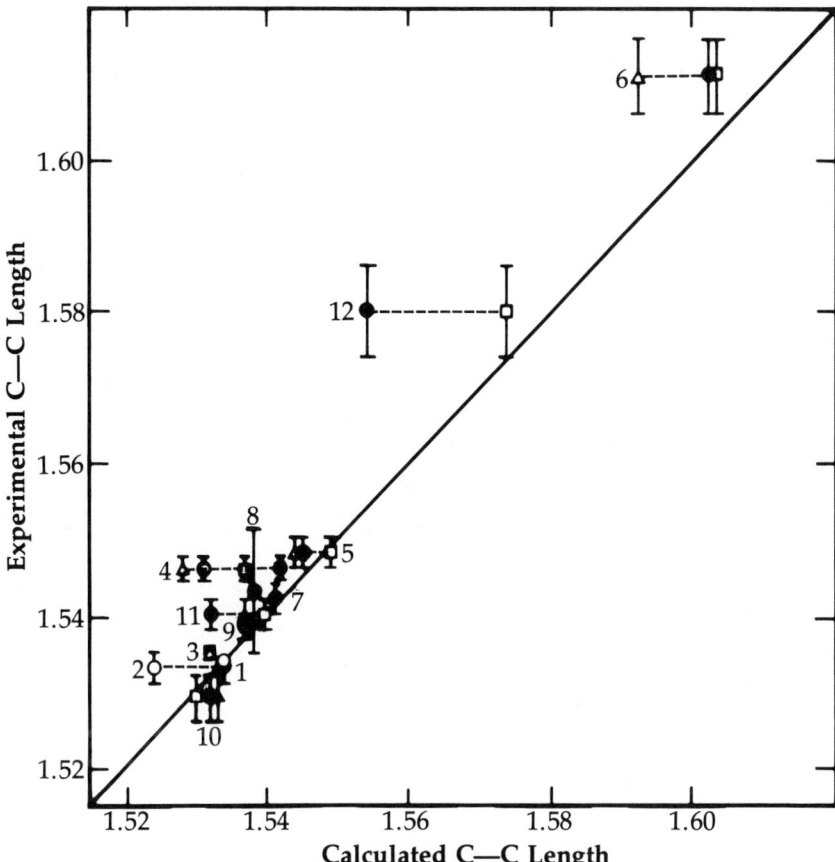

Figure 4.1. Comparison of experimental and calculated C–C bond lengths. The vertical error bars correspond to ± σ. Points that represent fits of the same bond length by different force fields are joined by dotted lines. The numbers refer to the following molecules: 1, ethane; 2, n-butane; 3, isobutane; 4, cyclopentane; 5, tri-t-butylmethane (C^q–Me); 6, tri-t-butylmethane (C^t–C^q); 7, hexamethylethane (end bond); 8, tetramethylethane (central bond); 9, tetramethylethane (end bond); 10, methylcyclohexane; 11, adamantane; 12, hexamethylethane (central bond). Symbol key: ●, MUB-2; □, MM2; ∧, Schleyer; ○, Warshel.

tion within the ab initio calculation itself. Although the equilibrium geometries differ somewhat from the experimental and molecular mechanics geometries, this seems like a good way to obtain ab initio geometry and energy *differences*. Molecular mechanics geometries have also been advocated for use with Extended Hückel (*30–31*) calculations, where the quantum mechanical method itself is too crude to furnish good geometries. A combination of semi-empirical quantum mechanics and force field calculations has also been used (*359*).

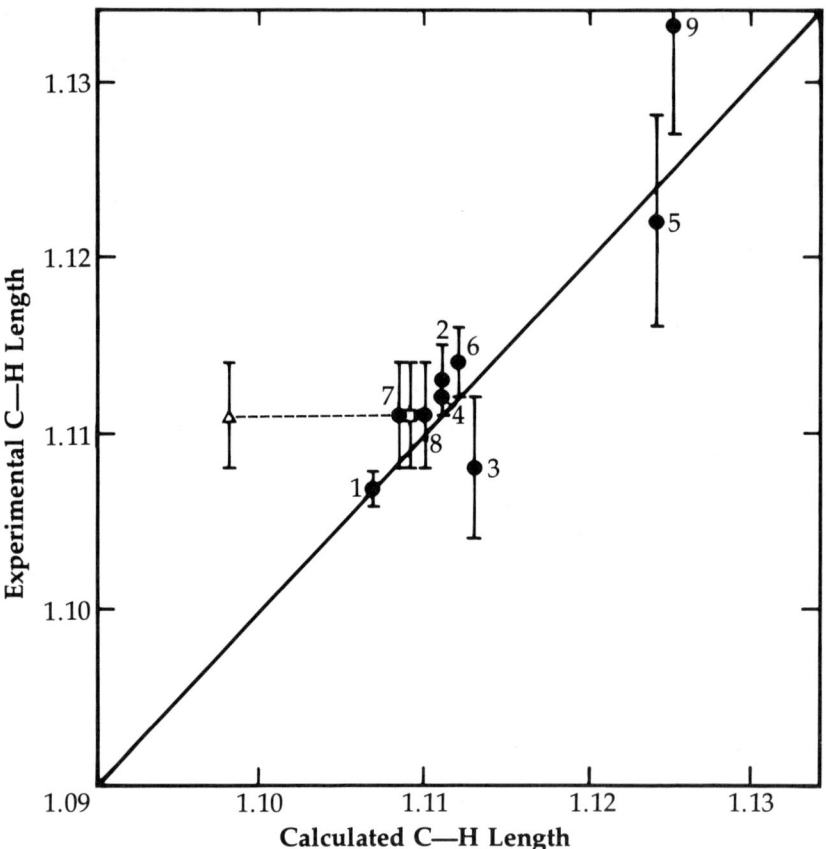

Figure 4.2. Comparison of experimental ($\pm\sigma$) and calculated C–H bond lengths. Points that represent fits of the same bond length by different force fields are joined by dotted lines. The numbers refer to the following molecules: 1, methane; 2, ethane; 3, n-butane; 4, isobutane (methyl), tetramethylethane (methyl); 5, isobutane (tertiary); 6, cyclopentane; 7, tri-t-butylmethane; 8, hexamethylethane; 9, tetramethylethane (tertiary). Symbol key: ●, *MUB-2;* △, *Schleyer;* □, *MM2.*

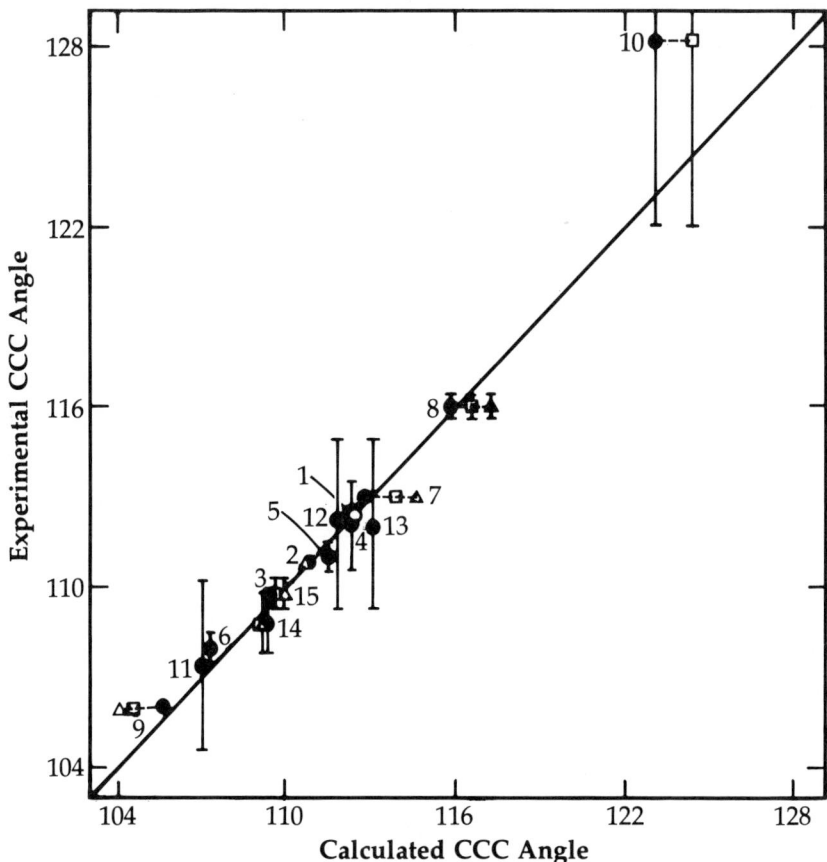

Figure 4.3. Comparison of experimental (±σ) and calculated CCC angles. Points that represent fits of the same angle by different force fields are joined by dotted lines. The numbers refer to the following molecules: 1, n-butane; 2, isobutane; 3, tetramethylethane (Me–C–Me); 4, tetramethylethane (C–C–Me); 5, hexamethylethane (C–C–Me); 6, hexamethylethane (Me–C–Me); 7, tri-t-butylmethane (C^t–C–Me); 8, tri-t-butylmethane (C–C^t–C); 9, tri-t-butylmethane (Me–C–Me); 10, di-t-butylmethane (C_2–C–C_2); 11, di-t-butylmethane (C–C_2–C_3); 12, di-t-butylmethane (C–C_2–C_4); 13, di-t-butylmethane (C–C_2–C_5); 14, adamantane (C^t–C^s–C^t); 15, adamantane (C^s–C^t–C^s). Symbol key: ●, MUB-2; △, Schleyer; □, MM2; ○, Warshel.

The precision and reliability of geometries calculated for hydrocarbons by molecular mechanics are now often better than those from ED. This is especially true for hydrogen positions, and for cases where more than one conformation is present. With high-quality geometries available from the calculations, new methods for the determination of conformational equilibria are feasible. For example, in a recent approach to the determination of structures and conformational equilibria from ED data, the geometries of possible conformations were calculated by molecular mechanics. These structures were used to reduce the number of parameters to be determined from the ED pattern. The conformational composition has been treated as a parameter (*14, 15*), or as the only parameter (*32*), in fitting the diffraction data. In the latter case, the fit of the calculated radial distribution function to experiment is not as good as when all parameters are optimized, but it is not bad, and it utilizes only one parameter.

As previously mentioned for tri-*t*-butylmethane, the occurrence of torsional deformations that destroy the potential symmetry planes of the

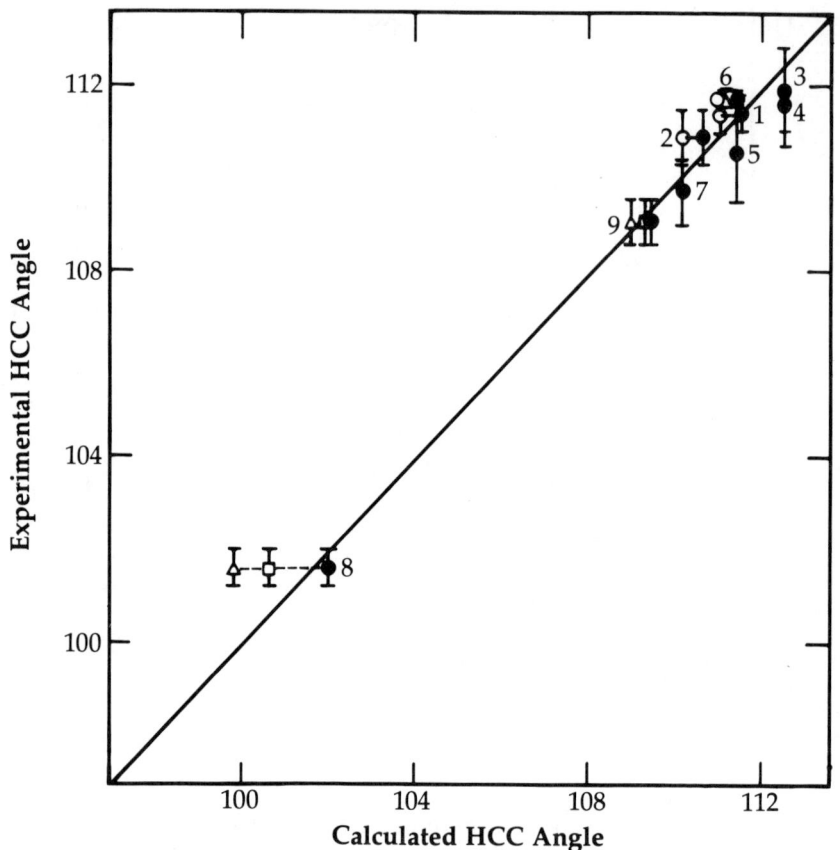

Figure 4.4. Comparison of experimental (±σ) and calculated HCC angles. Points that represent fits of the same angles by different force fields are joined by dotted lines. The numbers refer to the following molecules: 1, ethane; 2, n-butane; 3, di-t-butylmethane; 4, hexamethylethane; 5, tetramethylethane; 6, cyclopentane; 7, methylcyclohexane; 8, tri-t-butylmethane; 9, adamantane. Symbol key: ●, MUB-2; △, Schleyer; □, MM2; ○, Warshel.

molecule has been predicted for other hindered molecules. Examples are 2,2,4,4-tetramethylpentane (di-t-butylmethane) and di-1-adamantylmethane (326). Several examples were noted among dialkyl sulfides (33), where the approximately 90° bond angle at sulfur leads to serious interactions between the alkyl groups. In extensive studies of compounds of the type $M[L(CH_3)_3]_4$ derived from tetra-t-butylmethane with M = C, Si, Ge, Sn, Pb and L = C, Si, the geometries with T_d, T, D_{2d}, D_2, S_4, and C_2 symmetries, which depend on the twist of the $L(CH_3)_3$ group, were all examined (11, 12). The geometry with ideal T_d tetrahedral symmetry does not correspond to an energy minimum in any of these molecules, but the $L(CH_3)_3$ groups are twisted, usually by about 14°, and even more than in tri-t-butylmethane (8). The T symmetry structures with all $L(CH_3)_3$ groups twisted in the same direction are always the most stable, and in the case of tetra-t-butylmethane, the geometry with S_4 symmetry is calculated to be 59–67 kJ higher in energy than the T geometry if it was a minimum at all (335). The T conformation was indeed found in the crystal of tetra-t-butylphosphonium tetrafluoroborate (336). The calculated geometries exhibit the expected bond stretching (11, 12). A symmetrically related molecule is tetra-t-butyltetrahedrane, where M is the tetrahedrane ring system. Twisted geometries were found to be preferred over the T_d structure with several different force fields (327).

Another highly hindered molecule, in which molecular relaxation causes a distortion of the geometry from the 60° torsion angles, is hexamethylethane. The electron diffraction pattern indicated a twist

about the central bond of 5 ± 4°, but it could not be decided whether or not the terminal bonds were twisted (0 ± 7°) (34). Molecular mechanics studies revealed structures in which the central and the terminal bonds are twisted. Calculation and experiment agree that the geometry has D_3 symmetry, not D_{3h}, although the experimental data indicate smaller distortions than the calculations, where a torsional angle at 13° for the central C–C bond was found with Bartell's MUB-1 (34), 15° with MM2 (38), and the surprisingly high values of 41° and 44° with two older force fields (39).

Another example of a correlated geometry change by partial rotation about single bonds (a phenomenon occasionally denoted as a "gear effect") (40) concerns the coupled rotation of isopropyl groups in vicinal positions. Adjacent isopropyl groups behave like interlocking tetrahedral rotors in a synchronous rotation (40). Molecular mechanics calculations on methyl-N,N-diisopropyldithiocarbamate [CH$_3$SC(=S)N-(i-Pr)$_2$] showed that there is an interlocking with regard to the preferred

$$CH_3S-\overset{\overset{S}{\|}}{C}-N(iPr)_2$$

conformation, and the rotation of one isopropyl group induces a gear-type correlated motion of the second isopropyl group. But, instead of

following the movement of the driven isopropyl group over the torsional barrier to another staggered conformation, the second isopropyl group falls back into its first minimum. Consequently, full correlated rotation was not found in the calculation (40), which may, however, be due to a deficiency of the driver routine employed (see pp. 72-76). In another example (41), the rotation of the methyl groups in hexamethylbenzene was found to show a gear effect in some force fields, but not in others.

Many greatly hindered ethane derivatives have had their structures studied by molecular mechanics and by x-ray crystallography in recent years (348). Extreme deformations in bond lengths and bond angles have been found. The calculated and experimental structures generally agree very well, although at the time the force fields used were developed, the data available on highly congested molecules were quite limited.

Major advances in the *prediction* of conformational energies are possible, thanks to the molecular mechanics method. Most of classical conformational analysis is first-order in the sense that additivity of conformational energies is assumed. Conformational energies are then predicted by adding interaction increments. Discrepancies are often observed and explained by the action of some kind of "effect" (42). All of these effects are, of course, caused by the same fundamental types of atomic interactions, and the molecular mechanics calculation can be expected to give directly the experimental results if these interactions have been correctly accounted for in the force field. The large number of effects existing in conformational analysis show clearly that the additivity of conformational energies is only a rough approximation. One of the reasons for the nonadditivity of conformational energies is the nonideal geometry at carbon discussed above, namely the fact that the standard CCC bond angle of a methylene group should better be taken to be about 112° rather than the tetrahedral value. Ball-and-stick models usually work with tetrahedral values, and their use in conformational analysis work has caused some problems in the past. Effects should now only be designated as such when the results of a molecular mechanics calculation show a significant systematic deviation from the experimental results (42).

The additivity rule for the conformational analysis of alkanes is that each *gauche* C/C interaction contributes the same increment to the total conformational energy, while C/H and H/H interactions are neglected. The value of the butane *gauche/anti* equilibrium is approximately half the value of the conformational energy of methylcyclohexane, namely 3.6 kJ/mol in the liquid or 4.0 kJ/mol in the gas phase. This classical interaction model does not work at all well in some cases, a simple example of which is 2,3-dimethylbutane. This scheme predicts greater stability for the *anti* isomer (ΔH = 3.6 kJ), as this form has only two *gauche* C/C interactions while the *gauche* has three. But according to force field calculations (5, 9, 43), the methyl groups are forced into one another in the *anti* conformation as a result of the larger than tetrahedral CCC angles, which increases slightly the energy of this form. At the same time the *gauche* conformer evades much of the methyl/methyl repulsion by a small rotation of the two isopropyl groups. The result is that the two conformations end up with about the same energy, and this is confirmed experimentally (34–37, 44, 45).

For 2-methylbutane the conformational energy was again found to deviate from the simple additive value. One finds that the symmetrical isomer is not destabilized as much as the *n*-butane value would predict, [experimental ΔH: 2.47 kJ (45), 3.39 kJ (46)], and this result was found in calculations with different force fields [e.g., 2.80 kJ/mol (38), 3.54 kJ/mol (7)].

Tetra-*t*-butylethane exhibits much stronger repulsions between the alkyl groups than does 2,3-dimethylbutane. The compound might have been expected to be a conformational mixture of *anti* (**1**) and *gauche* (**2**) conformers, as pictured. Neither of these is an energy minimum according to molecular mechanics. Only a single conformation (**3**) is found, which has the unusual feature that its Newman projection does not

have alternation of front and back substituents (*47, 48*). The central CC bond is stretched to about 1.60 Å, and the angle between the *t*-butyl groups is 119–120.8° as found with several force fields (*38, 47, 48, 49*). The conformation calculated for 1,2-bis(2,6-dimethylphenyl)-1,2-di-*t*-butylethane also has a similar Newman projection (*48*).

The full conformational energy surface of butane is a fundamental problem, the understanding of which is crucial to the success of molecular mechanics. It was discussed in detail in pp. 36–52.

Common, Medium, and Large Rings

Cyclopentane. Aston and coworkers were the first to point out that the entropy of cyclopentane requires the molecule to have a symmetry number of one or two, which is compatible only with a nonplanar ring geometry (50, 51). Since Pitzer's classical paper, the conformation of cyclopentane has been discussed in terms of ten envelope (C_s) and ten twist or half-chair (C_2) conformations, all of equal energy, and interconverting through free pseudorotation (52).

The interconnection between these different conformations is shown in Figure 4.5. The five ring torsion angles are interrelated through Equation 4.1.

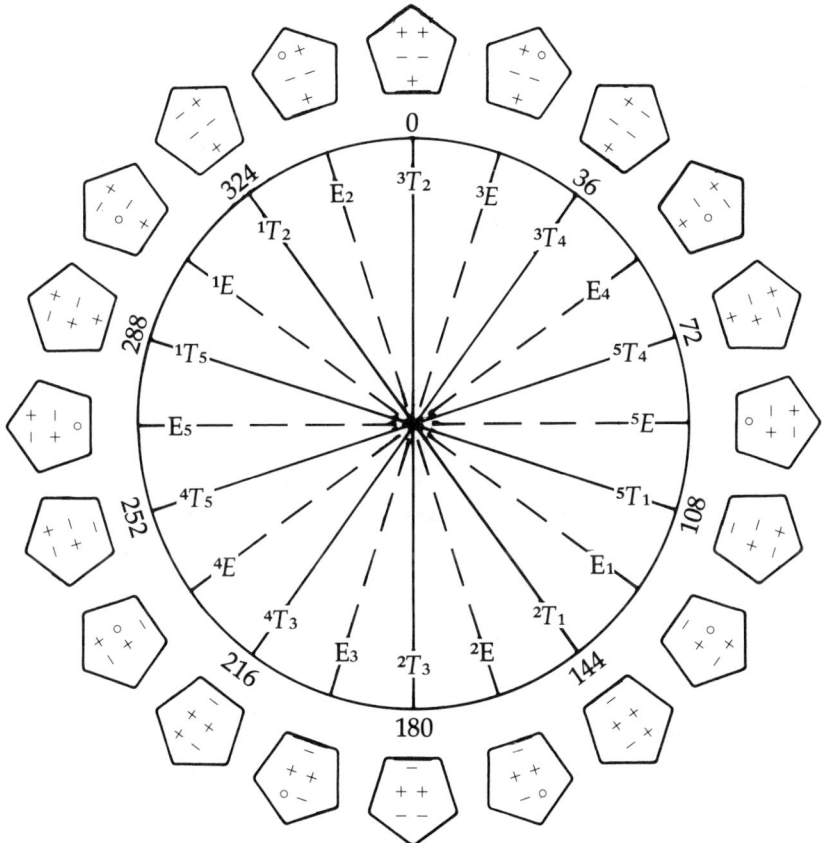

Figure 4.5. Pseudorotation pathway of the cyclopentane ring; each point on the circle represents a specific value of the phase angle of pseudorotation P.

$$\omega_j = \omega_{max} \cos(\Delta + j\delta) \qquad j = 0,1,2,3,4; \; \delta = 144° \qquad (4.1)$$

The variation of the ring torsion angles is shown in Figure 4.6. The value of the maximum torsion angle, ω_{max}, has been found to range from 42° to 50° in the D ring of a number of steroids (53). Several molecular mechanics calculations for cyclopentane have been reported, all of which agree that the half-chair and envelope have the same energy, and that pseudorotation is essentially free (activation barriers of less than 0.4 kJ), although the force fields applied differed considerably (54–58). The reason behind this is that the bending and torsional ener-

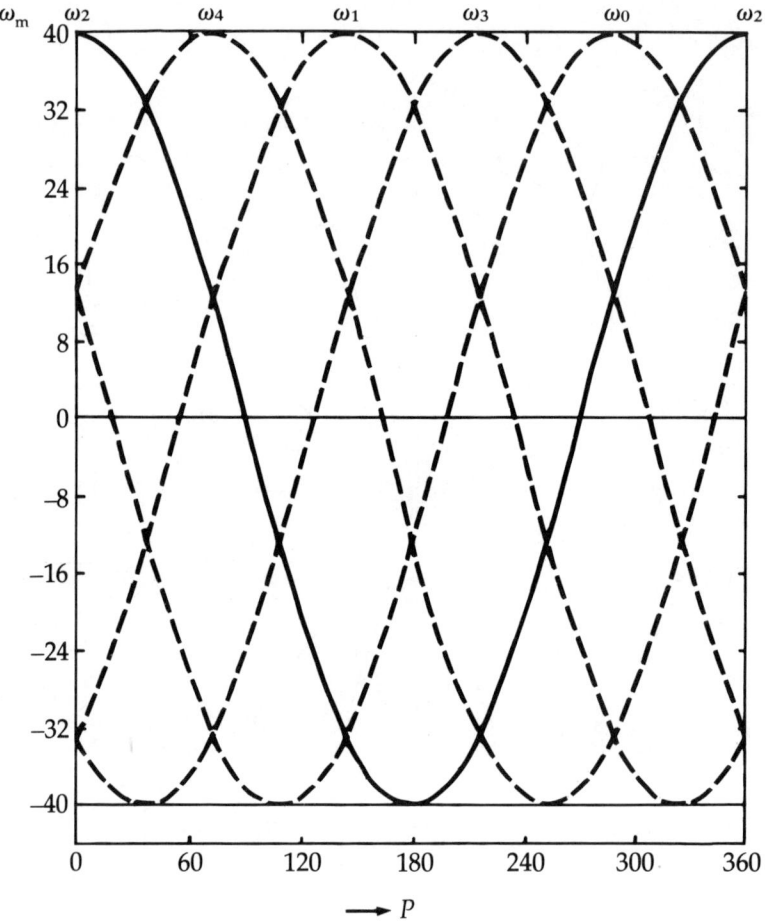

Figure 4.6. Change of the five torsion angles of a puckered five-membered ring during one full pseudorotational cycle $0 \leq P \leq 360°$.

gies are nearly independent of the phase angle (58). The calculated value of ω_{max} ranges from 42–48°, in agreement with the experimental results from the steroids and from vibrational studies. Lifson and Warshel were able to calculate all of the vibrational frequencies of cyclopentane to within 30 cm^{-1}, the lowest being the pseudorotation with zero frequency (58).

In alkylated cyclopentanes pseudorotation is more hindered, and different conformations appear as energy minima on the pseudorotation cycle. Methylcyclopentane was calculated to be most stable in the conformation with the methyl group at the "tip of the flap" of an envelope form. *trans-* and *cis-*1,3-Dimethylcyclopentane are each calculated to

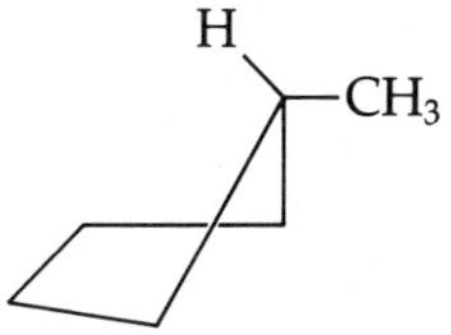

exist in two conformations of about equal energy, the *trans* isomer being about 1.2 kJ/mol less stable than the *cis* (9). Note that in all four of these conformations the methyl groups are in positions close to the molecular plane (pseudoequatorial).

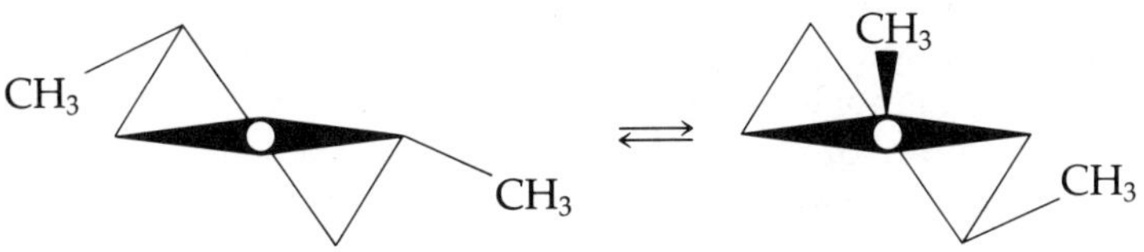

cis

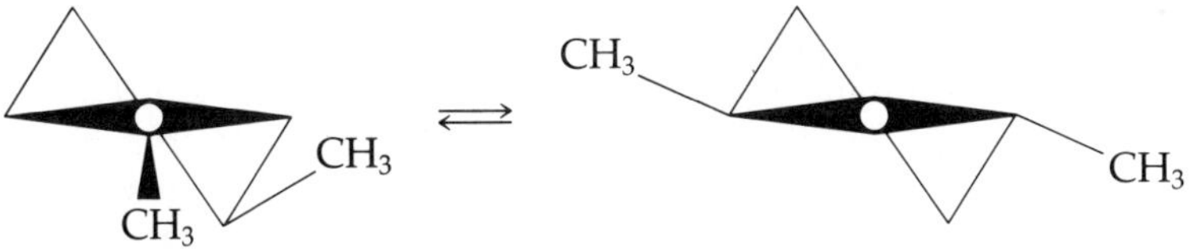

trans

Cyclohexane. The conformations of this most studied ring of organic chemistry, the chair and twist boat, were the topic of the work of Sachse (59–60) as early as 1890, and a rather good estimate was made of the height of the chair ⇌ boat barrier in 1946 (61). Molecular mechanics calculations on cyclohexane have been carried out by many different research groups, beginning with Hendrickson's pioneering machine computations (54). As stated elsewhere (62), a force field that cannot calculate a reasonable geometry for the cyclohexane chair with its bond angles of 111.4° and bond lengths of 1.536 Å [and consequently, as they are not independent coordinates in the cyclohexane ring, torsion angles of 54.5° (63)] must be rejected. Interestingly, cyclohexane was calculated to have this geometry and a C–C bond length of 1.530–1.534 Å (64) at a time when the best experimental data indicated a much shorter C–C bond distance [1.520 Å (65), later corrected to 1.528 Å (63)]. As Kitaigorodsky predicted in 1960 (66, 67), the cyclohexane ring is flattened in comparison with the ball-and-stick models (68–71); consequently, the axial substituents are bent out from the ring, not because of transannular repulsion, but simply because of the bond angles at carbon. Therefore, all of the torsion angles have slightly nonideal values. It is therefore to be expected that only a certain amount of additivity of conformational energies will occur within the cyclohexane system, but that the energy increments may not be transferable unaltered to noncyclic systems, to other ring sizes, or even to six-membered rings with a different ring pucker. Physical and chemical consequences of cyclohexane ring flattening have been discussed in the literature (70–72).

The twist-boat form is in a much shallower energy minimum than is the (rigid) chair, and interconversion of the different twist-boats by a pseudorotation requires such a small activation energy that the conformation is often regarded as flexible. Hendrickson (54) studied the process and found an activation energy of 6.70 kJ/mol, and similar values in the range from 0–7.5 kJ/mol have been found with other force fields (38, 71). The pseudorotation of the cyclohexane twist-boats through the classical boat forms, where small barriers occur, is somewhat different from the situation in cyclopentane. But since the twist-boat is not the ground state, it is not easily accessible to experiments. Nonetheless, the infrared vibrational spectrum of the twist-boat of cyclohexane was recently reported (73). So for studies on cyclohexane, the twist-boat form of molecular mechanics was fifteen years ahead of experiment.

The full interconversion pathway of the chair and twist-boat conformations is of fundamental interest and has been studied extensively by molecular mechanics. A cross section of the energy surface showing these cyclohexane conformations is given in Figure 4.7, and a contour map of the surface in Figure 4.8. Experimental studies can provide the height of the inversion barrier, and (through the activation entropy) some information about the symmetry of the transition state. Detailed discussions of the various pathways are given in the literature (71, 74, 75, 78, 349). Hendrickson's calculation scheme required that the inversion path have some kind of symmetry (54, 76). The inversion path retaining C_s symmetry has its barrier at a geometry where the ring has

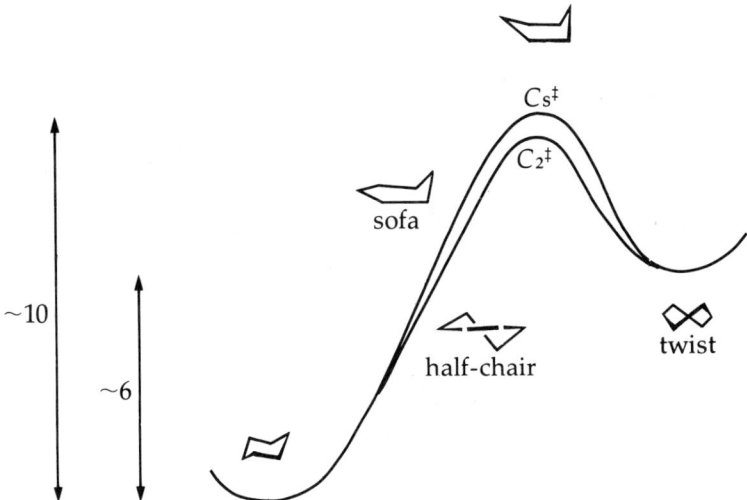

Figure 4.7. A cross section of the energy surface showing important conformations of cyclohexane.

just passed the "sofa" form (which has five coplanar carbon atoms), but Hendrickson calculated that the path retaining C_2 symmetry has a lower barrier (by 2.1 kJ) with a height of 45.2 kJ (76). Dropping all symmetry restrictions, and using their method of driving one torsion angle through a range of values (pp. 72–76), Wiberg and Boyd arrived at a transition state geometry close to Hendrickson's half-chair transition state, with four coplanar atoms (77). The driving method does not, however, pass over 0° torsion angles automatically, and when this is done by hand (with the MM1 and MM2 force fields), the energy increases further to a transition state in which the torsion angle has passed the value of zero (the half-chair), and which does not exhibit symmetry (10). With other force fields it has been found that all of these transition states have nearly the same energy (56, 71, 74, 75, 78), which means that the cyclohexane ring undergoes almost free pseudorotation at the transition state. Although it is not clear to what extent this result depends on the force field used, it seems certain that the barrier to pseudorotation in the transition state is quite small.

Alkylcyclohexanes. The early calculations of Hendrickson indicated that the equatorial methyl group in methylcyclohexane does not disturb the cyclohexane geometry, but that the axial methyl group will bend slightly away from the ring and from the symmetry plane, distorting the ring and moving the molecule a little along the ring inversion pathway (structures 1 and 2) (79). This conclusion is probably an artifact of Hendrickson's calculational method, which allowed relaxation in only a few

94 MOLECULAR MECHANICS

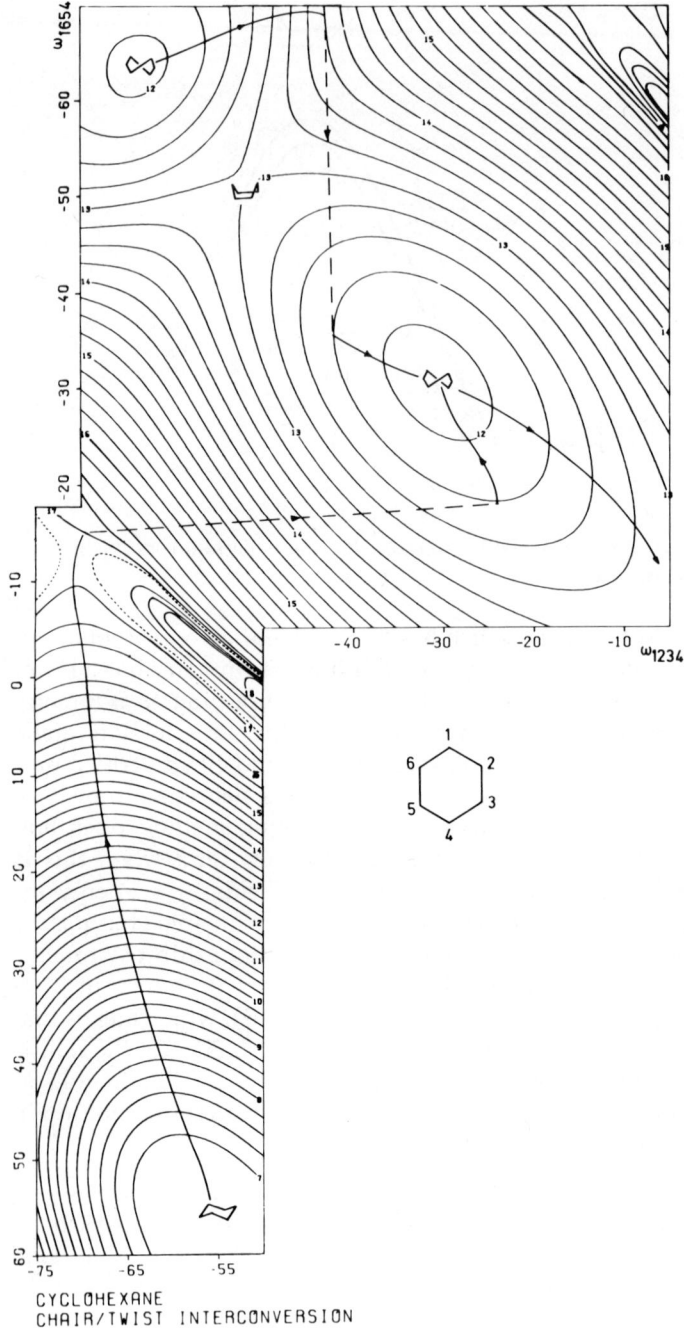

Figure 4.8. A contour map of the important part of the cyclohexane energy surface.

designated coordinates. Thus the angle α was allowed to open, but β was not; γ was not allowed to close when the symmetry plane was maintained. More recent calculations indicate these deformations are all of similar importance.

1   **2**

The equatorial methyl group was found to have some influence on the cyclohexane geometry (though not on the carbon skeleton), because it pushes the iso-hydrogen next to it into the ring (*68, 80*), which increases the energy of the interaction of the hydrogen with any syn-axial groups that might be present. This phenomenon has been referred to as "Allinger buttressing" (*81*), and experimentally it was confirmed in the conformational energies of dimethyl- and trimethylcyclohexanes (*80, 82*). The energy of the axial methyl group on a cyclohexane ring has been an important calibration point for force fields, most of which now give values of 6.7–7.5 kJ/mol for this quantity. The solution and gas phase experimental values lie around 7.1 and 7.9 kJ/mol, respectively. The buttressing of the iso-hydrogen found in molecular mechanics calculations can explain the variation in the conformational energies in the series: 1,4-dimethylcyclohexane (8.00 kJ), 1,3-dimethylcyclohexane (8.20 kJ), and 1,3,5-trimethylcyclohexane (8.58 kJ) (*80*), where the number of such hydrogens interacting with the axial methyl is, respectively, zero, one, and two.

ΔE (kJ/mol)   8.00   8.20   8.58

The slight twisting of the cyclohexane ring by axial substituents reported by Hendrickson has not been found in a number of other calculations, which might be caused by the tendency of some energy minimizing methods not to break symmetry, as Hendrickson criticized (*54*). Instead, two other effects have been found, a ring flattening (with a decrease in the ring torsion angles from 54.6° in cyclohexane to 52.3° in ax-methylcyclohexane) (*83*), and a decrease of the endocyclic bond angle at the carbon atom carrying the methyl group. In the experimental structures of a number of molecules for which structural data were available, Altona and Sundaralingam found a bond angle decrease in γ down to 104° and local flattening of the ring shape, but no notable decrease in the mean ring torsion angles (*68*). Their force field unfortunately failed to reproduce the bond angle decrease, but other force fields (MM2) show both effects. While it is clear that the repulsion between the axial methyl and the syn-axial hydrogens would lead to the angle α opening out, the angle β opens out a little more, the values being respectively 112.2° and 112.6° (MM2) (*38*).

In 1,1-dimethylcyclohexane, the ring is considerably flattened, with a torsion angle of 50.4° next to the quaternary carbon, and Schleyer's force field does, in fact, reproduce this further flattening in comparison with axial methylcyclohexane (*83*). This is mainly due to the decrease of the ring bond angle at the position of methyl substitution from 110.4° in ax-methylcyclohexane to about the tetrahedral value (109.9°). In this molecule the axial methyl group cannot bend out from the ring very much because it is buttressed by the equatorial methyl group (*83*). Bending out the whole isopropylidene group from the ring is the best way to separate the axial methyl from its syn-axial hydrogen atoms. This motion results in considerable bond and torsion angle changes.

For ethyl- and isopropylcyclohexane, relatively few of the poten-

tial axial conformations are available because the methyl-inside forms are extremely strained (*80*). Thermodynamic studies agree with this (*84, 85*).

The equatorial *t*-butyl group has been used extensively as a holding group in conformational studies (*86*), with the underlying assumption that it has no pronounced effect on the cyclohexane ring geometry. An early molecular mechanics study (*68*) concluded that the C–C bond linking the *t*-butyl group to the cyclohexane ring is stretched, the endocyclic bond angle is decreased (to minimize the repulsion between the methyl groups and the cyclohexane methylene groups), and the axial iso-hydrogen is pushed into the ring by the *t*-butyl group. A flattening of the cyclohexane ring was not observed in the calculated structure, but a surprising twist of the t-butyl group by 17° away from the staggered conformation, together with a twist of the whole ring was predicted. The energy difference between the twisted and staggered forms was found to be 0.8–1.2 kJ, and it was proposed that a torsional vibration occurred between two enantiomeric forms (*68*). (For the related effect in hexamethylethane, see p. 85.) Meanwhile, the structures of several cyclohexane and cyclohexanone rings with t-butyl substituents in equatorial positions were solved by x-ray crystallography, and only in a few cases were twisted t-butyl groups found (*87, 88*). Most of these had torsion angles smaller than 2°. The largest twist (10.9°) was found in one unsymmetrically substituted derivative (*88*). Other results of the molecular mechanics study were confirmed. The existence of an abnormally large torsional vibration in the crystal can be excluded in nearly every case on grounds of the temperature factors found. A calculation employing the Schleyer–Engler force field found a twisted geometry as the potential minimum with a torsion angle of 8.3° (*89, 333*). The more recent MM2 force field found the symmetrical geometry to be the most stable one (*38*), and it would appear that the twisted t-butyl groups were artifacts of the earlier force fields.

For the axial t-butyl group the symmetrical structure was (with the Schleyer–Engler force field) calculated to have an energy 4.6 kJ higher than a geometry where the t-butyl group was twisted from the symmetrical structure by 21.0° (*89*). The cyclohexane ring was twisted, which is a geometry similar to the one obtained by Hendrickson for axial methylcyclohexane (*79*). Indeed, an axial t-butyl derivative studied by crystallography has the group twisted by 10° (*328*).

From the variation of the infrared spectrum of *trans*-1,3-di-t-butylcyclohexane with temperature, it was concluded (*90*) that there are two conformations in equilibrium that differ in energy by 1.7 kJ/mol. They must be the chair (with an axial t-butyl) and the boat. It has been suggested (*90*) that the boat is of lower energy, but this is not certain. The calculated (MM2) conformational energies are 22.43 kJ for twist t-butylcyclohexane and 20.88 kJ/mol for ax-t-butylcyclohexane, which is again twisted as found by Baas (*89*). Assuming that the t-butyl groups which were not considered do not further destabilize these strained conformations, this energy difference is in accordance with the experimental study, and indicates that the chair form of ax-t-butylcyclohexane is more stable than the twist form. Calculations with the MUB-1 force field favored the chair over the twist form, but the Warshel–Lifson field gives a lower energy for the twist conformation, as does an ab initio (STO-3G) calculation (*91*).

Calculations of the *cis*- and *trans*-1,2- and 1,4-di-t-butylcyclohexanes with the MM1 and Schleyer force fields again gave twisted side chain conformations when the six-membered ring was in the chair form (333). The t-butyl groups of the trans-1,4 derivatives are independent of each other, and two conformations are possible. The cis-1,4 isomer is calculated to have the chair and twist form in equilibrium, with the latter predominating. The same qualitative result was also found for the trans-1,2 isomer, and this was considered to be in agreement with the NMR coupling constants reported for this compound (333). The conformations of the cis-1,2 isomer and their interconversions have been studied in some detail (334).

Cycloheptanes. Two families of conformations, the chair–twist-chair (**C** and **TC**) and the boat–twist-boat (**B** and **TB**), are possible for cycloheptane, and doubtlessly the **TC** with C_2 symmetry is the lowest energy minimum. The fourteen possible **TC** conformations can be easily interconverted by a pseudorotational process (just as with the cyclopentane conformations) with energy maxima at the fourteen **C** forms with C_s symmetry. The interrelationships between these conformations are shown in Figure 4.9, and the energies are given in Table 4.1.

One direct structure determination (96) of a cycloheptane derivative found a disordered mixture of **TC** and chair conformations in the crystal, with **TC** predominating. Another determination also showed a disordered structure that was taken as an indication of pseudorotation in cycloheptane (97). The **B/TB** family is also open to pseudorotation. Older calculations did not agree as to which of the two possible conformations was the energy minimum, while in a more recent calculation these conformations were found to pseudorotate more or less freely. The barrier to interconversion of the chair and boat families has a transition state with a geometry slightly closer to **TB** than **TC**, and was calculated by Hendrickson to have an energy of 33.9 kJ/mol (76), and, by Bocian et al., of 34.3 kJ/mol (95). A recent electron diffraction study (98) seems to indicate a mixture of twist-chair and chair forms. The latter almost certainly (from the calculations) corresponds to an energy maximum, although perhaps quite a small one. The treatment of the vibrational amplitudes, on which the structure derived from electron diffraction is quite dependent, thus becomes rather uncertain.

A methyl group hinders pseudorotation in both cycloheptane conformations to such an extent that the inversion is easier than the full pseudorotation. Hendrickson (54) suggested that in methylcycloheptane two sets of conformations out of the pseudorotational cycle of cycloheptane are present; these two sets interconvert by combined ring inversion, pseudorotation in the twist-boat, and reconversion to the other twist-chair family. A dynamic process with $\Delta G^{\ddagger}$ 41.0 kJ/mol was

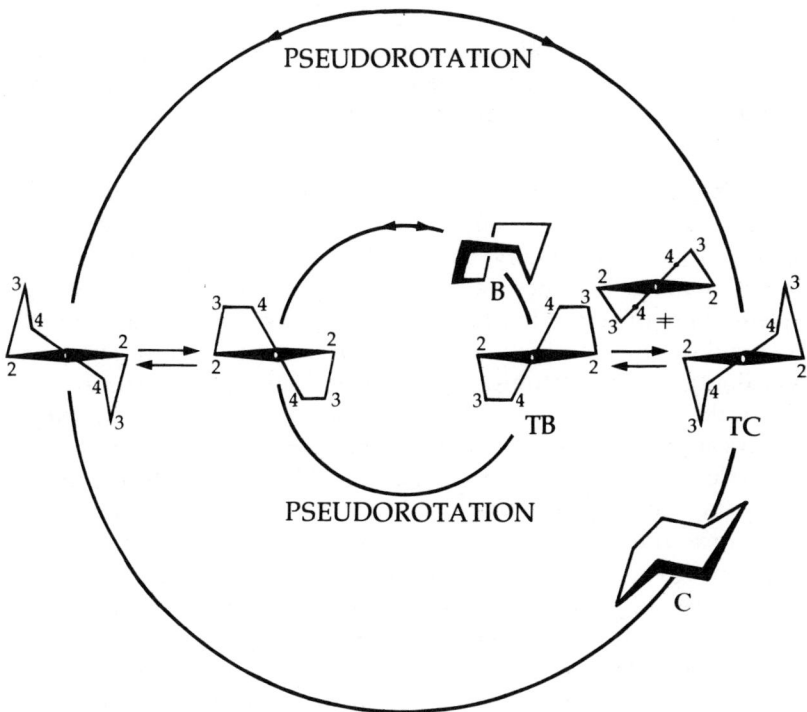

Figure 4.9. The interrelations between the cycloheptanone conformations.

Table 4.1

Relative Energies (kJ/mol) of Cycloheptane Conformations

	TC	C	TB	B
Hendrickson (1961) (54)	0.0	9.04	10.42	12.64
Lifson (1966) (92)	0.0	2.80	11.05	10.04
Hendrickson (1967) (93)	0.0	5.86	10.04	11.30
Allinger (1971) (94)	0.0	—	13.8	—
Schleyer (1973) (94)	0.0	—	16.7	—
Strauss (1975) (95)	0.0	4.52	2.51	5.82
Allinger (1977) (38)	0.0	4.23	13.18	13.18

found experimentally for 4,5-*trans*-dibromo-1,1-difluorocycloheptane (*99, 100*) that may belong either to the hindered pseudorotation or to the inversion of the ring. As usual, the positions calculated to be preferred by the methyl group are those close to the mean plane of the carbon atoms (*54*).

Cyclooctane. For rings of this size and larger, the molecular models become so flexible, and the number of conformations (which are usually designated by their symmetry) so large, that a complete analysis of all of the conformations becomes very laborious, and has usually been incomplete. As pointed out in Chapter 3, the results in such cases often do not depend so much on the force field used, but on the intuition of the person doing the calculations, and on which starting geometries were used for the energy minimizations. What were earlier discussed as symmetrical conformations may sometimes correspond to energy maxima (saddle points), but this cannot be decided as long as the full matrix Newton-Raphson criteria discussed on pp. 72-73 have not been applied. In studies of conformational interconversions, a complete analysis of all pathways is usually not necessary, as high energy pathways or routes between high energy conformations can often be excluded at the outset. To make the whole problem tractable, symmetry restrictions (sometimes unintentional) are commonly used. The torsion angle-driving routine can also be used to interconvert the conformations. In cyclooctane (*101*), and other large rings, use of the driving routine is more complicated than in smaller molecules. To reach all of the desired conformations may require studies in which a single torsion angle is first driven, and then two or more torsion angles are driven simultaneously (*101*).

Detailed discussions of molecular mechanics calculations on cyclooctane conformations in relation to NMR spectroscopic results and the crystal structures of cyclooctane derivatives have been given by Anet (*102, 103*). There are many conformations of cyclooctane that fall into three families (boat-chair, chair-chair, and boat-boat) separated by energy barriers of 33-46 kJ/mol (*104*). The members of a given family are interconverted by processes similar to pseudorotation, with activation energies of less than 12.6 kJ/mol. The more important conformations are shown in Figure 4.10 (*102*).

Only one half of the conformations in each family can be reached via pseudorotation; the other half are reached by ring inversion. This information is summarized in Figure 4.11. The lowest energy conformation is the boat-chair **(BC)** form according to most molecular mechanics calculations (*57, 76, 101, 104*) (Table 4.2). Most cyclooctane derivatives crystallize in this conformation. Wiberg's parameters indicate that a member of the chair-chair **(CC)** family (which suffers mainly from un-

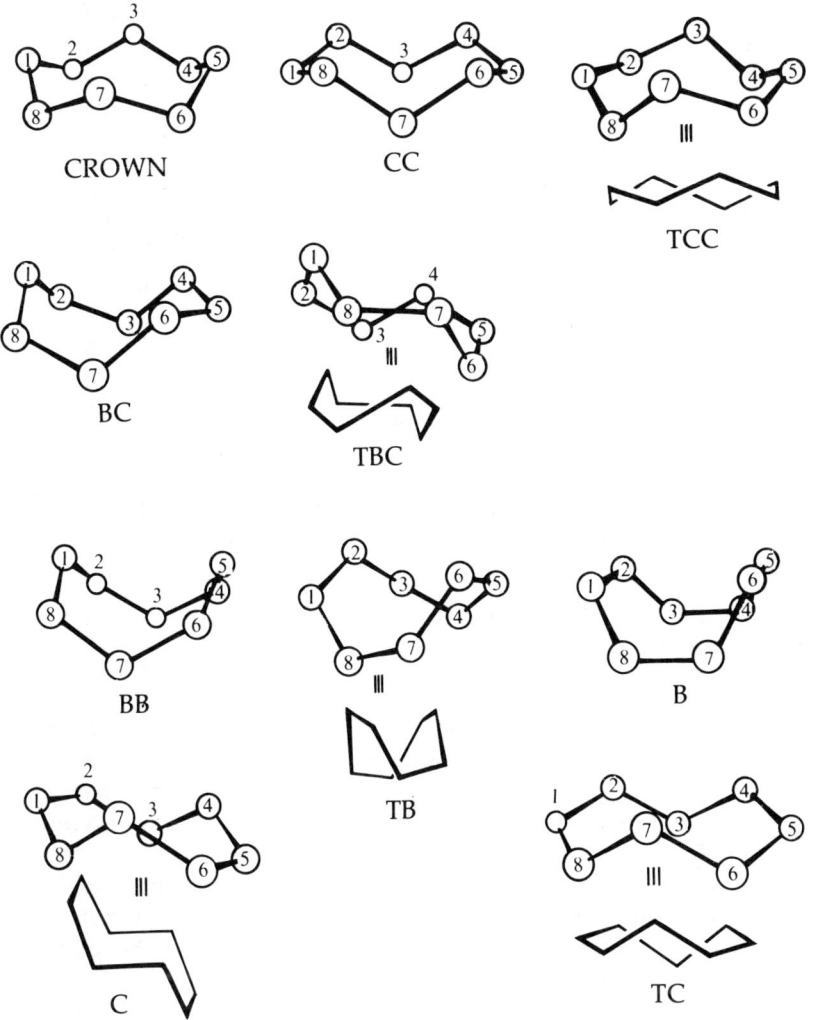

Figure 4.10. Symmetrical conformations of cyclooctane (102) according to the calculations of Hendrickson (104). Alternate views of some conformations are given.

favorable torsion angles), the twist-chair–chair (**TCC**), is more stable (105), while in most calculations it is found to be the second most stable. The Schleyer force field gives three conformations of nearly equal energy (94), with the crown being the most stable (Table 4.2). The third real energy minimum is in the boat family, but, as a result of nonbonded repulsions, it has an energy much higher than those of the other two families.

102 MOLECULAR MECHANICS

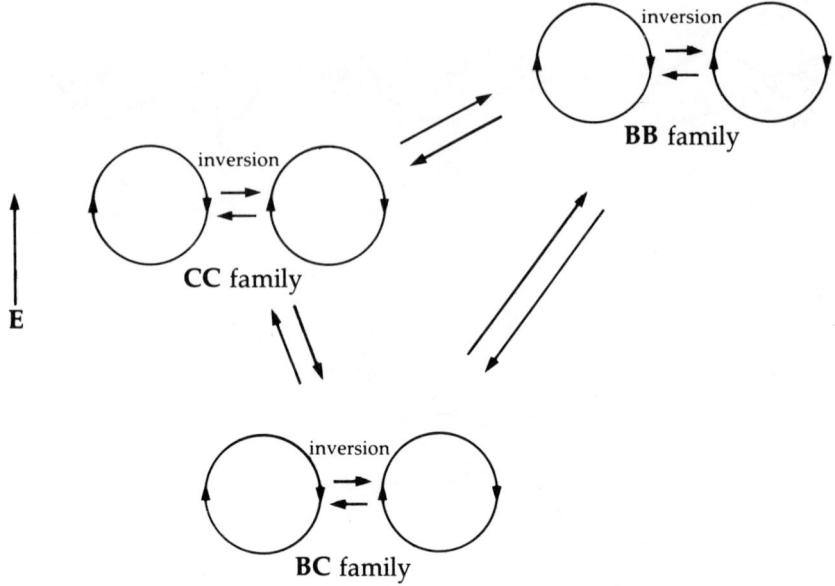

Figure 4.11. Pseudorotation, ring inversion, and other conformational interconversions in cyclooctane.

Pseudorotation of the BC conformations (107) has been simulated by driving the torsion angles ω_2 or ω_3 (101). The twist-boat–chair (TBC) conformation is an intermediate on the pseudorotation pathway, and the point of highest energy (13.8 kJ/mol) was found at an asymmetric geometry.

This pseudorotation pathway does not lead to ring inversion, and two possible pathways must be considered for inversion of the BC conformations. Anet found (with the Boyd force field) that the pathway described by Hendrickson (76, 104) passing through the TBC and the C conformations has the lowest barrier, and may best be simulated by simultaneously driving two torsion angles (ω_1 and ω_5) of the TBC form through zero (101). The C is the transition state with an energy in excellent agreement with the experimental value of 33.9 kJ/mol (Table 4.2) (364). A twist-chair conformation lying along the direct BC/C interconversion was calculated by Hendrickson to have a higher energy than C (104), and Anet, when driving the torsion angle ω_2 of the BC form only, found a higher energy transition state very close to the twist-chair (101). Therefore, ring inversion passes not through a twist-chair, but through the TBC conformation. The simulation of the inversion process by means of the torsion angle driving method proved to be especially difficult in this case because the lowest energy pathway is found only by driving two symmetry-related torsion angles simultaneously, starting

CALCULATIONS ON HYDROCARBONS 103

Table 4.2
Relative Energies (kJ/mol) of Cyclooctane Conformations

	BC	TBC	TCC	Crown	CC	BB	TB	B	C	TC
Hendrickson (1964) (104)	0	3.77	1.25	10.04	6.70	15.90	—	40.58	33.89	27.20
Wiberg (1965) (105)	1.05	—	0	2.10	—	19.62	—	38.53	—	26.53
Lifson (1966) (92)	0	—	7.91	15.15	—	—	12.18	—	—	—
Hendrickson (1967) (93)	0	8.4	7.1	11.7	7.9	5.9	3.8	45.2	34.7	36.4
Allinger (1967) (80)	0	—	9.20	8.74	9.41	—	—	—	—	—
Schleyer (1973) (94)	0.8	—	0.4	0	—	—	—	—	—	—
Anet (1973) (101)	0	7.1	3.3	6.3	7.5	11.7	11.7	46.9	31.4	32.2
Allinger (1977) (38)	0	7.1	4.2	5.0	5.0	15.1	13.0	42.7	31.4	—
Experimental (106)	0	—	7.1	—	—	>8.4	—	—	—	—

not with the **BC** conformation, but with the less stable **TBC** conformation.

The most stable conformation in the crown family with almost all force fields is the **TCC** form, and two symmetrical conformational transformation processes within the family have been proposed by Hendrickson (*104*): one passing through the (roughly circular) crown with a C_4 symmetry axis (an energy minimum); and the other, with slightly higher energy, passing through the chair–chair, with only a C_2 axis (an elongated crown), possibly another energy minimum. All conformations were calculated to be within 4 kJ of each other, so the crown family is very flexible.

In the boat–boat **(BB)** conformation excessive H···H repulsions are present, which can be partly relieved by twisting the ring. Hendrickson calculated that a twist-boat is even more stable than the **TCC**, and slightly more stable than the **BB** (*93*). More recent calculations showed that **TB** and **BB** are equal in energy and separated by an extremely small barrier (*101*). The tub shaped boat **(B)** is much more strained than the **BB** form and is not attractive, even as an intermediate in an interconversion.

The most important interfamily interconversion is that of the boat–chair and crown families. It can be simulated conveniently by driving the ω_5 torsion angle of the **TBC** through zero, retaining C_2 symmetry throughout the interconversion. The barrier is 43.1 kJ/mol, (experimentally 43.9 kJ), and the **TCC** form is obtained at the completion of driving (*101*).

For the **BC/TB** interconversion, two different transition state geometries with identical energies were found; the geometry was dependent on from which of the two conformations the driving was started (*101*). This has been interpreted as indicating two different pathways with identical energies, but it really is an artifact of the torsion angle driving method (*see* pp. 74–76). The lowest energy pathway involves driving two angles towards the **BB** conformation, which goes to the **BC** form over an extremely low barrier.

The effect of methyl substituents on the energies of cyclooctane conformations has been studied by Hendrickson (*108*), and by Ferro et al. (*109*).

Cyclononane. All force fields applied to cyclononane agree that the **TBC** conformation with perfect D_3 symmetry is the most stable form (*38, 92–94, 104, 114*). This form was also found in the crystal of cyclononanol-1-dimethylphosphonate (*111*). In the crystal of cyclononylamine hydrochloride, however, a twist-chair–boat **(TCB)** conformation with approximate C_2 symmetry was found (*112*), which was calculated to be only slightly higher in energy than the D_3 structure

Table 4.3
Relative Energies (kJ/mol) of Cyclononane Conformations

	TBC	TCC	TCB	BC	CC
Hendrickson (1964) (*104*)	0	10.0	22.2	31.0	43.5
Hendrickson (1967) (*76*)	0	9.2	—	—	—
Lifson (1966) (*92*)	0	13.01	16.36	—	—
Allinger (1971) (*94*)	0	4.2	4.6	—	—
Schleyer (1973) (*94*)	0	0	2.9	—	—
Allinger (1977) (*38*)	0	—	3.14	—	—
Anet (1980) (*345*)	0	7.11	5.02	—	—
Experiment (*345*)	0	9.4	4.0	—	—

(Table 4.3). ^{13}C NMR data show a structure with D_3 symmetry at low temperatures (*113, 114*), but because of the unfavorable entropy of the D_3 structure, it accounts for only 40% of the equilibrium mixture at room temperature, with the C_2 [255] conformation (*115*) predominating (*114*).

The interconversion map of cyclononane has been studied qualitatively by Dale (*115*) and by Hendrickson (*76*), and calculations have been reported (*114, 116*) that yield interconversion barriers in good agreement with the experimental value (*113*).

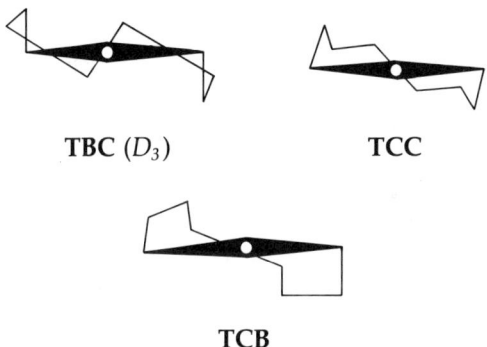

TBC (D_3)   TCC

TCB

Cyclodecane. The number of conformations that can be imagined for cyclodecane is so large that no systematic investigation of all of them has been reported. Therefore, experimental studies have led the way for the molecular mechanics calculations. From x-ray crystallography it was known that cyclodecane derivatives usually crystallize in the **BCB** conformation (*117*). The first calculations on cyclodecane, therefore, involved this conformation, together with a number of other reasonable ones (*92, 104, 105*). All of the latter were calculated to be several kilojoules less stable than the **BCB**, and cyclodecane was regarded as a very immobile molecule, conformationally as biased as cyclohexane. The preference for the **BCB** conformation has also been confirmed by electron diffraction (*14*). Problems in the x-ray analysis of a tetramethyl cyclodecane derivative in which the **BCB** conformation would suffer transannular repulsions from methyl groups pointing into the ring in-

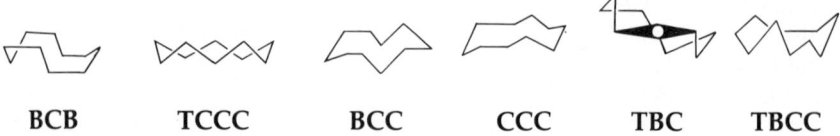

BCB   **TCCC**   **BCC**   **CCC**   **TBC**   **TBCC**

duced the Lifson group to study additional conformations. Two new cyclodecane conformations, a **TBC** and a **TBCC** conformation, which had previously not been considered, were found and these were calculated (for the parent cyclodecane) to be only 8.8 and 13.0 kJ/mol less stable than **BCB** when the starting geometry for energy minimization was obtained from the unrefined coordinates of the tetramethyl derivative (*118*). Still another low energy conformation, **TCCC**, was located by Schleyer in the crown family of the cyclodecane system (*94*); he found that with the Allinger 1971 force field and with his own, this form was calculated to be even more stable than the **BCB** form. The **BCB** form is disfavored by strong H···H repulsions inside the ring, which are not present in **TCCC**, and, consequently, force fields with strong H···H repulsions tend to disfavor **BCB** (*119*). Hendrickson did not find this conformation because of the symmetry restrictions in his approach. The force field of Lifson and Warshel (*58*) has been reported to yield several conformations of greater stability than **BCB**, including **BCC** and **CCC** conformations (*14*). More recent force fields were employed to study again the **TCCC** and **BCB** conformations, and with these the **BCB** was calculated to be of lower energy (*10, 14, 110, 120*) in White's force field (which was developed from Schleyer's earlier force field) by 2.1 kJ/mol. The two conformations may be quite close in energy, and it is surprising that in the crystals of many different derivatives, only the **BCB** conformation has been found. Electron diffraction suggests a Gibbs free energy difference of more than 3.3 kJ between the **BCB** form and the next most stable form (*14*). Interconversion pathways calculated under symmetry restrictions yield barriers that are too high (*76*), but good agreement between calculated and experimental barriers was obtained with the driver method (*110*).

Cycloundecane. In early calculations of this flexible ring only one (*92*) or two (*94*) conformations were considered. More extensive calculations

CALCULATIONS ON HYDROCARBONS 107

considered other conformations, too (121, 122). Dale reports two conformations that are isoenergetic, and two more that have relative energies of 5.0 kJ/mol and 6.3 kJ/mol, respectively (115). He also studied the conformational interconversion pathways, which consist of complicated series of steps (115).

Cyclododecane. The original structural work here was not totally straightforward. The crystal structure is disordered, and the average electron density at each lattice point corresponds to a superposition of two kinds of molecules randomly distributed with respect to the mirror plane. The observed arrangement of peaks in the disordered structure was compatible with two different conformations of the individual molecule, as illustrated in the idealized diagram (Figure 4.12), where the superposition C, of equal amounts of A and A', is indistinguishable from the superposition of B and B' in equal amounts. It was first decided (123) that conformation B could be rejected, since the bond lengths deduced from this model varied by an unreasonable amount relative to the estimated standard deviation in the measurements. Shortly after the publication of the preliminary communication (123) on the subject, conformation A was presented at an Organic Chemistry Colloquium held in early 1959 at the Swiss Federal Institute of Technology (E.T.H.) where in the discussion, Professor V. Prelog expressed his opinion that conformation A could not be the stable one. A critical reevaluation of the x-ray

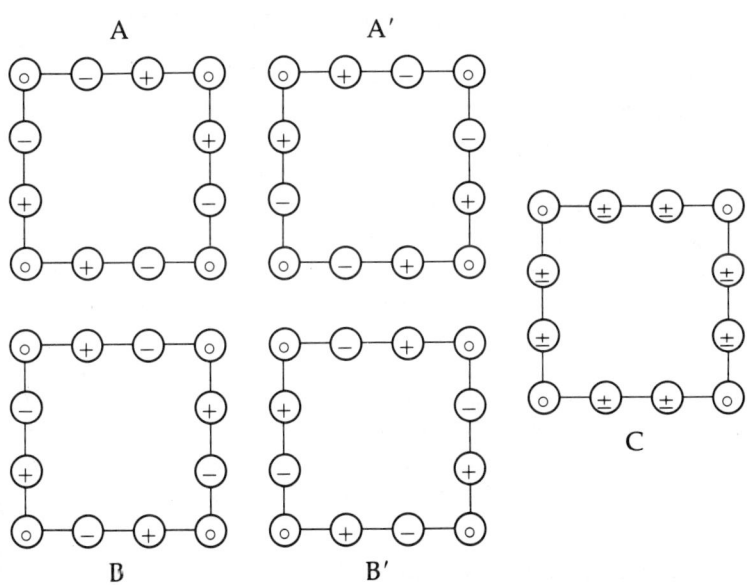

Figure 4.12. Two different conformations for cyclododecane, A and B, that are equally compatible with the observed arrangement of peaks C in the disordered crystal structure.

evidence (124, 125) showed that the alternative conformation B fitted the evidence as well as did conformation A, and was energetically preferable. A tempting conclusion was to draw the moral that a good theory is better than a bad experiment. However, it was also pointed out that the contrary is also true, namely that a good experiment is better than a bad theory.

The experimental structure B (125, 126) has been confirmed using several different force fields (38, 57, 58, 92, 94, 121, 128). The alternative structure A is calculated to be higher in energy by 43.9 kJ/mol (MM2). In the crystal of the 1,4,7,10-tetraoxa derivative, however, a different conformation has been detected (127). Dynamic processes in cyclododecane have been studied computationally and experimentally by Anet, and a multistep process accounts for the observed barrier [$\Delta G^{\ddagger}_{exp}$ = 30.5 kJ/mol (126), $\Delta H^{\ddagger}_{calc}$ = 35.1 kJ/mol (128)].

Large Rings. The conformational problem becomes more and more difficult as the ring size becomes larger, and there is a distinct possibility that important conformations may be overlooked when the starting geometries are generated (*see* pp. 59–61). An automatic procedure for the generation of all possible starting geometries was reported recently and should help in this regard (129). Cyclotridecane has been studied semiquantitatively, and four conformations of similar energies were obtained (121). However, full force field calculations (122) give results where the stability ordering differs from these simple calculations, and each minimum energy conformation proved to be unsymmetrical. For cyclotetradecane, the conformation found in the crystal (130, 131) and in solution (126) has a diamond-lattice type backbone, and was calculated to be the most stable (121, 132). The same conformation has been reported for 1,3,8,10-tetraoxacyclotetradecane (133), but in the 1,4,8,11-tetraoxa analog, a conformation with approximate C_s symmetry has been found (134). Semiquantitative calculations have been reported for cyclopentadecane (121), but no experimental information is available on this compound. Full calculations (122) agree with the earlier simple ones. The largest unsubstituted hydrocarbon ring studied to date is cyclohexadecane, and a strong preference for a square, diamond-lattice type conformation has been obtained (121, 135, 136). This conformation is also found in the crystals of the 1,1,9,9-tetramethyl derivative (137), and 1,5,9,13-tetraoxacyclohexadecane (138). Finally, some 16- and 18-membered diketones and ketals have been studied, by molecular mechanics and by x-ray crystallography (136, 139) (*see* pp. 216–217).

Fused Rings and Other Bicyclic Hydrocarbons

In this section we will discuss saturated carbocyclic fused ring systems in which each ring contains five or more members. Examples in which a small ring is present will be deferred to the next section. Even with this limitation the number of possible compounds to be considered is very large, and only the more important examples can be covered in detail. A large number of calculations on known and hypothetical molecules have been reported by Schleyer and coworkers (94), and many hydrocarbons falling in this group are mentioned in Chapters 5 and 8.

Bicyclo[3.3.0]octane. This compound exists as moderately strained *cis* and highly strained *trans* isomers. The structure of the *cis* isomer is perhaps unexpected, being the form with C_s symmetry, rather than the C_{2v} or a twisted form suggested earlier (57). The calculated energy difference favors the C_s form by 1.0 kJ/mol with MM2 (38), which is in good agreement with experiment (140). The *trans* isomer is highly

C_{2v} C_s

strained, mainly because of the very wide CCC inter-ring angle (124° by MM2). The *cis–trans* energy difference is quite large, with the *cis* isomer being more stable [by 29.5 kJ/mol (MM2) (38) and 54.8 kJ/mol (Boyd) (141) computationally, and by 28.5 kJ/mol (141) by heats of combustion].

Bicyclo[4.3.0]nonane (Hydrindane). This ring system occurs widely in natural products, and it has been studied in some detail, both experimentally and computationally (142). The heats of formation (143) and an accurate energy difference (144) between the *cis* and *trans* isomers are available. The bridgehead methyl derivative, 8-methylhydrindane, analogous to the C/D structure found in steroids, also has been studied (9, 142). In the parent system the *trans* form is more stable by 4.48 kJ/mol (144), while in the methyl derivative the *cis*-fused form is of lower energy (142). The five-membered ring is in an envelope conformation in the *cis* form, and in a half-chair with C_2 symmetry in the *trans* form. Other conformations are found in steroids, depending on the substitution pattern (53, 145).

Decalin. The earliest calculations on the decalins gave too small an energy difference between the chair and twist conformations of the six-membered rings, but for a long time now, all calculations have agreed on the strong preference for double chair conformations in both the *cis* and the *trans* isomers (9, 80, 146). The electron diffraction structures of the two isomers (146) are in good agreement with recent calculated structures, although the latter are considered to be more accurate. The *cis–trans* energy question was a fundamental one in the early days of conformational analysis (*147*), and the molecular mechanics calculations reproduce well the observed data, and also those for the bridgehead methyl derivative (9-methyldecalin) used as a model for the steroids (*9, 80, 141*).

The *cis* isomer of decalin has been studied in considerable detail (*360*). The various chair, twist, and boat forms, the pseudorotations they can undergo, and the interconversions of one to the other have all been examined.

Bicyclo[5.3.0]decane (Perhydroazulene). The *cis–trans* energy difference in this molecule is small. Direct equilibration experiments showed that the *trans* was more stable by 1.3 ± 0.8 kJ/mol (*148*). Early approximations (*54, 149*) led to the conclusion that conformations exist for five- and seven-membered rings such that they can be fused to each other in either *cis* or *trans* configurations without much added strain. Calculations were reported by Boyd who found the *cis* to be more stable by 0.8 kJ/mol (*141*).

Perhydroanthracene. There are five stereoisomers of this compound, all

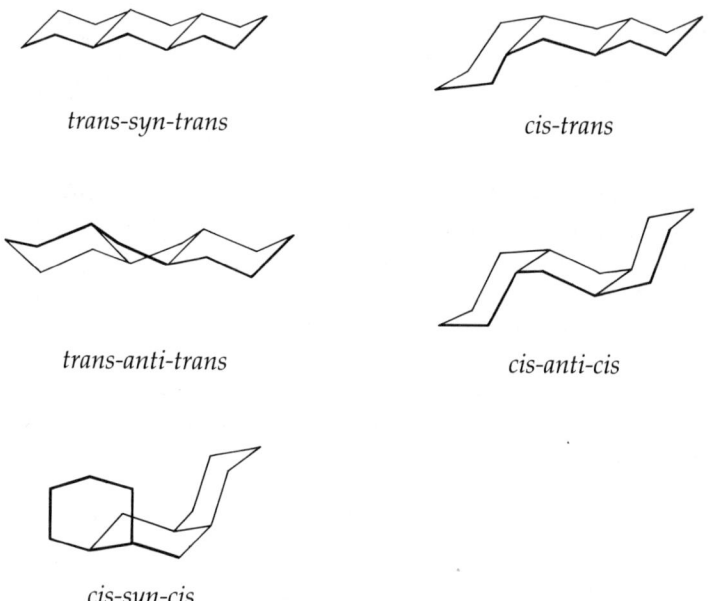

trans-syn-trans

cis-trans

trans-anti-trans

cis-anti-cis

cis-syn-cis

of which are known. Equilibration studies have given experimental values for the relative energies (*150*). Most of these were predicted long ago by additivity considerations (*151*). Molecular mechanics (1971 force field) gave results in good agreement with experiment (Table 4.4) (*150*).

Table 4.4

The Thermodynamic Quantities[a] for the Perhydroanthracenes (150)

Isomer	$\Delta H°_{exp}$	$\Delta S°_{exp}$	$\Delta G°_{exp}$[b]	$\Delta H°_{calc}$	$\Delta S°_{calc}$	$\Delta G°_{calc}$[b]	$\Delta H°_{est}$[c]
tst	0	0	0	0	0	0	0
ct	+11.55	+ 8.8	+ 6.78	10.96	+11.7	+ 4.60	10.0
tat	+17.36	− 6.7	+21.00	24.52	0	+24.52	23.4
cac	+23.35	+ 1.3	+22.68	23.26	+ 5.9	+20.08	20.1
csc	+36.57	+16.7	+27.45	34.02	+ 9.2	+29.00	26.8

[a] Units are kJ/mol for $\Delta H°$ and $\Delta G°$ and J/mol-deg for $\Delta S°$.
[b] 544 K.
[c] Reference 151.

Perhydrophenanthrene. Early predictions were also made for the six stereoisomers of perhydrophenanthrene (151), most of which have been borne out by molecular mechanics (MM1) and by equilibration experiments (152, 153). Two of the isomers are rather exceptional. The *trans-syn-trans* isomer has the central ring constrained to a boat conformation,

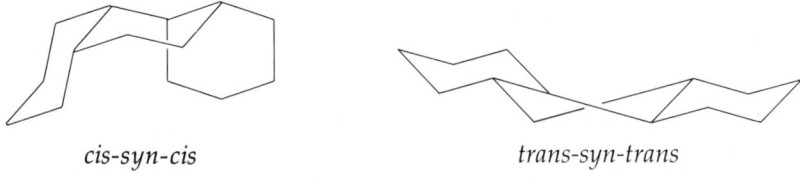

cis-syn-cis   trans-syn-trans

while the *cis-syn-cis* isomer contains a severe 1,3-diaxial interaction between methylene groups attached to the middle ring. Conflicts remain between theory and experiment, and there may be some experimental problems as well, since not all of the isomers have been identified by independent synthesis, and the final word on this system is not yet in.

Bicyclo[2.2.1]heptane. The norbornane molecule is a standard one that those working with molecular mechanics want to deal with correctly. The heat of formation is known (154, 155), and the structure moderately well characterized (156, 321). Since the compound is usually used in initial force-field parameterizations, its geometry and energy are gener-

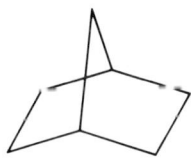

ally well reproduced. Of special interest is the small CCC bond angle at the bridging carbon, C_7. This angle compression is accompanied by bond stretching, as would be expected from the principles outlined above (pages 27 and 79), (*57, 66, 67, 78, 94, 158*), but this stretching is reproduced by the calculations only when stretch–bend cross terms (or Urey–Bradley terms) are included.

Bicyclo[2.2.2]octane. The structure of this molecule is known (*158–161*) to be one with D_3 symmetry, and not D_{3h}, both by crystallography (*159*) and by calculation (*158–161, 169*). The latter has the bridges eclipsed, and the calculated (MM2) energy can be reduced by allowing the molecule to twist to a shallow energy minimum (0.25 kJ) so that the torsional angle in each bridge increases from 0° to 12°. Since the calculated barrier is much smaller than RT, the molecule will presumably show a very low frequency, wide amplitude torsional vibration, which will carry across the barrier. This structure is unusual in that the bridgehead carbons are

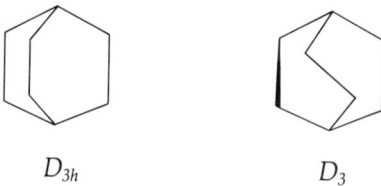

D_{3h} D_3

held very close together and exert considerable van der Waals repulsion on one another with most force fields. So much, in fact, that heats of formation are calculated to be too positive in the main (Chapter 5). The only force field that appears to have correctly calculated this quantity was MM1 (*64*), but the error with MM2 (which gives a good structure for the diamond) is only 4.10 kJ/mol.

Bicyclo[3.3.1]nonane. This molecule has long been of interest because of the interference between the methylene groups at C_3 and C_7, which tends to destabilize the chair–chair conformation. This interference is not present in the chair–boat conformation, which suggests that the two conformations will be similar in energy. It was anticipated that different examples of substituted derivatives might be found in which either conformer would predominate. This is, indeed, the case.

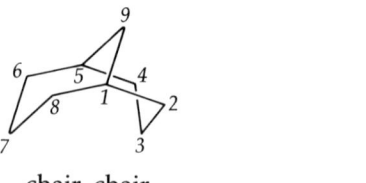

chair–chair   chair–boat

The parent compound has been found to exist preferentially in the chair–chair conformation. The chair–boat is calculated (162) (MM2) to be higher in energy by 9.67 kJ/mol. Electron diffraction studies at 65° and at 400° C have yielded 5 ± 4 and 25 ± 3% chair–boat, respectively, in agreement with the calculations (162, 163). The C_3–C_7 nonbonded distance in the double chair conformation is rather well known from x-ray and neutron diffraction measurements on various derivatives (164–166) (approximately 3.09 Å). The more interesting H/H distance is less well known, but is about 1.9–2.0 Å. The calculated values tend to be slightly larger (MM2, 3.18 and 2.02 Å).

The twist-boat form of cyclohexane is more stable than the symmetrical boat (C_{2v}) by 4.60 kJ/mol, and is less stable than the chair by 22.43 kJ/mol, while the transition state separating the chair from the boat manifold is 44.06 kJ above the former [MM2 (10)]. In the bicyclo[3.3.1]nonane system, the chair–boat can be taken as similar to the boat form in cyclohexane, as a reference point. From here, the chair–chair is only 9.67 kJ/mol lower in energy (MM1), and the transition state is only 20.92 kJ/mol above the latter (167). Interestingly, the twist-boat is not considerably more stable than the symmetrical boat here, as it is in cyclohexane. Rather, the energy is invarient to a sizable amount of twisting, and the potential surface is quite flat perpendicular to the inversion coordinate leading from the chair–chair to the chair–boat (167). The heat of formation was recently determined experimentally (168), and it agrees exactly with the previously calculated MM2 value (162, 163). The Bovill–White force field also gives a good structure and heat of formation for the double chair conformation (164–166).

Bicyclo[3.3.2]decane. This molecule has been studied theoretically in much detail (169, 170). The calculated structure of lowest energy was found to contain one seven-membered ring in the boat and the other in the chair form, with the ethano bridge eclipsed. The double chair conformation has a slightly higher energy, and it is twisted to avoid the eclipsed ethano bridge. The energy minima are much shallower than in the case of bicyclo[3.3.1]nonane (170).

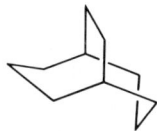

Perhydroquinacene. This molecule is of interest as a fragment of the elusive dodecahedrane, and its heat of formation has been determined (171). Perhydroquinacene has a rather different geometry than its counterpart in dodecahedrane, however. The five-membered rings in the

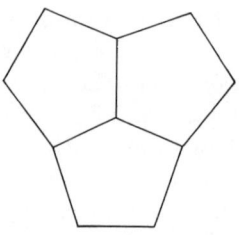

latter are all planar (172), whereas they are each twisted in perhydroquinacene in such a way as to give C_3, not C_{3v}, symmetry. The molecule can, in principle, go over to its enantiomer by a concerted twisting of all rings through planar forms, but torsion angle driving indicates that it twists one ring at a time (350), and NMR studies support these results. The experimental heat of formation is well reproduced by the Schleyer force field (94), but not so well by MM2 (38). A check of the experimental value would be desirable.

Dodecahedrane. The geometry of dodecahedrane was calculated to be fully symmetrical by several force fields. The vibrational frequencies were all calculated to be nonzero, confirming that the symmetrical structure represents a minimum of energy, and not a saddle point between minima (172). This molecule represents the outstanding difference between the MM1 and Schleyer force fields regarding their predicted heats of formation (94). The calculated values with the MM1, Schleyer, and MM2 force fields are −0.84 kJ/mol, +192.5 kJ/mol, and +92.7 kJ/mol, respectively. The reasons for these extreme discrepancies appear to be understood (see pp. 50–52 and Chapter 5), but an experimental value is not known. Since this quantity represents a sizable extrapolation by the force fields, the correct value is uncertain.

The parent dodecahedrane is itself an unknown compound at this writing. The 1,16-dimethyl derivative has recently been synthesized, however (322). A rather good crystal structure for the compound has also been reported (323), but the heat of formation is not yet known. The very interesting monosecodimethyldodecahedrane was also prepared in the course of the synthesis. This compound is highly strained indeed, since there are two hydrogens trying to fit into the space between two carbons that would, if the rest of the structure were to have minimum strain, be only 1.54 Å apart. The crystal structure of this compound has been obtained, but it is not highly accurate (323). The MM2 structures for both compounds are in good qualitative agreement with the crystal structures. The bond lengths calculated for dodecahedrane itself are all equal at 1.530 Å, but the presence of the methyl groups leads to quaternary carbons, which generally have longer bonds than tertiary carbons. The quaternary bond lengths here are calculated to be 1.536 Å, while the

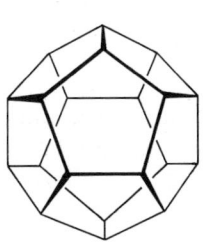
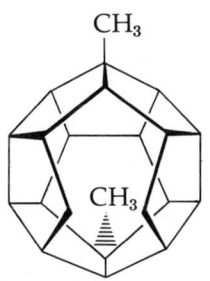

dodecahedrane   secodimethyldodecahedrane

other bonds in the ring system are about 1.530 Å. Experimentally (323), the bond lengths are 1.551 Å for the quaternary ring bonds, and 1.545 Å for the rest of the ring bonds. The calculated values are therefore systematically too short. Ermer calculated 1.538 Å for the bond length in dodecahedrane (172). With the monoseco compound, the calculated distance between the two carbons that were bonded in the parent is 3.03 Å, in exact agreement with the experimental value. The two closest hydrogens attached to those carbons are 1.862 Å apart (calculated, not located accurately by x-rays). The deformation of the molecule occurs with many expansions and contractions of bond lengths and angles, which are observed in the crystal, and these trends are faithfully reproduced by the calculation (38), but again the calculated bond lengths tend to be somewhat too short.

Diamantoid Compounds. Adamantane and related compounds are of special interest, because they contain many interactions between carbon atoms in the internal part of the molecule, which are few or lacking in most open chain or simple cyclic systems. Not much structural informa-

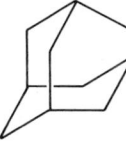 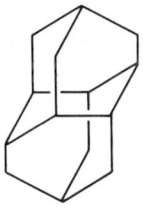

adamantane   diamantane

tion is available regarding these compounds. The structure of adamantane itself is known (173), and is reasonably well reproduced by most force fields. Derivatives such as the 1-methyl- and 2-methyladamantane and diamantane are known compounds, and have known heats of formation, but experimental structural information is lacking. Since the heats of formation are fairly well calculated (pp. 177–179), the theoretical structures are probably pretty good.

Compounds such as adamantane and diamantane are segments of a diamond lattice, with hydrogens around the periphery. It is easier to calculate a good structure for adamantane than it is for diamond, however, because, in the former, errors in the carbon van der Waals characteristics are diluted by the hydrogen van der Waals characteristics. Thus, the MM1 force field, with its very hard hydrogens and small carbons, gave a reasonable structure for adamantane; but the bond length for diamond was quite small (calculated 1.523 Å, experimental 1.544 Å). The MM2 force field gives a good structure for adamantane, and a much better structure for diamond (1.541 Å). The MUB-2 force field is also said to fit well to the diamond lattice, but the behavior of most force fields toward the diamond remains unreported. Since the diamond represents a limiting case for adamantane types of compounds, it is an important substance for a force field to deal with correctly.

Small Rings

Rings containing five or more ring members have usually been treated using the same parameters that are used for open chains. The three- and four-membered rings have sometimes been treated this way, (9, 94) but, alternatively, they have been given separate parameters (10, 94, 141). When the data are examined closely, it is found that only the latter approach works well, and quite justifiably (see pp. 24–27).

Cyclopropanes. It should be clear that cyclopropane is a special case. We cannot consider it as being simply an open chain, like propane, which is distorted down to a very small bond angle. Such a distorted system would differ from the actual cyclopropane in that it would have a severe repulsive interaction between the terminal carbons, which does not exist in cyclopropane because the two atoms are bonded together. In addition, the angle deformation is so great that the harmonic approximation is not expected to be satisfactory. Alternatively, one can think of the electronic structure of the molecule as quite different from that of an open chain molecule, just as an alkene has a different electronic structure from an alkane.

Some effort was expended over a period of several years to try to develop a molecular mechanics treatment for cyclopropane ring systems (174). This has not proved to be an easy task. There are many cyclopropane derivatives known that contain a great deal of strain apart from the cyclopropane ring, and to deal with this multitude of strained compounds simultaneously has been difficult. A force field for cyclopropanes now exists based on the MM2 force field; it is reasonably good, but is still not of the same quality as that for the alkanes (38).

There are a substantial number of relatively simple, cyclopropane-containing molecules with known structures, or known heats of formation, or sometimes both. We exclude from consideration here a number of relatively large molecules containing a cyclopropane ring (in steroids, for example), for which crystallographic structures are available. These are not really suitable for parameterization of the cyclopropane part of a molecule, because there are too many other kinds of interactions present to complicate matters. The compounds examined (174) are shown below.

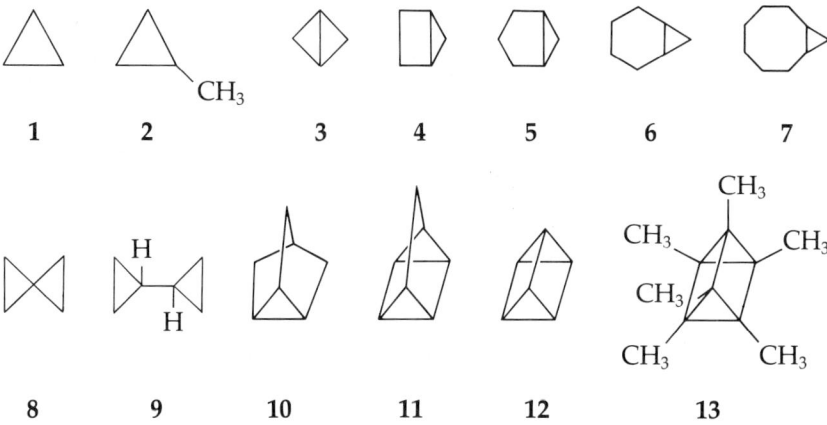

For compounds **1, 3–6,** and **8–11,** both the structures and the heats of formation are available. [The heats of formation of **10** and **11** were not known with certainty at the outset of this study (174), and the force field work depended on an experimental redetermination of these values (175).] For **13,** only the structure is known, and for **12** nothing is known. For the remaining compounds, the heats of formation are known, but not the structures.

The cyclopropane ring is relatively rigid, since any kind of angular deformation also requires a bond length deformation. The 60° angle internal to the cyclopropane, therefore, was taken as the standard, more or less ordinary force constants were used internally and externally to the cyclopropane ring, and natural bond lengths and angles were chosen in the usual way to optimize the fit between the calculated and experimental structures. It was crucial to also fit the energies of all of these compounds at the same time, and this was done with the introduction of only three heat of formation parameters. One parameter was for a C–C bond of the type in cyclopropane, a second was for the CC bond where one of the carbons is cyclopropyl in type, and the other is an ordinary sp^3 type of carbon. A third parameter was needed, because sometimes cyclopropane-type carbons are attached within a cyclopropane ring, and sometimes outside of such a ring (as in bicyclopropyl, for example). This third parameter was chosen to be simply the cyclopropane ring itself.

As might be supposed, it is not difficult to fit the data for molecules that are relatively simply substituted cyclopropanes. [See, for example, a study of 9 (37).] The difficulty comes when there are structures in which the cyclopropane ring is joined with the rest of the molecule to produce considerable additional strain (as in bicyclobutane 3, and spiropentane 8, for example). All available heats of formation were fit with a standard deviation of 2.10 kJ/mol (174) (compared with 1.92 kJ/mol with alkanes generally), but this result is probably not as good as it looks, because there are not as many data accumulated for the cyclopropanes as there are for the alkanes.

Most of the structures were fit to within three standard deviations of the reported values (174), except for a few of the most strained molecules, for which the results are shown. Experimental values are given in parentheses for comparison.

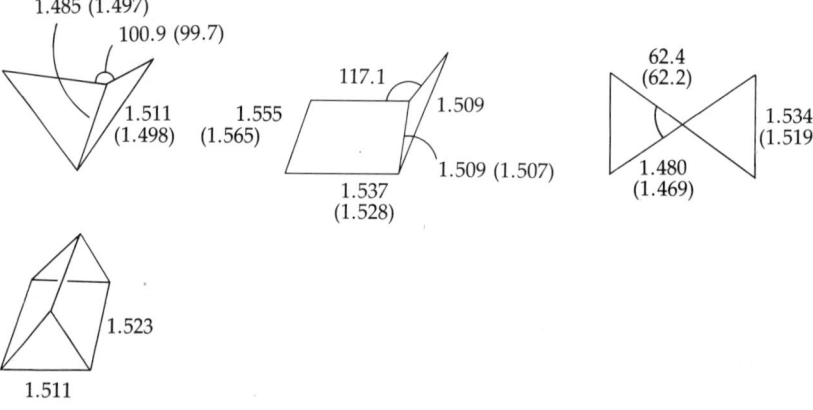

Bicyclo[3.1.0]hexane (5) is an interesting molecule. The six-membered ring here is known experimentally to occur in a boat conformation (176), as is correctly calculated (38, 174). The chair conformation is not known experimentally, and it has been estimated from the vibrational spectrum that it is not in fact an energy minimum (177). It was calculated to be an energy minimum, but 10.5 kJ above the chair, and only 3.3 kJ below the barrier that separates them. Hence, the calculations (38, 174) show it would exist as a separate species, but barely so. Structures have been reported for nortricyclene 10 (178) and quadricyclene 11 (179), but they are of limited accuracy. The calculated structures are reasonably consistent with the information available in the literature, and the predictions are believed to be pretty reliable.

Cyclobutanes. These molecules represent an intermediate between cyclopropane, where special parameters clearly are needed, and cyclopentane, where they are not. Treating cyclobutane with ordinary parameters using a harmonic bending potential results in a bending strain energy that is much too large. In an effort to retain the same parameter set, various adjustments may be made. A second term (cubic or other) can be added to the bending potential, with the sign chosen to reduce the bending energy of the cyclobutane ring. This was done by Schleyer (94), and also in the MM1 force field (64). But this does not in itself fully solve the problem. There is a tendency for the calculated cyclobutane ring to remain planar unless the bending potential becomes extremely low when the bond angle is near 90°. A way around this would be to increase the torsional energy, or, as was done with MM1, to introduce a torsion–bend interaction term (64). Again, the latter was not entirely satisfactory, as the torsion–bend interaction required for cyclobutane tended to give poor results for cyclopentane, and vice versa. Earlier force fields simply assigned different parameters as needed to fit cyclobutane (57, 141). While that works in simple cases, it introduces another problem, because the cyclobutane can be attached in various ways to other groupings, such as to other cyclobutane or cyclopropane rings, and the number of parameters required may become multitudinous. It is not immediately obvious how such cases are to be dealt with.

Whether or not the bending potential should be changed from a straight harmonic is uncertain. A harmonic potential is suitable when the bending is small, and is not expected to be suitable when it is large, but cyclobutane is somewhere in between. Ermer has implied (158) that bendings down to angles of about 90° are still within the harmonic approximation in his force field. In a five-membered ring (or, in general, in an n-membered ring) there are five (in general, n) van der Waals types of interactions between carbons 1,3 to each other; these are often not included explicitly in the calculation. These interactions have to be allowed for in a valence force field in the bending function and in the cross terms. In a Urey–Bradley force field they are included in part in the Urey–Bradley terms. In a four-membered ring, the valence force field does not include what here are very large interactions. There seems little hope that these terms can be ignored, because these repulsions occur only twice in cyclobutane, where there are four angles being deformed, as opposed to n times in an n-cycloalkane, where there are n angles being deformed. So theory tells us that the bending and torsional parameters should be different in a four-membered ring from those in larger rings (*see also* pp. 24–27). Hence, the decision was made with MM2 to simply introduce both a different bending constant and a different threefold torsional constant, specifically designed fit simultaneously both the energy and the pucker in cyclobutane (*10*). In view of what is outlined above, it was also decided to use this special set of parameters only for interactions that were completely internal to a four-membered ring. Any interaction that is partly or totally exo to the four-membered ring was calculated with the ordinary parameter set. Thus, fusing a four-membered ring to a five-membered ring, let us say, involves no indecision as to which parameters are to be used for which part of the molecule.

There is a great deal of experimental information available concerning structures with carbocyclic four-membered rings, but most of these rings contain one or more double bonds (pp. 125–127). Saturated four-membered rings that have been studied include cyclobutane and bicyclo[4.2.0]octane, for which both the heats of formation and the structures are known, and bicyclo[2.2.0]hexane, for which only the structure is known. Cyclobutane has several features of interest. The bond length is unusually long for a sp^3–sp^3 bond in an uncongested system (1.548 Å), and this lengthening has been attributed to the 1,3 van der Waals repulsions between carbons, as previously mentioned. The ring is quite puckered (the angle between the planes is around 35° in various derivatives), and the barrier to inversion is approximately 6.3 kJ/mol. Utilizing a smaller than usual CCC bending constant, and a larger than usual torsional V_3 constant, together with a substantial stretch–bend interaction term where the constant was given a sign to lengthen the bond when the angle is contracted, it was possible to fit accurately all of this information with the MM2 force field (*10, 38*).

Interestingly, it is known from careful ab initio calculations (*181*) that the bisector of the HCH angle is tilted with respect to the bisector of the attached CCC angle by a few degrees. This twisting has been said to result from a minimization of the "tortuosity" of the bonds in cyclobutane (*182*). The MM2 program reproduces this twisting, the angle between the bisectors being some 2° in the direction found by the ab initio calculations. In the force field calculation, the twisting results from the minimization of the torsional energy, even though syn hydrogens attached to carbons in the 1,3 positions come quite close together, and their van der Waals repulsion is increased by this twisting motion.

The structure of bicyclo[2.2.0]hexane is known from electron diffraction measurements (*183*). The four-membered rings are puckered both in the MM2 and in the experimental geometries, so the molecule has C_2 symmetry rather than C_{2v}, which would result if the rings were individually planar [as had been found with an older force field (*332*)]. With MM2 the C_{2v} structure was calculated to be higher in energy by 2.1 kJ/mol. The calculated and experimental structures are in reasonable agreement. The largest discrepancies concern the bridgehead bond angle and 1,2 bond lengths as shown. Note that the smaller bond angle and longer bond length in the experimental structure tends to make the 2,6 distance quite similar to the value found by calculation. This may

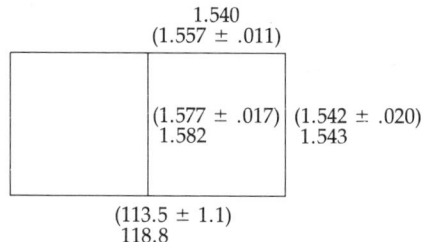

Calculated and experimental (in parentheses) structures for bicyclo[2.2.0]hexane

well be another case in which there simply is not enough information in the electron diffraction radial distribution function to arrive at a unique structure. The one chosen may be incorrect, although it may fit the diffraction data as well as, or perhaps slightly better than, the correct structure.

Bicyclo[4.2.0]octane. The *cis* and *trans* isomers both adopt a geometry with the six-membered ring in a flattened chair conformation, and there is good agreement between the calculated and electron diffraction structures (*19, 38*). The heat of formation of the *cis* isomer (the only one reported experimentally) is also correctly calculated.

Alkenes and Cycloalkenes

Small alkenes are used for the parameterization of the force field; therefore, their structural and energetic characteristics (*cis–trans* ratio, rotational barriers) are calculated in good agreement with experiment in every force field. This is true for the structure of ethylene, and both the structure and rotational barrier in propene. A hydrogen on the methyl in propene eclipses the double bond in the ground state. Contrary to what is observed with alkane types of structures, the ground states of alkenes and other compounds with double bonds will usually have the double bond eclipsed by a substituent.

1-Butene. This is an important compound in the parameterization of a force field, because it is the simplest example of a structure containing the C_{sp^2}—C_{sp^2}—C_{sp^3}—C_{sp^3} type of linkage; the torsional constants of this linkage are of general importance. There are three conformations in equilibrium, two of which are mirror images of C_1 symmetry and have a hydrogen eclipsing the double bond. The third is of C_s symmetry, and has the methyl group eclipsing the double bond. The energy difference

C_s C_1 (dl)

between these conformations now seems clear. Early work indicated that they were approximately equal in energy but, more recently, coupling constant studies in the NMR have been interpreted as indicating that the C_s conformation is somewhat less stable than the two enantiomeric ones (*184*), and good quality ab initio calculations bear this out (*185*). The exact position of rotameric equilibria in the butenes may be adjusted with the V_1 torsional parameter in molecular mechanics calculations.

The precise equilibrium position for the 2-butenes has been available for a long time, and *trans*-2-butene is correctly calculated to be more stable than the *cis* isomer. More interesting than the equilibrium position is the geometry of the *cis* isomer. The C—C=C angle opens because of the methyl···methyl repulsion, and angular values of 127–128° have been calculated (*38, 120, 186, 187*), while the values from ED (125.4°) and microwave spectra (126.7°) are slightly smaller (*188–190*). Three rotamers seem possible a priori for the *cis* isomer, and all current force fields agree that the C_{2v} (1) geometry with one close H···H contact

C_{2v} (1) C_s C_{2v} (2)

of 2.07 Å (*120*), 2.16 Å (*187*), or 2.12 Å (*38*) is the most stable. The favorable eclipsing of the hydrogen by the double bond outweighs the van der Waals repulsion. Earlier calculations had indicated that the methyls were rotated slightly from the C_{2v} (1) structure, to give a symmetry of C_2 (*186*). The experimental results are in agreement with either a C_2 or C_{2v} structure (*188–190*).

4,4-Dimethyl-2-pentene (methyl-*t*-butylethylene). The *trans* form has been calculated to be in the conformation shown (*187*) with eclipsed torsion angles at the double bond, as would be expected from the 1-butene data given earlier. In the *cis* isomer the eclipsed conformation is clearly not stable, and it is even questionable if the double bond

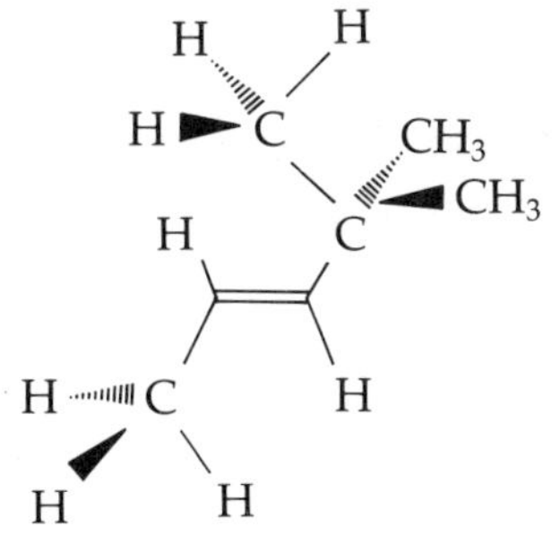

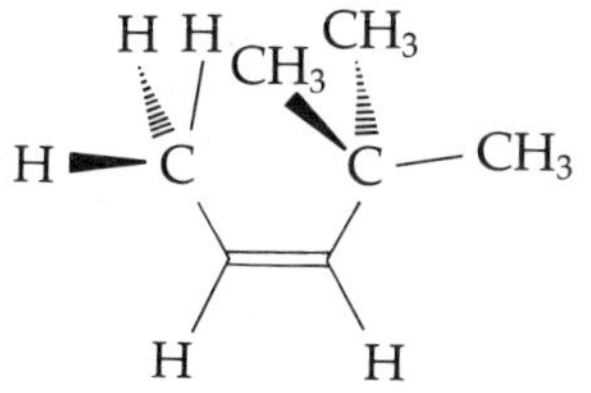

remains planar. The calculations indicate that the structure shown (C_s) is the most stable one, and the double bond is essentially planar (*187*). No experimental structural data are available.

1,2-Di-*t*-butylethylene. While the *trans* isomer is unexceptional, the *cis* isomer is extremely strained. The calculations indicate that the strain causes the bond angle C=C—C to open to 135.2°. The double bond is nearly planar [twist 0.5° (*38*) to 5° (*187*)], in agreement with the structures known for some 1,2-di-*t*-butyl olefins. The two isomers differ in their heats of hydrogenation by 40.6 kJ in acetic acid; in the gas phase this value is expected to be somewhat higher. The energy difference between the *cis* and *trans* structures was calculated (MM2) to be 32.59 kJ by Allinger and Sprague (*202*). Ermer and Lifson used the experimental quantity for the least squares optimization of their force field, but calculated an energy difference of only 29.87 kJ (*187*). The heats of hydrogenation of the individual isomers are calculated with each force field in only moderate agreement with experiment, so some errors seem to cancel out in this calculation of relative energies.

Calculations on several more highly substituted, and therefore more crowded, olefins have been carried out. These molecules attempt to escape their congested environments by stretching the double and adjacent single bonds, by bond-angle deformation, and interestingly, by a twisting of the double bond. This twist has been calculated to increase from 4.3° in tetramethylethylene (*191*) to 22° in *trans*-1,2-di-*t*-butyl-1,2-dimethylethylene (*191*) and 44–45° in tetra-*t*-butylethylene (*195, 196, 284, 335, 353*). [The early result of a torsion angle of 75° in the latter compound (*191*) seems unreasonable (*335*)]. The actual value of the calculated torsion angle depends strongly on the rotational barrier assumed for the tetraalkyl-substituted double bond (the CCCC torsional energy parameters). Since no useful experimental data are available on this point, the results are only tentative in cases of extreme torsion angles such as that of tetra-*t*-butylethylene. Structural data are available, however, for many related compounds, and show good agreement with the calculated results. The calculated double bond twist of *trans*-1,2-di-*t*-butyl-1,2-dimethylethylene agrees well with the value of 16° found for a related molecule in the crystal (*192*). Interestingly, tetraisopropylethylene is both calculated (*191*) and found experimentally (*354*) to have an essentially planar double bond with strongly interlocking isopropyl groups. A planar double bond is also present in adaman-

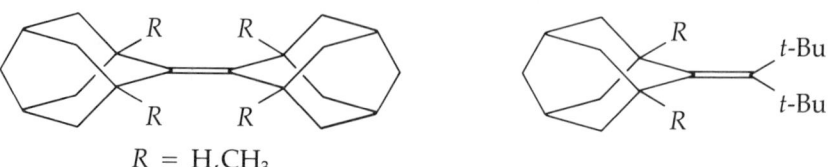

R = H,CH₃

tylideneadamantane (*191, 193*). An alkylated adamantylideneadamantane was found to prefer out-of-plane bending over a double bond torsion (*193–195*).

Several compounds that are structurally related to the elusive tetra-*t*-butylethylene were synthesized recently, and experimental as well as calculated structural data are available. In these compounds, at least two methyl groups on each side of the double bond are bound together to form a ring. The structure calculated for fenchylidenefenchane (*195, 196, 284*) is in excellent agreement with the x-ray structure

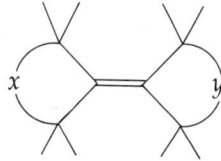

(*355*); the double bond is twisted by 7.4° (*284*). More hindered olefins result when only one bridge connects the substituents on one side of the double bond. The compounds containing one or two six-membered rings are nearly as strained as tetra-*t*-butylethylene itself, and they adopt twist conformations in the six-membered rings (*196, 284*). The double bonds in these compounds are predicted to be twisted by less than 15° (*196*), and still smaller double bond torsion angles were calculated for the compounds containing smaller rings (*196*). A valuable measure of the strain across the double bond in these compounds was found to be the difference in the steric energies of the hindered olefin and the two fragments that result when the two C_{sp^2} atoms are separated to infinite distance ("fragmentation strain") (*196*). Compounds containing four- and five-membered rings have been successfully synthesized, as could be anticipated from the low values of their calculated fragmentation strain energies (356, 357).

Cycloalkene Ring Systems. As was mentioned previously, the bending force constants and torsional constants for four-membered rings are necessarily different in a valence force field from those of the corresponding open-chain compounds or larger rings. For cyclobutene, after suitable adjustment of the bending and torsional constants, the calculated geometry and energy (197) were in good agreement with experiment (198). There are a few additional parameters needed for other unsaturated four-membered rings such as methylenecyclobutane. For this molecule the full structure is not known, but the moments of inertia are known [from the microwave spectrum (199)], and show that the molecule is nonplanar (barrier to planarity 1.92 kJ/mol). The heat of formation is also known (154, 155). Again, the parameters that reproduce these data can be found, and the force field then gives the entire structure (197).

In the equilibrium between methylcyclobutene and methylenecyclobutane, the compound with the endocyclic double bond is the more stable one. This is also true in the corresponding five and six-membered rings, but quantitatively the numbers differ substantially, as given below (38, 197, 200).

ΔH(kJ/mol)

6-membered ring: CH$_3$ (endocyclic) ⇌ CH$_2$ (exocyclic)   13.77

5-membered ring: CH$_3$ (endocyclic) ⇌ CH$_2$ (exocyclic)   18.95

4-membered ring: CH$_3$ (endocyclic) ⇌ CH$_2$ (exocyclic)   4.81

The interpretation of these numbers is straightforward. First, the compound with the most highly substituted double bond is the most stable one for bond energy reasons. If, for the discussion of the steric effects, we take the six-membered ring as our standard, torsion tends to be more serious than bending in the five-membered ring, and placing the second sp^2 carbon in the ring, rather than exocyclic, relieves more torsional strain than it increases bending strain. Accordingly, the endocyclic structure is stabilized, relative to the six-membered ring, increasing ΔH for the rearrangement as shown. On the other hand, the reverse is true with the four-membered ring. Here bending is more serious than torsion, and, relative to the six-membered ring, it is the exocyclic methylene that is stabilized. Hence, ΔH is reduced.

The structure of Dewar benzene has been calculated (*197*), and an electron diffraction structure is known (*201*). It has been suggested (*197*) that the disagreements, which are sizable, result from the radial distribution function being analyzable in two different ways. In other words, there is not enough information in the radial distribution function to arrive at a unique structure, and the one that has been arrived at is not in a good agreement with the molecular mechanics calculations, and is probably not a very accurate structure.

1,2-Dimethylenecyclobutane had its structure calculated by the MMP2 method (*202*) (pp. 52–55), because of the conjugated double bonds. The full structure is not known, but again, the moments of inertia are available from microwave spectroscopy (*203*). The structure is planar, but the bond angle between the methylene group and the ring differs by about 0.9° from the structure required by the moments of inertia. It has been suggested (*197*) that by using the calculations, together with the moments of inertia, a structure can be arrived at that is superior to that obtained by either method alone.

Cyclopentene. The puckered structure is calculated to be more stable than the planar transition state for ring inversion by 1.72 kJ (*186*) and 1.51 kJ (*187*), and vibrational spectra indicate a value of 2.76 kJ (*204*). The calculated ring pucker [19° (*186*), 21.4° (*187*), 19.1° (*120*)] also compares well with experimental values of between 22° and 29°.

Calculations on cyclopentene and several other olefins (for example, norbornene) indicate that the double bonds are bent out of plane, in the case of cyclopentene by 2.1°. This deformation in these molecules has the effect of diminishing torsional strain between the olefinic hydrogens and their more nearly eclipsed allylic neighbors (*205*) as shown in the drawing. This result was confirmed by ab initio calculations (*351*, *352*).

A similar deplanarization was also calculated to exist in Dewar benzene (284), but the electron diffraction data did not permit a conclusion on this point. The effect was definitely observed in the hexafluoro and hexamethyl derivatives of Dewar benzene (201).

Dewar benzene
(bicyclo[2.2.0]-2,5-hexadiene)

Cyclohexene. The two half-chair conformations of cyclohexene can be interconverted in a pseudorotation-like movement, passing through a symmetrical boat conformation as the transition state with a calculated energy of 25.48 kJ/mol (38), 27.6 kJ/mol (206), or 29.3 kJ/mol (207). These calculated energies are only slightly higher than the experimental barrier [$\Delta G^{\ddagger}$ = 22.6 kJ/mol (208), $\Delta H^{\ddagger}$ = 22.2 kJ/mol (206, 209)]. The relative energies of equatorial and axial methyl groups on cyclohexene rings, and the energies of the dimethylcyclohexenes have also been discussed (200).

Methylenecyclohexane. This compound is calculated to prefer the usual chair form over a pseudorotating twist-boat by about the same amount as does cyclohexane [18.58 kJ (200) vs. 21–25 kJ (206)]. The ring inversion barrier (experimental value: $\Delta G^{\ddagger}$ = 35.1 kJ) was calculated to be 33.9 kJ/mol (206). A somewhat smaller twist-chair energy difference was also found for 1,4-dimethylenecyclohexane (ΔH = 14.77 kJ, $\Delta H_{pseudorot}^{\ddagger}$ = 10.0 kJ), and this is in agreement with experimental results, which indicate the presence of the chair in both compounds (347).

1,3-Cyclohexadiene. This compound is a conjugated system, and force field calculations on it have been carried out (210) by the methods described in pp. 52–55. It will be discussed here, however, for comparison with the 1,4-isomer. The compound has been studied both experimentally and theoretically. The molecule has a twisted (C_2) geometry, so that the chromophore is almost but not quite planar (by 12–18° from different determinations). Interestingly, and contrary to the usual ideas about conjugation, this compound has almost exactly the same free energy as does the 1,4-isomer, and equilibration with base gives a mixture of the two compounds (294). If the symmetry effects are allowed for, the 1,4-isomer has a slightly more negative enthalpy (1.84 kJ/mol) than does the 1,3-isomer. The equilibration experiment is not well described, however, and the accuracy of the result is unknown. Calculations with MM2/MMP2 (210) indicate that the 1,3-isomer is of lower enthalpy by 3.93 kJ/mol.

1,4-Cyclohexadiene. The molecular mechanics calculations all agree that the planar structure is the energy minimum (120, 186, 187). It is, however, the bottom of a shallow potential well in which the molecule vibrates about the planar structure. A low frequency (wide amplitude) vibration corresponding to this process has been reported (211), and it seems clear that both the bottom of the potential well and the time average structure of the molecule occur at a planar geometry. Two independent electron diffraction structures have been published, one of which concludes (212) the molecule is planar, in agreement with the rest of the available experimental and theoretical evidence, and the other of which concludes (213) the molecule has a boat conformation (C_{2v}). The electron diffraction study that deduced a boat structure did not properly allow for this wide amplitude motion, which introduces shrinkage (*see* pp. 6–10). The fact that 1,4-cyclohexadiene really is planar was not widely appreciated until recently (214).

However, the dibenzo analog, 9,10-dihydroanthracene, is known to be puckered (215, 216). This is an example of the differences between olefins and their benzo analogs, which will be discussed in detail. In this case the known facts are correctly calculated (38, 186). The reason for the difference between the two systems is to be found in the angle bending deformations. The planar olefin system has the angle at the methylene group (Θ_1) opened above its natural value, and the internal angle at the olefinic carbon (Θ_2) is closed somewhat to spread the bend-

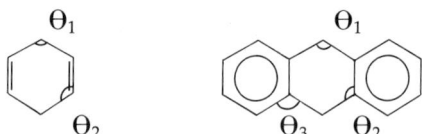

ing strain among all the angles. This angle strain can be relieved in the nonplanar system, but the torsional potentials are stronger and hold the molecule planar. For the benzo analog, the torsional potentials are the same as in the cyclohexadiene, and the angular forces are similar for the two angles Θ_1 and Θ_2. But in addition, there is a second angle at the olefinic carbon (Θ_3), which now is the other portion of the aromatic ring. The aromatic ring itself is rather resistant to deformation, so that when the methylene group is bent away from the aromatic ring, two angles (Θ_2 and Θ_3) must be deformed rather than one. The resistance to this deformation, therefore, is greater than in the case of cyclohexadiene, sufficiently great, in fact, that it overcomes the torsional forces and causes the center ring to pucker.

cis-Cycloheptene. All recent calculations agree that the chair conformation is between 2.1 kJ and 17.6 kJ more stable than the boat (*38, 120, 186, 187, 217*). As in cyclohexane, the interconversion of the two chairs goes through the boat, with an unsymmetrical twisted transition state. The barrier was calculated to be 21.6 kJ (*186*), and the experimental value [$\Delta G^{\ddagger}$ = 21.0 kJ (*218*)] is in excellent agreement with this. The boat is a true energy minimum, separated (calculationally) from the twist-boat transition state by only 0.4 kJ (Figure 4.13) (*120, 186*). At the time the calculations were performed, only the barrier for the 5,5-difluoro derivative was known, which is 31.0 kJ, so the calculation gave better results than the study of the derivative.

The benzo derivative has also been studied by low temperature NMR spectroscopy, and the inversion barrier found for this molecule is considerably higher than that for cycloheptene [$\Delta G^{\ddagger}$ = 45.6 kJ (*219*)]. To explain this high value, which shows again that benzo analogs may behave quite differently from alkenes, we must analyze the torsional situation at the sp^2 carbon. In the ground state of cycloheptene, unfavorable H/H eclipsing of the kind found in propene (rotational barrier 8.28 kJ) is observed, while in benzocycloheptene, the eclipsing is between H and C_{sp^2}, of the kind observed in toluene [rotational barrier 58 J (*280*)]. (For symmetry reasons the rotation barrier in toluene is approximately zero, regardless of the parameters actually chosen.) When going to the transition state, this torsional interaction is nearly completely relieved in cycloheptene, but in the benzo analog there is the same torsional interaction in all geometries. The transition state for ring inversion therefore has a much higher energy in the benzo derivatives.

trans-Cycloheptene. This highly strained and the least stable of all *trans*-cycloolefins prepared to date was calculated to be 84.9 kJ more strained than its *cis* isomer, and separated from it by a calculated barrier

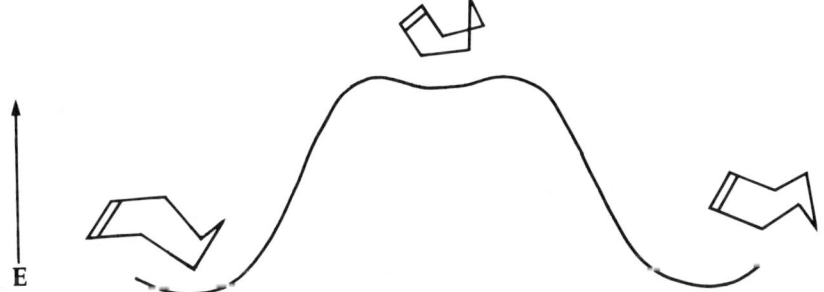

Figure 4.13. Qualitative energy diagram for the inversion of cycloheptene.

of only 24.98 kJ, indicating the serious instability of the *trans* double bond (*186*). The compound has been obtained in solution and is stable only at low temperatures. The double bond is considerably deplanarized and rehybridized, as is the case of *trans*-cyclooctene (*186*), which is discussed in more detail below.

1,3-Cycloheptadiene. Early calculations (MMP1) indicated that the C_s and C_2 geometries had very similar energies, with practically free pseudorotation between them (*220*). More recent calculations [MMP2 (*202*)] indicate that the C_s form is more stable than the C_2 by 8.66 kJ/mol, and that the barrier separating them is 7.82 kJ/mol above the C_2 form (*349*). This is consistent with the electron diffraction data, which detected only the C_s form (*221*).

C_s C_2

cis-**Cyclooctene.** The most stable conformation was calculated to be an unsymmetrical structure related to the cyclooctane boat–chair, not one of the symmetrical conformations (*186, 187, 217*). The x-ray structure of

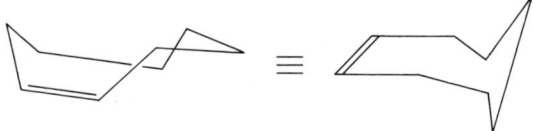

enantholactam hydrochloride (*222*), which has a *cis*-cyclooctene-like structure, shows the same conformation, and the geometric parameters compare well with those calculated (*187*).

trans-**Cyclooctene.** This molecule is sufficiently stable that its structure can be studied by the usual methods, so it is a good model for compounds with extremely strained (torsion and out-of-plane bending) double bonds. The twist (C_2) conformation is calculated to be 8.49 kJ (*38*) and 13.14 kJ (*187*) more stable than the chair (C_2), and the former structure is found in the x-ray structures of derivatives (*223–225*). An electron diffraction study (*226*) seemed contradictory to the calculations, and it was uncertain if the parent compound really assumed the geometry of the derivatives. However, a more recent ED study has confirmed the results of the calculations (*227*). This was one of the first cases where the disagreement between molecular mechanics and experiment was later resolved in favor of molecular mechanics. The torsion angle about the double bond [experimental (227) 137.7°; calculated 145.3° (38), 138.0° (225), 136° (206)] is far smaller than the usual 180°. The strain inherent in this strongly deformed double bond leads to an energy difference between the *cis* and *trans* isomers of 47.57 kJ [from heats of hydrogenation in hexane (228)], which is reproduced by calculations [48.70 kJ (38), 47.20 kJ (78)]. Pathways for the interconversion of twist and chair forms were calculated in detail by Ermer (78, 335).

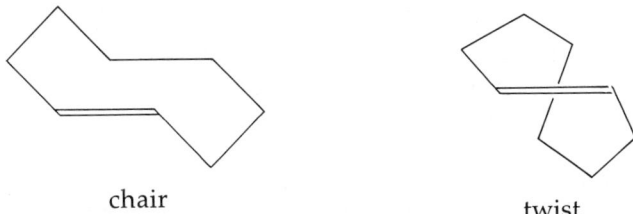

chair   twist

The deformations in *trans*-cyclooctene do not correspond to a simple twisting about the double bond, as might have been supposed from the above discussion. It was reported in 1958 that the molecule had a dipole moment of 0.8 D, and it was argued that this could not result from twisting about the double bond, but rather was from the pi bond taking on some s character (229). Thus, looking at the double bond end on, one does not see the projection **1**, but rather projection **2**, where all four of the atoms attached to the olefinic carbons lie on the same side of a plane that can be drawn containing the olefinic carbons.

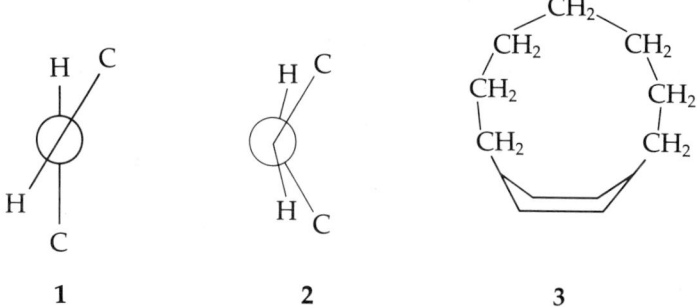

1   2   3

This kind of structure increases the electron density in the pi bond on the side of the plane opposite that containing the four atoms, and explains the observed dipole moment. The force field correctly predicts the same qualitative effect. The reason the force field calculation comes out this way is because all six of the atoms in question are trying to achieve coplanarity. They cannot do so, but they strive to come as close as they can, and deformation **2** comes closer than deformation **1**, for a given distortion of the carbon skeleton. While the structure of cyclooctene is not known with sufficient accuracy to position the olefinic hydrogens, a neutron diffraction study of a derivative was reported recently (*335*), and indeed the olefinic hydrogens are bent from the plane as indicated by 2. There are also known some paracyclophane structures in which the benzene ring is deformed into a shallow boat (structure 3) (*230, 231*). In these too, the hydrogens bend away from the plane in the same direction in which the ends are bent (up in structure 3), and for the same reason. Again, the calculations predict this very well (*232*).

1,2-Cyclooctadiene. The only calculations on allenes reported to date are those on one ten-membered and two nine-membered rings (*251*), and on 1,2-cyclooctadiene (*235*). The bending parameter for the central atom of the allene unit was chosen as similar to the value used for alkynes, while most other parameters could be taken over unchanged from alkenes. The terminal atoms of the allene simply have a torsional potential with a minimum at 90° instead of the potential with minimum at 0°, as used for alkenes. 1,2-Cyclooctadiene appears to have been prepared, although it has never been isolated in pure form (*233, 234*). A molecular mechanics study was recently reported by Yavari (*235*). Two conformations, twist-boat (C_2) and twist-chair (C_1) were found; they have nearly equal energies and are separated by a barrier of 36.8 kJ/mol. The molecule is appreciably strained, the central allene angle being 158° and 160° in the two conformations, respectively. Some of the other angles also show significant distortion.

cis,cis-**1,3-Cyclooctadiene.** Both the twist-boat (C_2) and an unsymmetrical conformation were calculated to have about the same energy [C_1 1.76 kJ higher than C_2 (*236*)], so a mixture of conformations is expected (*237*). The **TB** form was calculated to have a torsion angle between the double bonds of 57°, while that in the C_1 conformation is 47°. Both are

C_2 C_1

somewhat larger than a value of 41° that had been inferred from UV measurements (*238*). Low temperature NMR spectra showed both conformations were present in comparable amounts (*239*). Extensive calculations on this molecule have been reported recently (*240*).

cis,cis-**1,4-Cyclooctadiene.** In one study, the **BC** conformation was calculated to be more stable than the **BB** isomer, and it appeared that the

TB was not an energy minimum, because the geometry of a particular slightly twisted BB geometry returned to the C_s geometry during energy

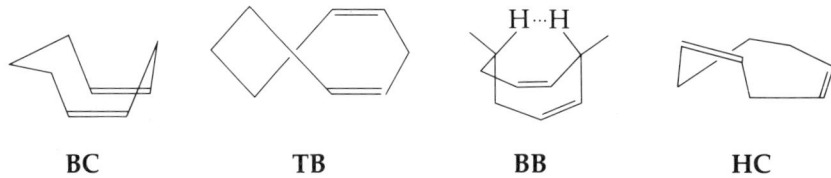

BC   TB   BB   HC

minimization (236). In another study BC and TB were compared, and BC was found to be more stable by 5.65 kJ (241). Anet, in a third study, found that BB is a transition state for the interconversion of the enantiomeric TBs, which he calculated to have the same energy as BC (242). The failure of the first study to find the TB as an energy minimum is one of two (243) examples known to the authors in which the block-diagonal Newton–Raphson method from a deformed starting structure ended upon a saddlepoint. Low temperature NMR revealed two conformations with an energy difference of 2.43 kJ (242), in agreement with Anet's calculations. Interconversion from the (rigid) BC to the (flexible) TB goes through a half-chair (HC) with a calculated energy of 37.6 kJ, which corresponds to a measured barrier of 33.5 kJ (242).

cis,cis-1,5-Cyclooctadiene. Several force fields were reported to yield the TB conformation as the energy minimum (120, 244–246) and the chair was calculated to be 6.07 kJ less stable (244). The various calculations differ somewhat in the ω_{2345} torsion angle of the TB because the minimum is very shallow, so the precise values depend on the force field and on the energy minimization procedure. The Newton–Raphson method usually locates the minimum more exactly in such cases. (See pp. 64–72). The conformational situation is summarized in Figure 4.14.

There has been some controversy about the various interconversion pathways (120, 245, 246); this was resolved by White (120) who showed that the method applied for the calculation of the interconversion pathway is important. On a symmetrical pathway, achieved by driving two torsion angles simultaneously, higher transition state energies are obtained than when only one angle is driven (120). The interconversion going through the skew form is easiest, followed by the one through the C_{2v} boat. The boat–boat interconversion via the chair is highest in energy, and is unimportant. (Table 4.5).

Dibenzo-1,5-cyclooctadiene was calculated to prefer the chair conformation (78), for reasons outlined earlier for the interconversion of cycloheptene chairs.

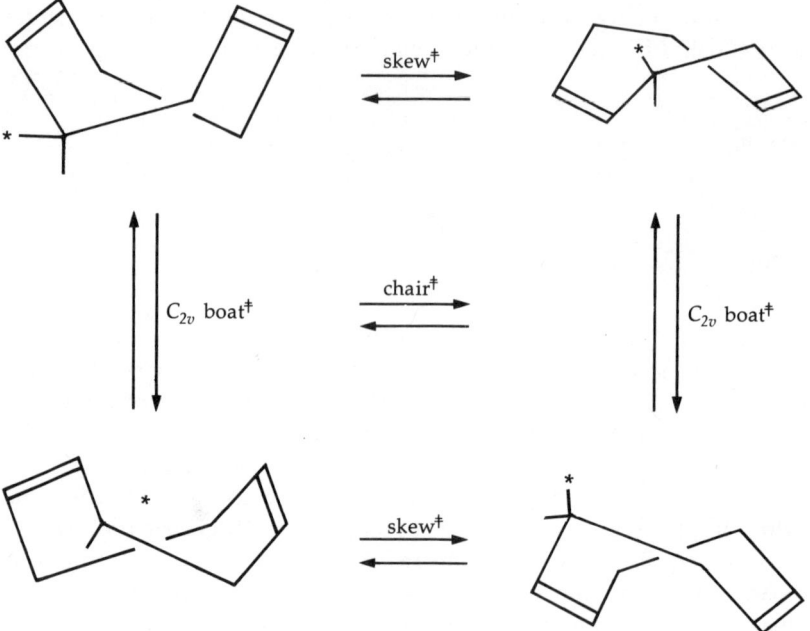

Figure 4.14. *The conformational analysis of* cis, cis*-1,5-cyclooctadiene.*

The relative energies of the 1,3-, 1,4- and 1,5-*cis,cis*-cyclooctadienes obtained from the MM1 force field were compared with equilibration values, and the calculated stabilities (1,3 > 1,5 > 1,4) are in agreement with experiment (236).

It was calculated that *cis, trans*-1,5-cyclooctadiene resembles *trans*-cyclooctene with respect to the *trans* double bond, and shows a torsion angle of 136° (330).

Table 4.5

Transition State Energies (kJ/mol) for 1,5-COD Interconversion in Different Force Fields

	Skew‡	Boat‡	Half-chair‡
Allinger (1975) (244)	24.7[a]	13.0[a]	21.8[a]
Ermer (1976) (246)	17.6	23.8	24.7
White (1977) (120)	9.6	18.8	19.2
Anet (1978) (206)	17.6	27.6	30.1
Experimental (245)	18.4	20.5	

[a] These calculations were not for the transition states themselves, but for structures of a specified symmetry that were approximations to the transition states.

cis,cis,cis-1,3,5-Cyclooctatriene. Calculations reported by Anet were based on a simple model for the conjugated double bonds, assuming a weak V_2 torsional term for the single bonds between conjugated double bonds (*206, 247*). With this force field the **TB** conformation was found to be the most stable one, in agreement with low temperature NMR data (*247*). The boat is the transition state for the interconversion of the **TB**s, with an energy of 20.1 kJ. Ring inversion is calculated to pass through

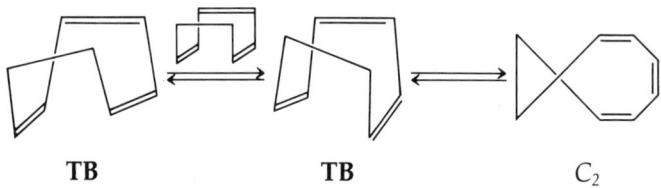

TB   **TB** C_2

the half-chair with a barrier of 27.6 kJ, and this was substantiated by the NMR experiment (*247*). Buemi et al. carried out calculations for a large number of conformations, but apparently missed the **TB**. Their energy minimum was the C_s boat (*248*).

cis,cis,cis-1,3,6-Cyclooctatriene. The **TBC** with C_2 symmetry is calculated to be most stable (*206, 247, 248*). Interconversion of the enantiomers goes through a high energy half-chair conformation (7.5 kJ) with a barrier of 17.2 kJ (*247*).

Cyclononenes. Little experimental information about the *cis* isomer is available that might be used to guide the calculations among the extremely large number of possible conformations. Calculations were carried out by Ermer and Lifson (*187*) on a conformation similar to that in the crystal of caproyllactam, which is cisoid at the amide linkage, and a number of conformations were studied by Favini et al. (*241*), but the conformation in solution is uncertain. The *trans* form was calculated to be only 12.1 kJ less stable than the *cis,* although the double bond in the former was predicted to have a torsion angle of only 150° (*187*). The experimental enthalpy of isomerization is just 12.1 kJ in acetic acid (*249*), and the *trans* isomer does have an experimental dipole moment (*250*) of 0.6 D (compared with 0.8 D for *trans*-cyclooctene where the torsional angle is 138°–145°).

cis   trans

1,2-Cyclononadiene. This cyclic allene was calculated to have a C_1 **BC** conformation and a C_2 conformation of similar energy (0.4 kJ), a **CC**

C_1 form with 2.5 kJ, and several high-energy minima. The transformations have been simulated by torsion angle driving; the calculated barriers (36.4 kJ and 29.3 kJ) are in agreement with the NMR data (251).

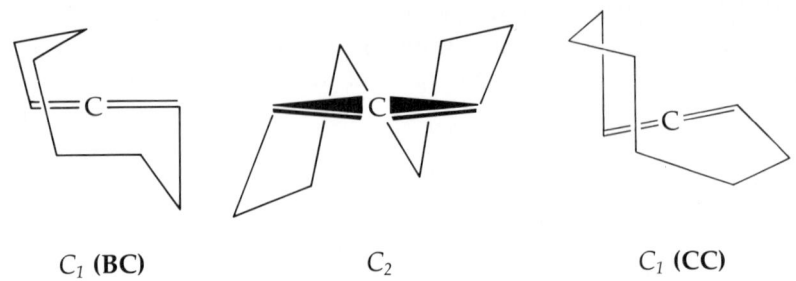

C_1 (BC) C_2 C_1 (CC)

Cyclonona-1,5-dienes. The most stable of the three *cis/trans* isomers is the *cis,cis* isomer, which fact is known from heat of hydrogenation data (252). It was calculated to prefer a twisted, flexible conformation with C_2 symmetry, designated as chair (120, 253, 254). In the crystal of a derivative, byssochlamic acid, a **BC** conformation with C_s symmetry was found, which in the calculations has a higher energy (120, 254). More detailed calculations revealed that the **BC** is, in fact, an energy maximum separating two enantiomeric **TBC**s, with a barrier height of 4.6 kJ, and that the **TBC** forms are 9.6 kJ above the C_2 chair (254). The calcula-

C_2 C_s

tions are confirmed by the energy barrier for the interconversion between the two stable conformations of 47.7 kJ, which compares with the experimental barrier of $\Delta G^{\ddagger} = 43.1 \pm 1.2$ kJ (254).

White also carried out calculations for different conformations of the *cis,trans* and *trans,trans* isomers. He reports that the most stable of his *cis,trans* conformations is 23.4 kJ less stable than the C_2 conformation of the *cis,cis* compound (120), which is in reasonable agreement with the heat of hydrogenation data (252). A lower energy conformation for the *cis,trans* isomer has not been found, but if one exists, the energy difference would be brought into even better agreement with experiment.

1,2,6-Cyclononatriene. The most stable conformation of this quite rigid molecule is the chiral **BC**. Ring inversion is a high energy process with an experimental barrier of 54.8 kJ. For such a strained transition state, the results of the force field calculations are not expected to be as precise as for lower barriers, and two possible pathways with barriers of 63.2 kJ

and 69.5 kJ have been found. The first of these two pathways goes through a high energy **TB**, which is an energy minimum (251).

1,4,7-Cyclononatrienes. The conformations of the cis,*cis*,cis-isomer (tris-homobenzene) and their interconversion pathways have been studied. A crown and a saddle form were found. The transition state is very unusual, as it has extremely large CCC bond angles. It has been proposed that this molecule be used as a model for parameterization of CCC bending potential functions (331).

Cyclodecene. Of the two conformations of cis-cyclodecene that were studied by Ermer and Lifson (187), the one existing in the crystal of the AgNO$_3$ adduct was calculated to be more stable. Allinger and Sprague (186) considered only one conformation, which appears to differ from those of Ermer and Lifson, so the results of the two calculations cannot be compared.

Ermer/Lifson   Allinger/Sprague

trans-Cyclodecene was studied in some detail. The lowest energy conformation found was the twist form, which is also found in the AgNO$_3$ adduct (186, 187). The torsion angle about the double bond calculated with different force fields (186, 187) is near 160°, while that in the crystal is 138.3° (255). The difference probably results from the silver ion complexation, and the molecular mechanics structures are a better description of the isolated molecule than they are of the complex. The calculated results are further supported by the dipole moments of *trans*-cyclooctene and *trans*-cyclodecene. The double bond in *trans*-cyclooctene is rehybridized due to the strong double bond twist/out-of-plane-bend, and its considerable dipole moment comes from the asymmetry of charge above and below the double bond (page 131). The dipole moment of *trans*-cyclodecene is small (indistinguishable from zero), pointing to a less deformed double bond (*186, 229, 256*).

1,5-Cyclodecadienes. The *cis,trans* isomer was calculated to be more stable than the other two isomers (*120, 258*), but the only conformation for which calculations were carried out was the one found in the crystal of a derivative (*259*). The two *cis,cis* conformations shown and the half-crown *trans-trans* isomer are all within 8 kJ of the most stable isomer. The structures of natural products containing *cis,cis* and *trans,trans* double bonds compare well with the calculated geometries (*120*).

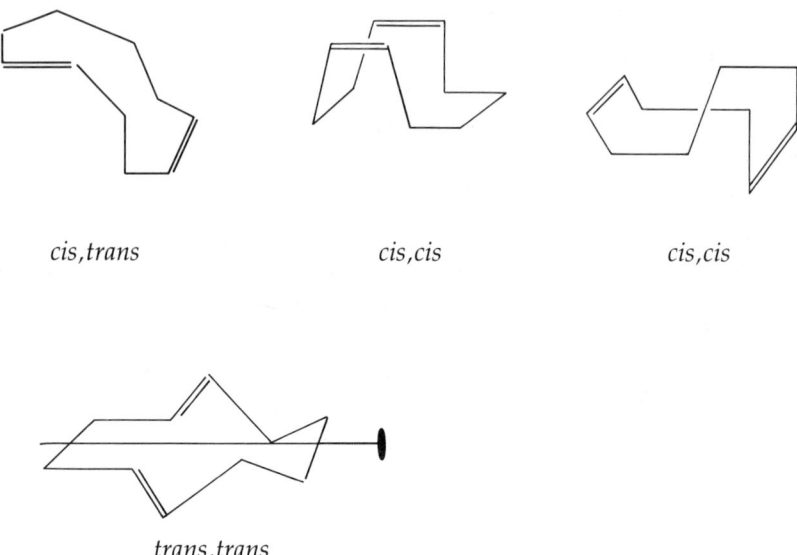

cis,trans   cis,cis   cis,cis

trans,trans

Cyclodeca-1,6-dienes. The most stable isomer is the *cis,cis*, which prefers the chair form over the boat (*120, 187, 260*). The exact energy difference obtained with different force fields varies from 0.7 to 5 kJ, and the boat has been detected as a minor constituent in an electron diffraction study.

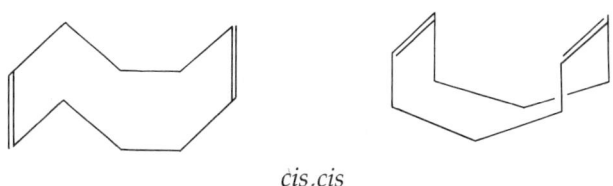

cis,cis

From an equilibration study it is known that the *cis,trans* isomer coexists in trace amount with the *cis,cis* isomer at room temperature (261). White calculated that the most stable conformation of *cis,trans*-1,5-cyclodecadiene lies 13.05 kJ above the most stable form of *cis,cis* (120). He also carried out calculations for six conformations of the *trans,trans* isomer and found the CC (crown) to be most stable.

cis,trans   trans,trans

meso-1,2,6,7-**Cyclodecatetraene.** Two conformations are important, and the one that was calculated to be more stable (8.8 kJ) was also found in the crystal (251, 262). This is the centrosymmetric structure corresponding to the extended cyclohexane chair. The extended TB can be reached over an activation barrier, which has a calculated value of 28.5 kJ, and an experimental (low temperature NMR) value of 29.7 kJ (251).

C   TB

dl-1,2,6,7-**Cyclodecatetraene.** This molecule was studied by molecular mechanics (251), together with the other allenes mentioned earlier, although it has not been prepared. Inspection of Dreiding models indicates that the molecule is very flexible (263), but the calculations do not bear this out. The conformation of lowest energy, a crown, is converted to a **TB** with an energy of 7.5 kJ, and to another boat with an energy of 5.4 kJ, via energy barriers that are both 38.5 kJ (*251*).

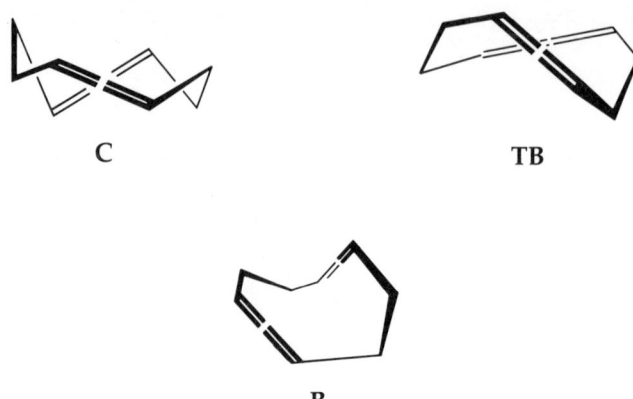

cis-**Cyclododecene.** Of the seven conformations that were calculated for this olefin, two were found to be much more stable than the others. The more stable of these (**1**, by 2.5 kJ/mol) resembles closely the stable conformation of cyclododecane (pp. 107–108).

Cyclododeca-1,5,9-triene. The *trans,trans,trans* isomer of this molecule, for which good structural data are available, was used by White for the parameterization of his force field. Calculations on conformations other than the experimental one were not reported, however (*120*). Calculations and dynamic NMR studies on the *cis,cis,cis* isomer have been reported by Anet (*343*).

Calculations on many of the above olefins have recently been reported by Ermer (*335*), using a modified version of his earlier force field (*187*).

Polycyclic Olefins. The molecular mechanics method has been used to study the relative energies of the double bond isomers of various polycyclic hydrocarbons, although experimental data are available for only a few of them. [For a large number of calculations employing the Ermer/Lifson force field, see (335)].

A well-studied compound is octalin (octahydronaphthalene), for which the energies of the several isomers were calculated in excellent agreement with the experimental equilibrium composition (200). The isomer with a central double bond (the most highly substituted isomer) is the most stable. Similarly, all of the isomers of dodecahydrophenanthrene were compared, and agreement with the few available experimental data was found (264).

An electron diffraction study on Δ^6-bicyclo[3.2.0]heptene was interpreted as being best reconciled with a chair conformation (361). A study of the rotational spectrum of the molecule, however, indicated that the conformation was a boat (362). A molecular mechanics calculation (MM1) was recently reported (363), and concludes that the chair–boat equilibrium contains a negligible amount of the chair conformation at ordinary temperatures, the calculated energy for that conformation being 8.90 kJ/mol above that of the boat.

Caryophyllene is a somewhat complex natural product, which consists of a *trans*-cyclononene, *trans* fused to a cyclobutane ring, and containing a few other complications. CMR measurements showed that the compound consisted of two major conformers in the ratio of 76:24 at room temperature. The four possible conformers suggested by models were studied by molecular mechanics (MM1), and it was concluded that the $\beta\alpha$ is the major conformer, with $\beta\beta$ being the minor one (346). The remaining two are much higher in energy.

$\beta\alpha$ $\alpha\alpha$ $\beta\beta$ $\alpha\beta$

Bridged olefins may be divided into the Bredt olefins, those having the double bond at one of the bridgeheads, and the regular olefins. Many of the latter have been studied by molecular mechanics. Dewar benzene has been mentioned above. Bicyclo[2.2.2]oct-2-ene and the 2,5-diene, as well as norbornadiene and pinene (for which structural information is available) were used by White for the construction of his force field (120). Some of these compounds were studied earlier also, and their heats of hydrogenation have been discussed (186).

Tetramethylbicyclo[4.2.2]deca-3,7,9-triene and its dibenzo derivative, polycyclic olefins with extremely large C=C—C bond angles, were studied by Ermer (265). His study showed that a harmonic bending potential function can be used for this kind of angle for deformations of up to at least 15°. The calculated angle in the dibenzo derivative (138.9°)

agrees very well with that found experimentally (138.7°). It is also interesting that both the calculated and the experimental structures have C_{2v} symmetry, although the Dreiding model shows a strong preference for C_s symmetry. Molecular models usually give information about angle strain, but are deficient when torsional and nonbonded interactions are important. In the C_s structure, bond angle bending is minimized. It is, however, less favorable for torsional reasons; the C_{2v} geometry minimizes torsional energy at moderate angle strain.

Bridgehead olefins violating Bredt's rule are often too unstable for isolation, and their geometries are accessible only by computations. Double bond torsion and out-of-plane bending are the main sources of strain, but torsional and other kinds of strain can be high in the saturated parts of these molecules. The Bredt olefins are structurally related to the *trans*-cycloalkenes. For example, the geometry calculated for the *trans*-cyclooctene ring unit in bicyclo[3.3.1]non-1-ene is very similar to the twist conformation calculated for *trans*-cyclooctene. The geometries and energies of bridgehead olefins were calculated with three different force fields (266–268, 344, 358). For bicyclic systems the calculations agree that, following Wiseman (269), strain is determined mainly by the size of the smallest ring in which the double bond is *trans*. Bridged *trans*-cyclooctenes are found to be stable at room temperature, cycloheptenes at low temperatures, and cyclohexenes occur as transient intermediates only. The size of the bridge across this ring is of secondary importance.

The relative stabilities calculated for the bicyclo[4.2.1]- and -[5.1.1]nonenes (next page) agree very well with the ratio of isomers found in the preparation by Hoffman elimination from the bridgehead ammonium salts. The isomer with the double bond in the smaller ring is the more stable one, contrary to what was concluded from an examination of Dreiding models (266, 267).

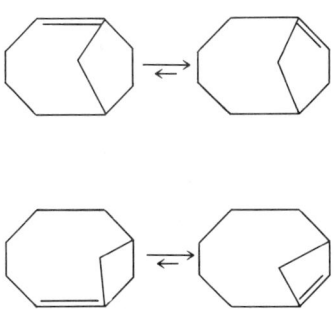

In polycyclic Bredt olefins the simple analogy with *trans*-cycloalkenes is no longer very useful. Compounds A and B, both of

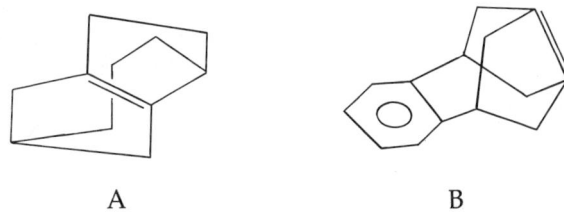

A   B

them bridged *trans*-cyclooctenes, are transient intermediates only. Simple rules break down in such cases, and only the calculated strain energy can provide a measure of the compounds' stabilities (268, 344). It was found, however, that the total value of the strain energy is misleading in the understanding of reactivity, because only part of this quantity is usually due to a strained double bond. Ermer proposed that the sum of the out-of-plane bending and torsional energy at the double bond would be a better measure of reactivity (267). A more general possibility is to use the difference between the total strain energies of the strained olefin and its hydrogenation product as a measure of the tendency of the molecule to undergo dimerization and addition reactions. Thus threshold values for the calculated strain energy difference have been determined and may be used to predict if a bridgehead olefin will be stable under different conditions (268, 344).

Alkynes

The amount of information available on these compounds is relatively limited. Structures have been determined by electron diffraction or microwave methods for acetylene, propyne, 2-butyne, 1-pentyne, 3-methyl-1-butyne, 3,3-dimethyl-1-butyne, and cyclooctyne. Some data on force constants are available (270). The heats of formation have been reported (271) for all of the linear alkynes containing the triple bond in all possible positions up to C_{10}, but little information is available regarding branched chain compounds.

From these data it was possible to extend the MM1 (272) and MM2 (271) force fields to include alkynes. The cycloalkynes containing from seven to ten carbons were examined briefly (271). The calculated strain energy for cycloheptyne is 130.1 kJ/mol, and the larger ring cycloalkynes have decreasing strains, down to 41.4 kJ/mol for cyclodecyne. The heats of hydrogenation calculated for the cycloalkynes are systematically larger than the experimental ones by 8–17 kJ/mol. Since the latter were determined in acetic acid (324), solvation would tend to yield a discrepancy in this direction.

The structure of cyclooctyne that is calculated (272) differs appreciably from the ED structure (273–275). The calculated bond lengths between saturated carbons are all in the range of 1.54–1.56 Å, whereas the electron diffraction values range from 1.49–1.58 Å. The calculated CCC bond angles are quite distorted, ranging from 104–119°, while the electron diffraction values range from 109–111°. The calculated geometries may be somewhat in error. They are probably more accurate than the experimental ones, however. With a molecule of this size, a number of possible structures might fit the electron diffraction data about equally well, and without additional information it seems unlikely that the one arrived at will be the correct one. The calculated structure puts most of the distortion in the bond angles rather than in the bond lengths, as anticipated, while the ED structure has most of the distortion in the bond lengths, and the angles are nearly normal.

The structures and energies of cycloocta-1-en-5-yne and of cycloocta-1,5-diyne have been calculated with two force fields (276).

Allinger and Meyer carried out force field calculations on a C_s conformation for cyclononyne (272). Subsequently Anet and Yavari published (277) an extensive study on the molecule, and concluded that there were 3 low-energy conformations (within 3 kJ/mol of one another), C_1 being most stable, followed by C_s and C_2. Calculations by Garrett (275) showed the C_s conformation to be much more stable than the C_1. An ED study was subsequently published (274) that was interpreted as being inconsistent with anything other than the C_1 or C_s conformations (if a single conformation were present). The structure, which was deduced assuming that only the C_s conformation was present, showed some suspiciously abnormal geometric quantities, and it seems likely that a conformational mixture of at least the C_s and C_1 forms is present.

The largest cycloalkyne studied to date is cyclododecyne (278, 279).

Molecules with Cyclic Conjugated Pi-Electronic Systems

Conjugated systems can be treated by the standard methods used for olefins only if the compound can be well described by a single

Kékulé form. Thus linear polyenes, for example, can be so treated by picking different parameters for the long and short bonds, and proceeding as before. Benzene is a special case that can also be treated by the standard method; a set of parameters is chosen for benzene such that the bond lengths are equal. But if we try to calculate the structure of naphthalene, using either of the above sets of parameters, we will get a structure that has either all equal bond lengths, or strongly alternating bond lengths—neither of which is correct. The correct structure of naphthalene is obtained by a molecular orbital calculation of some kind on the pi system. If we want to have a general treatment for conjugated systems, molecular orbital calculations (or equivalent valence bond calculations) on the pi system have to be included (pp. 52–55). Examples of the application of such a general treatment will be discussed below. First, however, we discuss the special case of the phenyl group.

The Phenyl Group

The rotameric behavior of aryl groups has found much recent interest. Among molecules containing a C_{sp^2}–C_{sp^3} bond, phenyl derivatives are unique because of the C_{2v} symmetry of the phenyl ring. The 2-propenyl group, in comparison, has only a symmetry plane.

The extremely low rotational barrier of toluene (280) (which is six-fold) is a consequence of these symmetry properties. For isopropylmesitylene the rotational barrier is much higher. The conformation with the methine hydrogen atom of the isopropyl group eclipsing the benzene ring was calculated to be the preferred one, in agreement with experiment (281–283). For 2,2-dimethyl-3-phenylbutane, calculations carried out using the rigid rotor approximation showed two energy minima corresponding to 1 and 2 (page 146), but rotation with full relaxation of all degrees of freedom except the torsion (driving) showed that only one of these minima (1) is real (284). A similar situation was noted with nucleoside calculations, where rigid rotation about the glycosyl bond also yielded four energy minima, only two of which appear to be real (284).

146 MOLECULAR MECHANICS

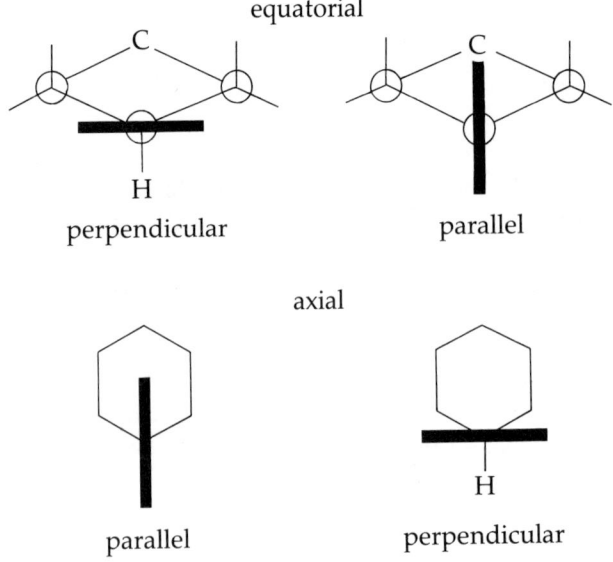

A phenyl group on a cyclohexane ring usually prefers to be equatorial by 12.6 kJ/mol, a much stronger preference than that found for a methyl group; yet the phenyl prefers the axial position in 1-methyl-1-phenylcyclohexane. This nonadditivity of group energies results from the rotational characteristics of the phenyl group (285). In phenylcyclohexane the equatorial phenyl is calculated to prefer a conformation in which its plane contains the bond to the iso-hydrogen, analogous to the situation in isopropylmesitylene discussed above. When the phenyl is axial, the conformation with the phenyl plane perpendicular to the CH bond is the better one, because of transannular repulsion. Replacement of the iso-hydrogen by a methyl group makes the equatorial parallel orientation less favored, because of interference between the methyl and an ortho hydrogen, and accordingly for

1-methyl-1-phenylcyclohexane, the equatorial phenyl group was calculated to prefer the perpendicular conformation, whereas for the axial conformation, the additional methyl group does not introduce additional strain. The energy difference between the equatorial and axial orientations of the phenyl group is therefore reduced because of a destabilization of the equatorial form.

In 2-phenyl-1,3-dioxane the atoms next to C2 do not carry equatorial hydrogens, which were mainly responsible for the high energy of the perpendicular conformation of the equatorial isomer of phenylcyclohexane. In fact, molecular mechanics calculations show a very low rotational barrier for the equatorial 2-phenyl group in 1,3-dioxane (*286, 287*). This is in good agreement with the lack of a preference for any one orientation in x-ray structures of 2-aryl-1,3-dioxanes, and in entropy measurements (*288, 289*).

When more than one phenyl group is attached to a carbon center, the conformation is determined by cooperative effects. Several studies have dealt with tri- and tetra-arylmethanes and -silanes. The preferred conformation of triphenyl- and trimesitylmethane was found to be propeller-shaped (*290*). The chirality (helicity) of the propeller can be inverted by four different mechanisms involving simultaneous flips of one, two, or all three rings, or, with boranes and amines, by inversion at the central atom (Figure 4.15). The process has been studied by molecular mechanics in trimesitylmethane (*290*). Simulation of the intercon-

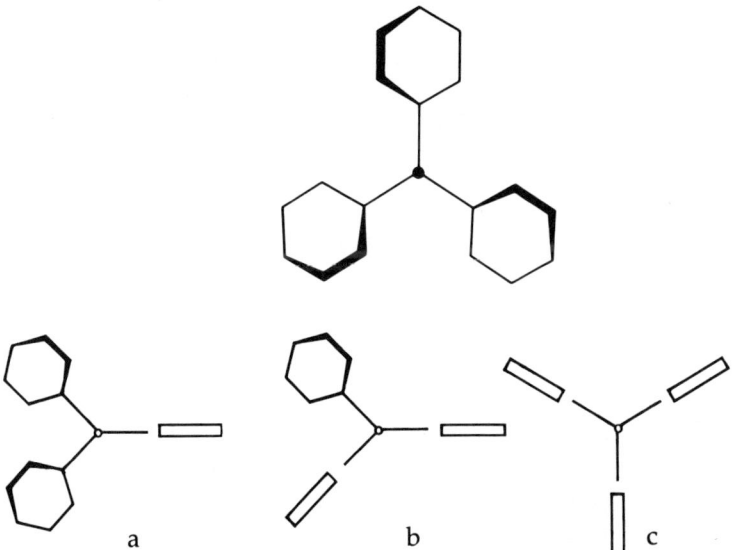

Figure 4.15 Transition states for chirality inversion by the one-ring (a), two-ring (b), and three-ring (c) flip mechanisms.

version allowing for only partial relaxation of the coordinates resulted in the prediction that the two-ring flip has the lowest transition state energy. Full-relaxation driving of a torsion angle of one mesityl ring in a clockwise direction from the energy minimum leads to a correlated counterclockwise motion of the two other mesityl rings. The transition state reached in this way is that of the two-ring flip, with a calculated barrier (83.7 kJ/mol) in agreement with the experimental barrier (91.6 kJ/mol) (*290*). The same preference for the two-ring flip mechanism has been found with trimesityl derivatives of several group IIIa, IVa, and Va elements (*291, 292*).

For the ground state geometry of tetraphenylmethane, one S_4 structure and one $°D_{2d}$ structure (° for open) appear the most favorable on

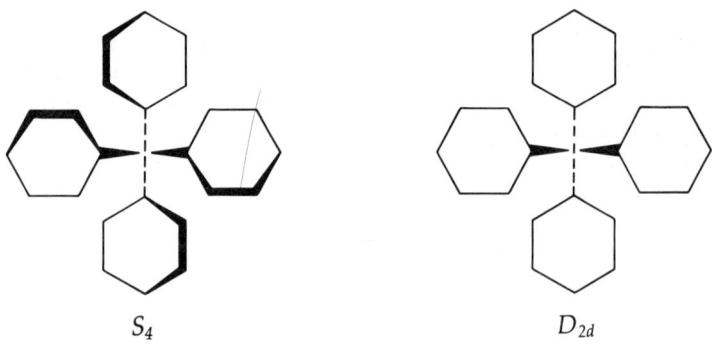

S_4 D_{2d}

qualitative grounds. From rough molecular mechanics calculations it was concluded that the S_4 structure was more stable, both in the gas phase and in the crystal (*293*). More refined calculations have shown that while the S_4 geometry is more stable for tetra-*o*-tolylmethane (with the methyl groups on the outside of the molecule) and for tetraphenylsilane (*12*), the $°D_{2d}$ structure is the preferred one for tetraphenylmethane. The exchange of diastereotopic positions in the aryl rings can again be studied by driving a phenyl ring through a range of torsion angles from the D_{2d} geometry across the barrier. In this motion the three other rings follow the driven ring at first to structures with approximate S_4 symmetry, but as the transition state is approached, two of the rings reverse their movement, and a two-ring flip mechanism follows for this molecule also (Figure 4.16) (*295*).

Most force fields had problems reproducing the conformational preferences in compounds with phenyl groups on adjacent carbon atoms (1,2-diphenylethane, for example). Ivanov et al. (*296*) and Jacobus (*297*) calculated with various force fields (including the Schleyer and the MM1 force field) that the *gauche* isomers of 1,2-diphenylpropane

C₆H₅ above, H & CH₃ & H & H	C₆H₅ above, C₆H₅ & CH₃ & H & H	C₆H₅ above, H & CH₃ & H & C₆H₅
0.0	3.35	5.10

Energies calculated with the
Schleyer force field (kJ/mol)

and of 1,2-diphenylethane are more stable than the *anti* forms, in contradiction to experimental evidence. This deficiency does not appear with the MM2 force field, which, without any special precautions, gave the correct equilibrium position (*38*).

1,1,2,2-Tetraphenylethane and 1,1,2,2-tetrakis(2,6-xylyl)ethane were calculated to have only one *anti* conformation exhibiting C_2 symmetry and resembling a four-blade propeller, due to cooperative effects. This form is calculated to be 20.5 kJ/mol more stable than the most stable of three *gauche* conformers (*298*), in agreement with NMR and x-ray data (*299*). The contrast between this molecule and tetramethylethane, where the *gauche* and *anti* conformations have identical energies, is

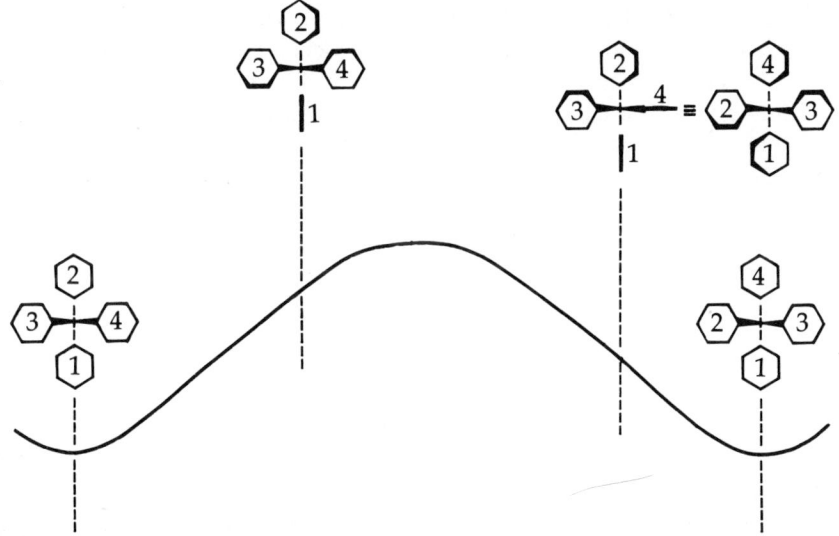

Figure 4.16. The two-ring flip for tetraphenylmethane.

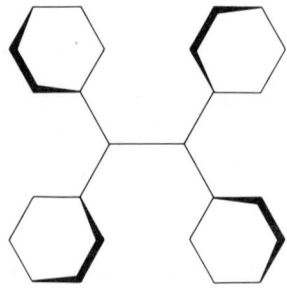

noteworthy. The contrast between these two cases and the tetra-t-butylethane, which exists exclusively in a *gauche* conformation, is striking.

For pentaphenylethane and hexaphenylethane, molecular mechanics predicted propeller conformations with very long central bonds (*300, 301*). In the case of pentaphenylethane, the calculated value [1.595 Å (*302*)] agrees with the experimental value in the crystal [1.606 Å (*303*)], but for the hexaphenyl derivative a much shorter bond (1.47 Å) was found in the crystal of a derivative (*304, 305*) than was calculated [1.64 Å (*301*)]. (It is difficult to imagine an electronic effect that would yield such a short bond length. While there really is no reason to suspect that the x-ray value is wrong, it would certainly be desirable to check the crystal structure of a different hexaphenylethane derivative, and see if a similar bond length could be found.) Such a discrepancy was also found for 9,9'-bitriptycyl, where experiment gives a much shorter bond (1.558 Å) than does molecular mechanics (1.589 Å). This discrepancy was explained by a rehybridization of the quaternary carbons giving the C_{sp^2}–C_{sp^3} bonds more p character and the C_{sp^3}–C_{sp^3} bond more s character (*301*). (*See also* the related discussion in pp. 155–156).

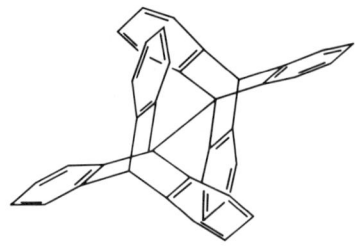

General Treatment of Conjugated Pi-Electronic Systems

The two methods that have been applied to conjugated hydrocarbons were described in pp. 52–55. The Warshel–Karplus approach

(*306*) assumes the validity of the pi–sigma separation regardless of planarity. This assumption is hard to justify for nonplanar systems, but the method does seem to work reasonably well in the few such cases studied. For the ethylene molecule, as an example, the bond lengths were calculated to within a few thousandths of an Angstrom, the bond angles to within 0.5°, the energy of formation to within 6 kJ/mol, the ionization potential and first electronic excitation energy to within 0.1 eV, the torsional barrier to within 8 kJ/mol, and the vibrational frequencies to within about 50 cm^{-1}. For the ground states of benzene and butadiene, the corresponding properties were calculated with similar accuracies. For the excited states of ethylene, the accuracy was less by a factor of two or three.

In the Allinger and Sprague approach (*202, 210, 220, 307*), [also developed independently by Lindner (*342*)] it is necessary to establish the relationship between bond order and natural bond length (r_0) on the one hand, and between bond order and force constant on the other. This was done by picking some standard compounds, the bond lengths of which are well established, namely ethylene, butadiene (both the long and the short bond), and benzene. As is well known (*308*), an approximately linear relationship between bond order and bond length exists. The constants required for this relationship were picked so that these bond lengths were all well reproduced. For the force constant, sufficient experimental data were available to permit a linear relationship with bond order to be established.

Of course, bond-order–bond-length relationships have been used for a long time to establish the structures of relatively unstrained planar conjugated systems. The advantage of the force field calculation is that one can extend the area of coverage to include molecules that are quite strained, such as *ortho*-di-*t*-butylbenzene.

These calculations give very good results with respect to structure for planar conjugated systems. There are two further points of interest. The first point concerns nonplanar systems, and the second concerns the heats of formation of these compounds (*307*). Here we will discuss the question of nonplanar systems, and heats of formation will be taken up in Chapter 5.

The initial supposition was made that the SCF calculation can be carried out on the planar system, and the molecule is then allowed to distort to find the energy minimum. This is not quite right, however, as can be seen by considering the butadiene molecule. The planar system calculation gives pi bond orders of 0.97 and 0.26 (MMP2) for the formally double and single bonds, respectively. As a rotation occurs about the single bond, the pi bond orders tend in the direction of 1.0 and 0.0. The difficulty is that the double bonds become stronger as the rotation occurs, so this must also be taken into account in dealing with the nonplanarity of the system. This effect usually has a minor influence with respect to the bond length, and probably could be neglected if that were the only problem. However, the torsional barrier about the double bond would also be increased by the increasing strength of the bond as rotation occurred about the single bond. For more distorted, particularly cyclic, systems, this effect is not negligible. Consequently, this reduction in the pi bond torsional force constants was calculated by a perturbation method, averaged over the whole molecule, and the torsional constants of the molecule were then appropriately adjusted. The method was applied to about fifty molecules, most of which have available data either in terms of structure or heat of formation. In general the results were in good agreement with experiment (202). The calculations have been used to provide structures for small polyenes, both planar and nonplanar, for the purposes of carrying out SCF calculations on electronic spectra (337). The calculations have also been extended to large polyenes (338), aimed at understanding better the UV spectra of carotenoids. Further extensions have been made to include conjugated carbonyl compounds (339, 340).

Some compounds studied that are of special interest are hexahelicene, bicyclo[5.5.0]dodecahexaene, phenanthrene, bicyclo[4.4.1]-undecapentaene, and 18-annulene. These will be explicitly discussed here.

Hexahelicene was of particular interest because of its nonplanar character, and it was desirable to check that the pitch of the helix was correctly obtained, since this depends upon a balance between the van der Waals repulsions between carbons and the various force constants that are attempting to maintain planarity. A good structure was obtained (210). Rather extensive calculations on helicenes have been reported by Lindner (309).

hexahelicene

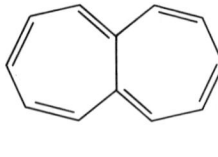
bicyclo[5.5.0]dodecahexaene

Bicyclo[5.5.0]dodecahexaene, a $4n$-pi-system, is only stable at low temperatures in dilute solution. While the compound could not be isolated, its UV spectrum was obtained (*310*). The calculations indicate that the molecule has a strong alternation of bond lengths, which is consistent with the Hückel rule. The UV spectrum calculated from this geometry is in good agreement with experiment (*202, 311*).

The phenanthrene molecule was interesting, because the difference between the calculated and experimental structures earlier led to the conclusion that the experimental values for certain bond lengths were incorrect (*210*), and indeed, a subsequent x-ray study agreed with the calculations (*312*).

Bicyclo[4.4.1]undecapentaene (1,6-methano-10-annulene) has had its structure determined (as a carboxy derivative); the bond lengths of the peripheral bonds are much more constant than in naphthalene, which suggests considerable aromatic character (*313*). A similar trend is found in the calculated bond lengths (Figure 4.17). The observed dihedral angles of the conjugated part of the molecule are distorted from planarity by up to about 30°, and the calculations reproduce this well (*210*).

It can be noted that the bond lengths calculated for the conjugated part of the molecule show a slight alternation in the naphthalene sense, where 2–3 is shorter than 1–2 or 3–4 (Figure 4.17). This is partly due to the fact that the 1–6 resonance integral, which is fairly large because the two pi orbitals are tilted toward one another, introduces a naphthalene kind of character in the pi system. (Resonance integrals between all pi orbitals were included in MMP1 calculations). But for consistency, this 1,6 resonance should be allowed to pull the two atoms closer together. But since they are not bonded together, the bond-order–bond-length relationship was not applied here (*210*). Recently Lindner (*341*) has employed a special bond-order–bond-length relationship, applied as a perturbation to his own force field (*342*), to this problem, and the results are very good indeed. For the 1–2, 2–3, and 3–4 bonds, his calculations

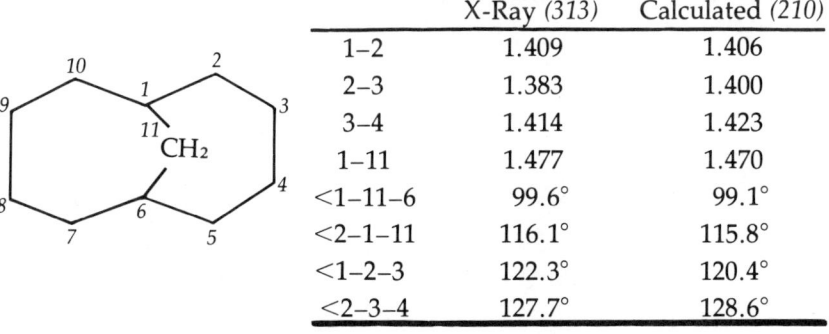

	X-Ray (*313*)	Calculated (*210*)
1–2	1.409	1.406
2–3	1.383	1.400
3–4	1.414	1.423
1–11	1.477	1.470
<1–11–6	99.6°	99.1°
<2–1–11	116.1°	115.8°
<1–2–3	122.3°	120.4°
<2–3–4	127.7°	128.6°

Figure 4.17. The geometry of bicyclo[4.4.1]undecapentaene.

gave 1.400, 1.406 and 1.415 Å without the perturbation, and 1.412, 1.391, 1.427 Å with it (cf. Figure 4.17).

This problem of a large non-neighbor resonance interaction is also important in a few other cases, 1,3,5-cycloheptatriene being one (*341*). Here the overlap (and hence the resonance integral) between the

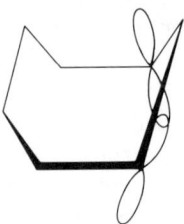

p-orbitals on atoms 1 and 6 is much larger when the ring is in the stable boat form than when it is planar. This interaction greatly stabilizes the boat form, so that it is not possible to correctly calculate the inversion barrier of this ring unless this interaction is included (*341*).

The less symmetrical bridged 10-annulene, bicyclo[5.3.1]undecapentaene (1,5-methano-10-annulene), is calculated (*210*) to have complete bond alternation (Fig. 4.18). The sigma system constraints force several dihedral angles to differ up to 54° from planarity, with an average distortion of 23°, effectively generating a pi system composed of isolated segments, each with an even number of pi-atomic centers. Thus, the compound is predicted to be fundamentally quite different from the isomeric [4.4.1] system, even though their average planarities are rather similar. Recent NMR studies have been interpreted as indicating that this conclusion may not be correct (*314*). It might be that the neglect of electron correlation in the SCF calculation leads to an erroneous result here, as is discussed below for 18-annulene, and a crystal structure is really needed.

The 18-annulene molecule still presents something of a mystery. The x-ray structure indicates that the bond lengths are all nearly equal (*315*) (approximately D_{6h} symmetry), while the calculations (*202, 316*)

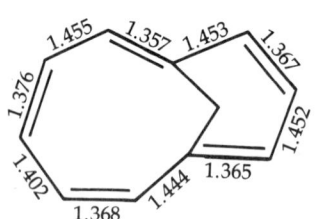
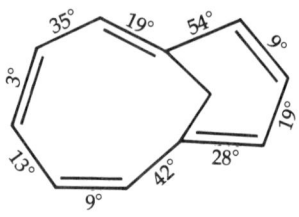

Figure 4.18. Calculated bond lengths and torsional angles in bicyclo[5.3.1]undecapentaene.

favor an alternating nonplanar structure (D_3). The temperature factors are normal, which indicates that rotational disorder in the crystal leading to an apparent D_6 symmetry is unlikely. The difference in energy between these structures was substantial with the MM1 (210) force field (42.7 kJ/mol), but is much reduced with the MM2 force field. The experimental UV spectrum is in much better agreement with that calculated (316) for an alternating structure than with that calculated for the x-ray structure. MINDO/3 calculations (317) also indicate bond alternation, but a planar (D_{3h}) structure. It was remarked that the energy increase on deforming the molecule from planarity was very small (317).

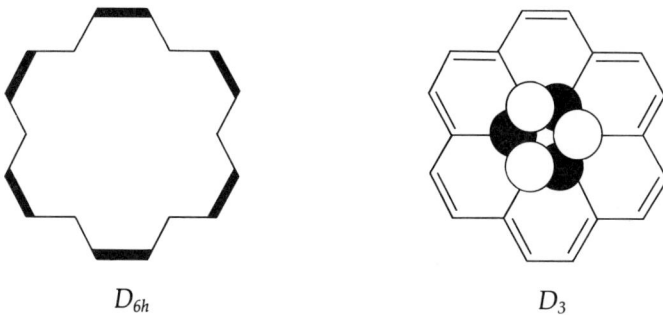

D_{6h} D_3

Baumann has recently done a very approximate calculation of the correlation energy of the pi electrons for the D_{3h} and D_{6h} structures (318), and has concluded that it is much larger for the latter. If correct, this might explain the crystallographic structure, and would suggest that other cases will be found where the difference in correlation energies between two structures will be large enough that SCF methods will give wrong answers.

The 2,2-paracyclophane (**1**), and molecules **2** and **3** (page 156) illustrate an interesting limitation in the molecular mechanics type calculation. In each of these molecules, the bonds indicated with an asterisk have bond lengths that are calculated to be too short. This effect has been studied in some detail theoretically by Mislow and coworkers (319). In still earlier studies, Gleiter (320) showed in the case of molecule **1**, that the parallel alignment of the p-orbitals of the pi systems and the sigma bonds indicated by asterisks led to a large overlap, and large pi–sigma interaction effects, which include the lengthening of the bond. This effect is expected to occur whenever similar geometrical conditions prevail, and since it is not allowed for in molecular mechanics calculations, at least up to this point, the latter will be systematically in error. The magnitude of this effect can be conveniently determined by looking at the magnitude of the deviation of the calculated bond length (where the effect is absent) from that observed experimentally (where the effect is present).

CH₂—⬡—CH₂
  *   *
CH₂—⬡—CH₂

1

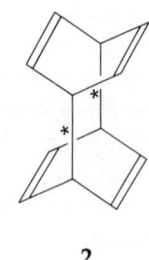

2

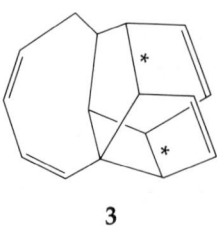

3

Hydrocarbons Containing Deuterium

The equilibrium geometry of a deuterated hydrocarbon is exactly that of the same compound without the deuterium. But because the mass of deuterium is greater than that of hydrogen, the vibrational energy levels for those vibrations involving deuterium are set down lower in the potential well than the analogous hydrocarbon vibrations, and the amplitude of the vibration is correspondingly smaller. When we consider experimental structures and properties, these vibrational differences between deuterium and hydrogen are expected to show up in some specific ways.

The C–H vibration is anharmonic, and generally the motion is described by a Morse potential. As the vibrational level gets lower, a weighted average of the position of the proton over the vibrational motion (the center of gravity of the wave function) moves in closer to the attached carbon. Because these frequencies are so high relative to other frequencies in the molecule, they usually do not couple very much with the other vibrational frequencies, and only the zero-point level is significantly populated at room temperature. So it is clear that the C–D bond length will be shorter than the C–H analog, and indeed this has been found to be true with small molecules such as perdeuteromethane (329) and perdeuteroethane (257). It is less clear in general how the bending vibrations that involve deuterium will differ from those involving hydrogen, but on the whole they will be of smaller amplitude.

In deciding how to deal with deuterium in a molecular mechanics calculation, it is clear that, besides the difference in mass and the difference in the value of r_o (in Equation 2.7), it would seem that the difference in the vibrational amplitude could best be represented by picking a smaller van der Waal's radius for deuterium than for hydrogen. The difference between deuterium and hydrogen is sufficient to shift conformational equilibria by small amounts (usually of the order of 0.04–0.50 kJ/mol). Such small shifts are difficult to detect by most experimental techniques, but they show up very clearly in experiments involving optical rotatory dispersion or circular dichroism. Exploratory force field calculations were therefore carried out by Djerassi and his group (157) using some ad hoc parameters for the van der Waal's radius of deuterium and for the C–D bond length. Indeed, several experiments on deuterium-containing compounds could thus be qualitatively interpreted.

More recently, a detailed study of deuterium in molecular mechanics calculations was carried out using the MM2 program (180). Many things can be reproduced quite well. For example, the conformational energy of a trideuteromethyl group relative to a methyl group on cyclohexane, the geometry of ethane, and the barrier to rotation of 1,8-bis(trideuteromethyl)phenanthrene, relative to its hydrogen analog, are reproduced quite well. On the other hand, the axial/equatorial equilibrium of the deuterium in 2-deuterocyclopentanone has a calculated value of zero, as opposed to an experimental value of 0.04 kJ/mol. Thus it was possible to choose parameters for the deuterium that in this force field gave generally good results, especially if the interaction of the deuterium with the opposing group was not back along the D–C bond, but was out away from this bond. In 2-deutereocyclopentanone the crucial interactions are back towards the bond, and the results are poor. It would seem that the MM2 force field is insufficiently refined in the basic hydrocarbon portion to deal as well as one would like with the very small interactions in question, unless they involve a more or less head-on interaction of the deuterium with an interacting group.

Literature Cited

1. Bartell, L. S. *J. Chem. Phys.* **1960**, *32*, 827.
2. Bartell, L. S. *Tetrahedron* **1962**, *17*, 177.
3. Bartell, L. S. *J. Chem. Ed.* **1968**, *45*, 754.
4. Gillespie, R. J. "Molecular Geometry"; Van Nostrand: London, 1972.
5. Jacob, E. J.; Thompson, H. B.; Bartell, L. S. *J. Chem. Phys.* **1967**, *47*, 3736.
6. Fitzwater, S.; Bartell, L. S. *J. Am. Chem. Soc.* **1976**, *98*, 5107.
7. Bartell, L. S. *J. Am. Chem. Soc.* **1977**, *99*, 3279.
8. Bartell, L. S.; Bürgi, H. B. *J. Am. Chem. Soc.* **1972**, *94*, 5239.

9. Allinger, N. L.; Hirsch, J. A.; Miller, M. A.; Tyminski, I. J.; Van-Catledge, F. A. *J. Am. Chem. Soc.* **1968**, *90*, 1199.
10. Allinger, N. L. *J. Am. Chem. Soc.* **1977**, *99*, 8127.
11. Iroff, L. D.; Mislow, K. *J. Am. Chem. Soc.* **1978**, *100*, 2121.
12. Hutchings, M. G.; Andose, J. D.; Mislow, K. *J. Am. Chem. Soc.* **1975**, *97*, 4553.
13. Traetteberg, M. *Acta Chem. Scand. Ser. B* **1975**, *29*, 29.
14. Hilderbrandt, R. L.; Wieser, J. D.; Montgomery, L. K. *J. Am. Chem. Soc.* **1973**, *95*, 8598.
15. Schubert, W.; Southern, J. F.; Schäfer, L. *J. Mol. Struct.* **1973**, *16*, 403.
16. Schubert, W.; Schäfer, L.; Pauli, G. H. *J. Mol. Struct.* **1974**, *21*, 53.
17. Askari, M.; Schäfer, L.; Seip, R. *J. Mol. Struct.* **1976**, *32*, 153.
18. Pauli, G. H.; Askari, M.; Schubert, W.; Schäfer, L. *J. Mol. Struct.* **1976**, *32*, 145.
19. Spelbos, A.; Mijlhoff, F. C.; Bakker, W. H.; Baden, R.; van den Enden, L. *J. Mol. Struct.* **1977**, *38*, 155.
20. Bellott, E. M., Jr. *Tetrahedron* **1977**, *33*, 1707.
21. Allinger, N. L.; Yuh, Y.; Sprague, J. T. *J. Comput. Chem.* **1980**, *1*, 30.
22. Philip, T.; Cook, R. L.; Malloy, T. B., Jr.; Allinger, N. L.; Chang, S.; Yuh, Y. *J. Am. Chem. Soc.* **1981**, *103*, 2151.
23. Allinger, N. L.; Hickey, M. J. *J. Mol. Struct.* **1973**, *17*, 233.
24. Allinger, N. L.; Hickey, M. J. *Tetrahedron* **1972**, *28*, 2157.
25. Allinger, N. L.; Kao, J.; Chang, H.-M.; Boyd, D. B. *Tetrahedron* **1976**, *32*, 2867.
26. Askari, M.; Ostlund, N. S.; Schäfer, L. *J. Am. Chem. Soc.* **1977**, *99*, 5246.
27. Allinger, N. L.; Profeta, S., Jr. *J. Comput. Chem.* **1980**, *1*, 181.
28. Allinger, N. L.; Burkert, U.; Profeta, S., Jr. *J. Comput. Chem.* **1980**, *1*, 281.
29. Burkert, U. *J. Comput. Chem.* **1980**, *1*, 285.
30. Dougherty, D. A.; Mislow, K. *J. Am. Chem. Soc.* **1979**, *101*, 1401.
31. Dougherty, D. A.; Mislow, K.; Huffman, J. W.; Jacobus, J. *J. Org. Chem.* **1979**, *44*, 1585.
32. Mastryukov, V. S.; Popik, M. V.; Dorofeeva, O. V.; Golubinskii, A. V.; Vilkor, L. V.; Belikova, N. A.; Allinger, N. L. *Tetrahedron Lett.* **1979**, 4339.
33. Allinger, N. L.; Hickey, M. J. *J. Am. Chem. Soc.* **1975**, *97*, 5167.
34. Bartell, L. S.; Boates, T. L. *J. Mol. Struct.* **1976**, *32*, 379.
35. Heinrich, F.; Lüttke, W. *Chem. Ber.* **1977**, *110*, 1246.
36. Lunazzi, L.; Macchiantelli, D.; Bernardi, F.; Ingold, K. U. *J. Am. Chem. Soc.* **1977**, *99*, 4573.
37. Braun, H.; Lüttke, W. *J. Mol. Struct.* **1976**, *31*, 97.
38. Allinger, N. L., unpublished calculations employing the MM2 force field (Ref. *10*).
39. Baxter, S., cited by Iroff and Mislow (Ref. *11*).
40. Lidén, A.; Roussel, C.; Liljefors, T.; Chanon, M.; Carter, R. E.; Mezger, J.; Sandström, J. *J. Am. Chem. Soc.* **1976**, *98*, 2853.
41. Iroff, L. D. *J. Comput. Chem.* **1980**, *1*, 76.
42. Zefirov, N. S. *Tetrahedron* **1977**, *33*, 3193.
43. Wertz, D. H.; Allinger, N. L. *Tetrahedron* **1974**, *30*, 1579.
44. Ritter, W.; Hull, W.; Cantow, H.-J. *Tetrahedron Lett.* **1978**, 3093.

45. Boyd, R. H. *J. Am. Chem. Soc.* **1975,** *97,* 5353.
46. Verma, A. L.; Murphy, W. F.; Bernstein, H. J. *J. Chem. Phys.* **1974,** *60,* 1540.
47. Beckhaus, H.-D.; Hellmann, G.; Rüchardt, C. *Chem. Ber.* **1978,** *111,* 72.
48. Hounshell, W. D.; Dougherty, D. A.; Mislow, K. *J. Am. Chem. Soc.* **1978,** *100,* 3149.
49. Osawa, E.; Shirahama, H.; Matsumoto, T. *J. Am. Chem. Soc.* **1979,** *101,* 4824.
50. Aston, J. G.; Schumann, S. C.; Fink, H. L.; Doty, P. M. *J. Am. Chem. Soc.* **1941,** *63,* 2029.
51. Aston, J. G.; Fink, H. L.; Schumann, S. C. *J. Am. Chem. Soc.* **1943,** *65,* 341.
52. Kilpatrick, J. E.; Pitzer, K. S.; Spitzer, R. *J. Am. Chem. Soc.* **1947,** *69,* 2483.
53. Altona, C.; Geise, H. J.; Romers, C. *Tetrahedron* **1968,** *24,* 13.
54. Hendrickson, J. B. *J. Am. Chem. Soc.* **1961,** *83,* 5537.
55. Pitzer, K. S.; Donath, W. E. *J. Am. Chem. Soc.* **1959,** *81,* 3213.
56. Pickett, H. M.; Strauss, H. L. *J. Chem. Phys.* **1970,** *53,* 376.
57. Allinger, N. L.; Tribble, M. T.; Miller, M. A.; Wertz, D. H. *J. Am. Chem. Soc.* **1971,** *93,* 1637.
58. Lifson, S.; Warshel, A. *J. Chem. Phys.* **1968,** *49,* 5116.
59. Sachse, H. *Ber. Dtsch. Chem. Ges.* **1890,** *23,* 1363.
60. Sachse, H. *Z. Phys. Chem.* **1892,** *10,* 203.
61. Shoppee, C. W. *J. Chem. Soc.* **1946,** 1138.
62. Dunitz, J. D.; Bürgi, H. B. In MTP Int. Rev. Sci. Phys. Chem. Ser. 2, Vol. 11 (Chem. Crystallography); Butterworths: London, 1975; p. 81.
63. Bastiansen, O.; Fernholt, L.; Seip, H. M.; Kambara, H.; Kuchitsu, K. *J. Mol. Struct.* **1973,** *18,* 163.
64. Allinger, N. L. *Adv. Phys. Org. Chem.* **1976,** *13,* 1.
65. Buys, H. R.; Geise, H. J. *Tetrahedron Lett.* **1970,** 2991.
66. Kitaigorodsky, A. I. *Tetrahedron* **1960,** *9,* 183.
67. Ibid., **1961,** *14,* 230.
68. Altona, C.; Sundaralingam, M. *Tetrahedron* **1970,** *26,* 925.
69. Davis, M.; Hassel, O. *Acta Chem. Scand.* **1963,** *17,* 1181.
70. Wohl, R. A. *Chimia* **1964,** *18,* 219.
71. Bucourt, R. *Top. Stereochem.* **1974,** *8,* 159.
72. Bucourt, R.; Hainault, D. *C. R. Acad. Sci.* **1964,** *258,* 3305.
73. Squillacote, M.; Sheridan, R. S.; Chapman, O. L.; Anet, F. A. L. *J. Am. Chem. Soc.* **1975,** *97,* 3244.
74. Strauss, H. L. *J. Chem. Ed.* **1971,** *48,* 221.
75. Pickett, H. M.; Strauss, H. L. *J. Am. Chem. Soc.* **1970,** *92,* 7281.
76. Hendrickson, J. B. *J. Am. Chem. Soc.* **1967,** *89,* 7047.
77. Wiberg, K. B.; Boyd, R. H. *J. Am. Chem. Soc.* **1972,** *94,* 8426.
78. Ermer, O. *Structure Bond.* **1976,** *27,* 161.
79. Hendrickson, J. B. *J. Am. Chem. Soc.* **1962,** *84,* 3355.
80. Allinger, N. L.; Miller, M. A.; Van-Catledge, F. A.; Hirsch, J. A. *J. Am. Chem. Soc.* **1967,** *89,* 4345.
81. Eliel, E. L.; Powers, J. R.; Nader, J. W. *Tetrahedron* **1974,** *30,* 515.
82. Eliel, E. L.; Brett, T. J. *J. Am. Chem. Soc.* **1965,** *87,* 5039.
83. Osawa, E.; Collins, J. B.; Schleyer, P. v. R. *Tetrahedron* **1977,** *33,* 2667.
84. Allinger, N. L.; Hu, S. *J. Am. Chem. Soc.* **1962,** *84,* 370.

85. Allinger, N. L.; Hu, S. *J. Org. Chem.* **1962**, *27*, 3417.
86. Winstein, S.; Holness, N. J. *J. Am. Chem. Soc.* **1955**, *77*, 5562.
87. Lectard, A.; Lichanot, A.; Metras, F.; Gaulties, J.; Hauw, C. *J. Mol. Struct.* **1976**, *34*, 113.
88. Faber, D. H.; Altona, C. *Chem. Commun.* **1971**, 1210.
89. Baas, J. M. A.; van de Graaf, B.; van Veen, A.; Wepster, B. M. *Tetrahedron Lett.* **1978**, 819.
90. Allinger, N. L.; Hirsch, J. A.; Miller, M. A.; Tyminski, I. J.; Van-Catledge, F. A. *J. Am. Chem. Soc.* **1968**, *90*, 1199.
91. Askari, M.; Merrifield, D. L.; Schäfer, L. *Tetrahedron Lett.* **1976**, 3497.
92. Bixon, M.; Lifson, S. *Tetrahedron* **1966**, *23*, 769.
93. Hendrickson, J. B. *J. Am. Chem. Soc.* **1967**, *89*, 7036.
94. Engler, E. M.; Andose, J. D.; Schleyer, P. v. R. *J. Am. Chem. Soc.* **1973**, *95*, 8005.
95. Bocian, D. F.; Pickett, H. M.; Rounds, T. C.; Strauss, H. L. *J. Am. Chem. Soc.* **1975**, *97*, 687.
96. Birnbaum, G. I.; Buchanan, G. W.; Morin, F. G. *J. Am. Chem. Soc.* **1977**, *99*, 6652.
97. Flapper, W. M.; Verschoor, G. C.; Rutten, E. W. M.; Romers, C. *Acta Crystallogr. Sect. B* **1977**, *33*, 5.
98. Dillen, J.; Geise, H. *J. Chem. Phys.* **1979**, *70*, 425.
99. Knorr, R.; Ganter, C.; Roberts, J. D. *Angew. Chem. Int. Ed. Engl.* **1967**, *6*, 556.
100. Knorr, R.; Ganter, C.; Roberts, J. D. *Angew. Chem.* **1967**, *79*, 577.
101. Anet, F. A. L.; Krane, F. *Tetrahedron Lett.* **1973**, 5029.
102. Anet, F. A. L. *Top. Curr. Chem.* **1974**, *45*, 169.
103. Anet, F. A. L.; Anet, R. "Dynamic Nuclear Magnetic Resonance Spectroscopy"; Jackman, L. M.; Cotton, F. A., Eds.; Academic: New York, 1975; p. 543.
104. Hendrickson, J. B. *J. Am. Chem. Soc.* **1964**, *86*, 4854.
105. Wiberg, K. B. *J. Am. Chem. Soc.* **1965**, *87*, 1070.
106. Anet, F. A. L.; Basus, V. J. *J. Am. Chem. Soc.* **1973**, *95*, 4424.
107. Anet, F. A. L.; St. Jacques, M. *J. Am. Chem. Soc.* **1966**, *88*, 2585, 2586.
108. Hendrickson, J. B. *J. Am. Chem. Soc.* **1967**, *89*, 7043.
109. Ferro, D. R.; Heatley, F.; Zambelli, A. *Macromolecules* **1974**, *7*, 480.
110. Anet, F. A. L.; Krane, J., cited by Dale (Ref. *115*).
111. Samuel, G.; Weiss, R. *Tetrahedron Lett.* **1969**, 3529.
112. Bryan, R. F.; Dunitz, J. D. *Helv. Chim. Acta* **1960**, *43*, 3.
113. Anet, F. A. L.; Wagner, J. J. *J. Am. Chem. Soc.* **1971**, *93*, 5266.
114. Anet, F. A. L.; Krane, J. *Isr. J. Chem.* **1980**, *20*, 72.
115. Dale, J. *Top. Stereochem.* **1976**, *9*, 199.
116. Rustad, S.; Seip, H. M. *Acta Chem. Scand. Ser. A* **1975**, *29*, 378.
117. Dunitz, J. D. *Pure Appl. Chem.* **1971**, *25*, 495.
118. Dunitz, J. D.; Eser, H.; Bixon, M.; Lifson, S. *Helv. Chim. Acta* **1967**, *50*, 1572.
119. White, D. N. J.; Bovill, M. J. *J. Mol. Struct.* **1976**, *33*, 273.
120. White, D. N. J.; Bovill, M. J. *J. Chem. Soc. Perkin Trans. 2* **1977**, 1610.
121. Dale, J. *Acta Chem. Scand.* **1973**, *27*, 1115.

122. Anet, F. A. L.; Rawdah, T. *J. Am. Chem. Soc.* **1978**, *100*, 7810.
123. Dunitz, J. D.; Shearer, H. M. M. *Proc. Chem. Soc.* **1958**, 348.
124. Ibid., **1959**, 268.
125. Dunitz, J. D.; Shearer, H. M. M. *Helv. Chim. Acta* **1960**, *43*, 18.
126. Anet, F. A. L.; Cheng, A. K.; Wagner, J. J. *J. Am. Chem. Soc.* **1972**, *94*, 9250.
127. Groth, P. *Acta Chem. Scand. Ser. A* **1978**, *32*, 279.
128. Anet, F. A. L.; Rawdah, T. *J. Am. Chem. Soc.* **1978**, *100*, 7166.
129. White, D. N. J.; Morrow, C. *Comput. Chem.* **1979**, *3*, 33.
130. Newman, B. A., unpublished data.
131. Groth, P. *Acta Chem. Scand. Ser. A* **1976**, *30*, 155.
132. Anet, F. A. L.; Rawdah, T., unpublished data, quoted by Dale (Ref. 115).
133. Bassi, W. I.; Scordamaglia, R.; Fiore, L. *J. Chem. Soc. Perkin Trans. 2* **1972**, 1726.
134. Groth, P. *Acta Chem. Scand. Ser. A* **1978**, *32*, 91.
135. Anet, F. A. L.; Cheng, A. K. *J. Am. Chem. Soc.* **1975**, *97*, 2420.
136. Allinger, N. L.; Gordon, B.; Profeta, S., Jr. *Tetrahedron* **1980**, *36*, 859.
137. Groth, P. *Acta Chem. Scand. Ser. A* **1974**, *28*, 808.
138. Groth, P. *Acta Chem. Scand.* **1971**, *25*, 725.
139. Allinger, N. L.; Gordon, B. J.; Newton, M. G.; Norskov-Lauritsen, L.; Profeta, S. *Tetrahedron* **1982**, *38*, (in press).
140. Granger, R.; Bardet, L.; Sablayrolles, C.; Girard, J.-P. *C. R. Acad. Sci., Ser. C* **1970**, *270*, 1326.
141. Chang, S.; McNally, D.; Shary-Tehrany, S.; Hickey, M. J.; Boyd, R. H. *J. Am. Chem. Soc.*, **1970**, *92*, 3109.
142. Allinger, N. L.; Tribble, M. T. *Tetrahedron* **1972**, *28*, 1191.
143. Browne, C. C.; Rossini, F. D. *J. Phys. Chem.* **1960**, *64*, 927.
144. Allinger, N. L.; Coke, J. L. *J. Am. Chem. Soc.* **1960**, *82*, 2553.
145. Duax, W. L.; Weeks, C. M.; Rohrer, D. C. *Top. Stereochem.* **1976**, *9*.
146. van den Enden, L.; Geise, H. J.; Spelbos, A. *J. Mol. Struct.* **1978**, *44*, 177.
147. Eliel, E. L.; Allinger, N. L.; Morrison, G. A.; Angyal, S. J. "Conformational Analysis," Wiley-Interscience, 1965.
148. Allinger, N. L.; Zalkow, V. B. *J. Am. Chem. Soc.* **1961**, *83*, 1144.
149. Hendrickson, J. B. *Tetrahedron* **1963**, *19*, 1387.
150. Allinger, N. L.; Wuesthoff, M. T. *J. Org. Chem.* **1971**, *36*, 2051.
151. Johnson, W. S. *J. Am. Chem. Soc.* **1953**, *75*, 1498.
152. Allinger, N. L.; Gorden, B. J.; Tyminski, I. J.; Wuesthoff, M. T. *J. Org. Chem.* **1971**, *36*, 739.
153. Hönig, H., unpublished data.
154. Pedley, U. B.; Rylance, J. "N.P.L. Computer Analyzed Thermochemical Data: Organic and Organometallic Compounds"; Univ. of Sussex: Sussex, 1977.
155. Cox, J. D.; Pilcher, G. "Thermochemistry of Organic and Organometallic Compounds"; Academic: London, 1970.
156. Newton, M. G.; Pantaleo, N. S.; Kirbawy, S.; Allinger, N. L. *J. Am. Chem. Soc.* **1978**, *100*, 2176.
157. Sing, L. Y.; Lindley, M.; Sundararaman, P.; Barth, G.; Djerassi, C. *Tetrahedron* **1981**, *Suppl. No. 1, 37*, 183.

158. Ermer, O. *Tetrahedron* **1974**, *30*, 3103.
159. Ermer, O.; Dunitz, J. D. *Helv. Chim. Acta* **1969**, *52*, 1861.
160. Gleicher, G. J.; Schleyer, P. v. R. *J. Am. Chem. Soc.* **1967**, *89*, 582.
161. Fournier, J.; Waegell, B. *Bull. Soc. Chim. Fr.* **1973**, 436.
162. Mastryukov, V. S.; Popik, M. V.; Dorofeeva, O. V.; Golubinskii, A. V.; Vilkov, L. V.; Belikova, N. A.; Allinger, N. L. *J. Am. Chem. Soc.* **1981**, *103*, 1333.
163. Mastryukov, V. S.; Osina, E. L.; Dorofeeva, O. V.; Popik, M. V.; Vilkov, L. V.; Belikova, N. A. *J. Mol. Struct.* **1979**, *52*, 211.
164. Bovill, M. J.; Cox, P. J.; Flitman, H. P.; Guy, H. P.; Hardy, A. D. U.; McCabe, P. H.; MacDonald, M. A.; Sim, G. A.; White, D. N. J. *Acta Crystallogr. Sect. B* **1979**, *35*, 669.
165. Sim, G. A. *Acta Crystallogr. Sect. B* **1979**, *35*, 2455.
166. Bhattacharjee, S. K.; Chako, K. K. *Tetrahedron* **1979**, *35*, 1999.
167. Schneider, H.-J.; Gschwendtner, W.; Weigand, E. F. *J. Am. Chem. Soc.* **1979**, *101*, 7195.
168. Parker, W.; Steele, W. V.; Watt, I. *J. Chem. Thermodyn.* **1977**, *9*, 307.
169. Engler, E. M.; Chang, L.; Schleyer, P. v. R. *Tetrahedron Lett.* **1972**, 2525.
170. Osawa, E.; Aigami, K.; Inamoto, Y. *J. Chem. Soc., Perkin Trans. 2* **1979**, 172.
171. Clark, T.; Knox, T. M. O.; McKervey, M. A.; Mackle, H.; Rooney, J. A. *J. Am. Chem. Soc.* **1979**, *101*, 2404.
172. Ermer, O. *Angew. Chem.* **1977**, *89*, 431; *Agnew. Chem. Int. Ed.* **1977**, *16*, 411.
173. Hargittai, I.; Hedberg, K. *Chem. Commun.* **1971**, 1499.
174. Greengard, R.; Chang, S.; Yuh, Y.; Allinger, N. L., unpublished data.
175. Rogers, D. W.; Choi, L. S.; Girellini, R. S.; Holmes, T. J.; Allinger, N. L. *J. Phys. Chem.* **1980**, *84*, 1810.
176. Mastryukov, V. S.; Osina, E. L.; Vilkov, L. V.; Hilderbrandt, R. L. *J. Am. Chem. Soc.* **1977**, *99*, 6855.
177. Lewis, J. D.; Laane, J.; Malloy, T. B., Jr. *J. Chem. Phys.* **1974**, *61*, 2342.
178. Bohn, R. K.; Mizuno, K.; Fukuyamo, T.; Kuchitsu, K. *Bull. Chem. Soc. Jpn.* **1973**, *46*, 1395.
179. Mizuno, K.; Fukuyama, T.; Kuchitsu, K. *Chem. Lett.* **1972**, 249.
180. Flanagan, H. L., unpublished data.
181. Wright, J. S.; Salem, L. *J. Am. Chem. Soc.* **1972**, *94*, 322.
182. Bartell, L. S. *Chem. Commun.* **1973**, 786.
183. Anderson, B.; Srinivasan, R. *Acta Chem. Scand.* **1972**, *26*, 3468.
184. Karabatsos, G. J.; Fenoglio, D. J. *Top. Stereochem.* **1970**, *5*, 188.
185. Van Hemelrijk, D.; van den Enden, L.; Geise, H. J.; Sellers, H. L.; Schäfer, L. *J. Am. Chem. Soc.* **1980**, *102*, 2189.
186. Allinger, N. L.; Sprague, J. T. *J. Am. Chem. Soc.* **1972**, *94*, 5734.
187. Ermer, O.; Lifson, S. *J. Am. Chem. Soc.* **1973**, *95*, 4121.
188. Almeningen, A.; Anfinsen, I. M.; Haaland, A. *Acta Chem. Scand.* **1970**, *24*, 43.
189. Kondo, S.; Sakurai, Y.; Hirota, E.; Morino, Y. *J. Mol. Spectrosc.* **1970**, *34*, 231.
190. Tokue, I.; Fukuyama, T.; Kuchitsu, K. *J. Mol. Struct.* **1974**, *23*, 33.
191. Ermer, O.; Lifson, S. *Tetrahedron* **1974**, *30*, 2425.
192. Mootz, D. *Acta Crystallogr. Sect. B* **1968**, *24*, 839.
193. Lenoir, D.; Frank, R. *Tetrahedron Lett.* **1978**, 53.

194. Cordt, F.; Frank, R. M.; Lenoir, D. *Tetrahedron Lett.* **1979**, 505.
195. Lenoir, D.; Frank, R. M.; Cordt, F.; Gieren A.; Lamm, V. *Chem. Ber.* **1980**, *113*, 739.
196. Burkert, U. *Tetrahedron* **1981**, *37*, 333.
197. Allinger, N. L.; Yuh, Y.; Sprague, J. T. *J. Comp. Chem.* **1980**, *1*, 30.
198. Bak, B.; Led, J. J.; Nygaard, L.; Rastrup-Andersen, J.; Sorensen, G. O. *J. Mol. Struct.* **1969**, *3*, 369.
199. Sharpen, L. H.; Laurie, V. W. *J. Chem. Phys.* **1968**, *49*, 3041.
200. Allinger, N. L.; Hirsch, J. A.; Miller, M. A.; Tyminski, I. J. *J. Am. Chem. Soc.* **1968**, *90*, 5773.
201. McNeill, E. A.; Scholer, F. R. *J. Mol. Struct.* **1976**, *31*, 65.
202. Allinger, N. L.; Sprague, J. T., unpublished data.
203. Avirah, T. K.; Cook, R. L.; Malloy, T. B., Jr. *J. Mol. Spectrosc.* **1975**, *54*, 231.
204. Laane, J.; Lord, R. C. *J. Chem. Phys.* **1967**, *47*, 4941.
205. Burkert, U. *Angew. Chem.* **1981**, *93*, 602; *Angew. Chem. Int. Ed.* **1981**, *20*, 572.
206. Anet, F. A. L.; Yavari, I. *Tetrahedron* **1978**, *34*, 2879.
207. Bucourt, R.; Hainault, D. *Bull. Soc. Chim. Fr.* **1965**, 1366.
208. Anet, F. A. L.; Haq, M. Z. *J. Am. Chem. Soc.* **1965**, *87*, 3147.
209. Jensen, F. R.; Bushweller, C. H. *J. Am. Chem. Soc.* **1969**, *91*, 5774.
210. Allinger, N. L.; Sprague, J. T. *J. Am. Chem. Soc.* **1973**, *95*, 3893, and unpublished data.
211. Laane, R.; Lord, R. C. *J. Mol. Spectrosc.* **1971**, *39*, 340.
212. Dallinga, G.; Toneman, L. H. *J. Mol. Struct.* **1967**, *1*, 117.
213. Oberhammer, H.; Bauer, S. *J. Am. Chem. Soc.* **1969**, *91*, 10.
214. Rabideau, P. W. *Acc. Chem. Res.* **1978**, *11*, 141.
215. Ferrier, W. G.; Iball, J. *Chem. Ind.* (London) **1954**, 1296.
216. Rabideau, P. W.; Paschal, J. W.; Marshal, J. L. *J. Chem. Soc., Perkin Trans. 2* **1977**, 842.
217. Favini, G.; Buemi, G.; Raimondi, M. *J. Mol. Struct.* **1968**, *2*, 137.
218. St. Jaques, M.; Vasiri, C. *Can. J. Chem.* **1971**, *49*, 1256.
219. Kabuss, S.; Friebolin, H.; Schmid, H. *Tetrahedron Lett.* **1965**, 469.
220. Allinger, N. L.; Sprague, J. T. *Tetrahedron* **1973**, *29*, 3811.
221. Hagen, K.; Traetteberg, M. *Acta Chem. Scand.* **1972**, *26*, 3643.
222. Winkler, F. K.; Dunitz, J. *J. Mol. Biol.* **1971**, *59*, 169.
223. Ganis, P.; Lepore, U.; Martuscelli, E. *J. Phys. Chem.* **1970**, *74*, 2439.
224. Manor, P. C.; Shomaker, D. P.; Parkes, A. S. *J. Am. Chem. Soc.* **1970**, *92*, 5260.
225. Ermer, O. *Angew. Chem.* **1974**, *86*, 672.
226. Gavin, R. M., Jr.; Wang, Z. F. *J. Am. Chem. Soc.* **1973**, *95*, 1425.
227. Traetteberg, M. *Acta Chem. Scand. Sect. B* **1975**, *29*, 29.
228. Rogers, D. W.; von Voithenberg, H.; Allinger, N. L. *J. Org. Chem.* **1978**, *43*, 360.
229. Allinger, N. L. *J. Am. Chem. Soc.* **1958**, *80*, 1953.
230. Allinger, N. L.; Walter, T. J.; Newton, M. G. *J. Am. Chem. Soc.* **1974**, *96*, 4588.
231. Newton, M. G.; Walter, T. J.; Allinger, N. L. *J. Am. Chem. Soc.* **1973**, *95*, 5652.
232. Allinger, N. L.; Sprague, J. T.; Liljefors, T. *J. Am. Chem. Soc.* **1974**, *96*, 5100.

233. Marquis, E. T.; Gardner, P. D. *Tetrahedron Lett.* **1966**, 2793.
234. Wittig, G.; Dorsch, H. L.; Meske-Schüller, J. *Justus Liebigs Ann. Chem.* **1968**, *711*, 55.
235. Yavari, I. *J. Mol. Struct.* **1980**, *65*, 169.
236. Allinger, N. L.; Viskosil, J. F., Jr.; Burkert, U.; Yuh, Y. *Tetrahedron* **1976**, *32*, 33.
237. Traetteberg, M. *Acta Chem. Scand.* **1970**, *24*, 2285.
238. Braude, E. A. *Chem. Ind. (London)* **1954**, 1557.
239. Anet, F. A. L.; Yavari, I. *Tetrahedron Lett.* **1975**, 1567.
240. Anet, F. A. L.; Yavari, I. *J. Am. Chem. Soc.* **1978**, *100*, 7814.
241. Favini, G.; Zuccharello, F.; Buemi, G. *J. Mol. Struct.* **1969**, *3*, 385.
242. Anet, F. A. L.; Yavari, I. *J. Am. Chem. Soc.* **1977**, *99*, 6986.
243. Osawa, E.; Shirahama, H.; Matsumoto, T. *J. Am. Chem. Soc.* **1979**, *101*, 4824.
244. Allinger, N. L.; Sprague, J. T. *Tetrahedron* **1975**, *31*, 21.
245. Anet, F. A. L.; Kozerski, L. *J. Am. Chem. Soc.* **1973**, *95*, 3407.
246. Ermer, O. *J. Am. Chem. Soc.* **1976**, *98*, 3964.
247. Anet, F. A. L.; Yavari, I. *Tetrahedron Lett.* **1975**, 4221.
248. Buemi, G.; Zuccarello, F.; Grosso, D. *J. Mol. Struct.* **1977**, *42*, 195.
249. Cope, A. C.; Moore, P. T.; Moore, W. R. *J. Am. Chem. Soc.* **1959**, *81*, 3153.
250. Neumann, C. L., unpublished data.
251. Anet, F. A. L.; Yavari, I. *J. Am. Chem. Soc.* **1977**, *99*, 7640.
252. Turner, R. B.; Mallon, B. J.; Tichy, M.; Doering, W. v. E.; Roth, W. R.; Schröder, G. *J. Am. Chem. Soc.* **1973**, *95*, 8605.
253. Zuccarello, F.; Buemi, G.; Favini, G. *J. Mol. Struct.* **1971**, *8*, 459.
254. Anet, F. A. L.; Yavari, I. *J. Am. Chem. Soc.* **1977**, *99*, 6496.
255. Ganis, P.; Dunitz, J. D. *Helv. Chim. Acta* **1967**, *50*, 2379.
256. Allinger, N. L. *J. Am. Chem. Soc.* **1957**, *79*, 3443.
257. Bartell, L. S.; Higginbotham, H. K. *J. Chem. Phys.* **1965**, *42*, 851.
258. White, D. N. J.; Bovill, M. J. *Tetrahedron Lett.* **1975**, 2239.
259. McPhail, A. T.; Onan, K. D. *J. Chem. Soc., Perkin Trans. 2* **1976**, 578.
260. Allinger, N. L.; Tribble, M. T.; Sprague, J. T. *J. Org. Chem.* **1972**, *37*, 2423.
261. Dale, J. *Angew. Chem.* **1966**, *78*, 1069; *Angew. Chem. Int. Ed. Engl.* **1966**, *5*, 1000.
262. Irngartinger, H.; Jäger, H.-U. *Tetrahedron Lett.* **1976**, 3595.
263. Dunitz, J. D.; Waser, J. *J. Am. Chem. Soc.* **1972**, *94*, 5645.
264. Hönig, H.; Allinger, N. L. *J. Org. Chem.* **1977**, *42*, 2330.
265. Ermer, O. *Angew. Chem.* **1977**, *89*, 665.
266. Burkert, U. *Chem. Ber.* **1977**, *110*, 773.
267. Ermer, O. *Z. Naturforsch., Teil B* **1977**, *32*, 837.
268. Martella, D. J.; Jones, M., Jr.; Schleyer, P. v. R.; Maier, W. F. *J. Am. Chem. Soc.* **1979**, *101*, 7634.
269. Wiseman, J. R.; Pletcher, W. A. *J. Am. Chem. Soc.* **1970**, *92*, 956.
270. Herzberg, G. "Infrared and Raman Spectra of Polyatomic Molecules"; Van Nostrand: New York, 1945.
271. Rogers, D. W.; Dagdagan, O. A.; Allinger, N. L. *J. Am. Chem. Soc.* **1979**, *101*, 671.
272. Allinger, N. L.; Meyer, A. Y. *Tetrahedron* **1975**, *31*, 1807.

273. Haase, J.; Krebs, A. Z. Naturforsch., Teil A 1971, 26, 1190.
274. Typke, V.; Haase, J.; Krebs, A. J. Mol. Struct. 1979, 56, 77.
275. Garrett, D. W., Ph.D. Thesis, Indiana Univ., Bloomington, 1974, cited in Ref. 274.
276. Leupin, W.; Wirz, J. Helv. Chim. Acta 1978, 61, 1663.
277. Anet, F. A. L.; Yavari, I. Tetrahedron 1978, 34, 2879.
278. Anet, F. A. L.; Rawdah, T. N. J. Am. Chem. Soc. 1979, 101, 1887.
279. Holmes, T. J., M.S. Thesis, Univ. of Georgia, 1980.
280. Rudolph, H. D.; Dreizler, H.; Jaeschke, A.; Wendling, P. Z. Naturforsch. Teil A 1967, 22, 940.
281. Mannschreck, A.; Ernst, L.; Keck, E. Angew. Chem. 1970, 82, 840; Angew. Chem. Int. Ed. 1970, 9, 806.
282. Mannschreck, A.; Ernst, L. Chem. Ber. 1971, 104, 228.
283. Peeling, J.; Ernst, L.; Schafer, T. Can. J. Chem. 1974, 52, 849.
284. Burkert, U., unpublished data.
285. Allinger, N. L.; Tribble, M. T. Tetrahedron Lett. 1971, 3259.
286. Allinger, N. L.; Chung, D. Y. J. Am. Chem. Soc. 1976, 98, 6798.
287. Burkert, U. Tetrahedron 1977, 33, 2237.
288. Eliel, E. L.; Bailey, W. F.; Wiberg, K. B.; Connon, H.; Nader, F. W. Justus Liebigs Ann. Chem. 1976, 2240.
289. Bailey, W. F.; Connon, H.; Eliel, E. L.; Wiberg, K. B. J. Am. Chem. Soc. 1978, 100, 2202.
290. Andose, J. D.; Mislow, K. J. Am. Chem. Soc. 1974, 96, 2168.
291. Kates, M. R.; Andose, J. D.; Finocchiaro, P.; Gust, D.; Mislow, K. J. Am. Chem. Soc. 1975, 97, 1772.
292. Hummel, J. P.; Zurbach, E. P.; DiCarlo, E. N.; Mislow, K. J. Am. Chem. Soc. 1976, 98, 7480.
293. Ahmed, N. A.; Kitaigorodski, A. I.; Mirskaya, M. V. Acta Crystallogr., Sect. B 1971, 27, 867.
294. Bates, R. B.; Carnighan, R. H.; Staples, C. E. J. Am. Chem. Soc. 1963, 85, 3032.
295. Hutchings, M. G.; Andose, J. D.; Mislow, K. J. Am. Chem. Soc. 1975, 97, 4562.
296. Ivanov, P.; Pojarlieff, I.; Tyutyulkov, N. Tetrahedron Lett. 1976, 775.
297. Jacobus, J. Tetrahedron Lett. 1976, 2927.
298. Finocchiaro, P.; Gust, D.; Hounshell, W. D.; Hummel, J. P.; Maravigna, P.; Mislow, K. J. Am. Chem. Soc. 1976, 98, 4945.
299. Dougherty, D. A.; Mislow, K.; Blount, J. F.; Wooten, J. B.; Jacobus, J. J. Am. Chem. Soc. 1977, 99, 6149.
300. Mislow, K. Acc. Chem. Res. 1976, 9, 26.
301. Mislow, K.; Dougherty, D. A.; Hounshell, W. D. Bull. Soc. Chim. Belg. 1978, 87, 555.
302. Hounshell, W. D.; Dougherty, D. A.; Hummel, J. P.; Mislow, K. J. Am. Chem. Soc. 1977, 99, 1916.
303. Destro, R.; Pilati, T.; Simonetta, M. J. Am. Chem. Soc. 1978, 100, 6507.
304. Stein, M.; Winter, W.; Rieker, A. Angew. Chem. 1978, 90, 737.
305. Stein, M.; Winter, W., Rieker, A. Angew. Chem. Int. Ed. Engl. 1978, 17, 692.
306. Warshel, A.; Karplus, M. J. Am. Chem. Soc. 1972, 94, 5612.

307. Kao, J.; Allinger, N. L. *J. Am. Chem. Soc.* **1977,** *99,* 975.
308. Allinger, N. L.; Graham, J. C. *J. Am. Chem. Soc.* **1973,** *95,* 2523.
309. Lindner, H. J. *Tetrahedron* **1975,** *31,* 281.
310. Dauben, H. J., Jr.; Bertelli, D. J. *J. Am. Chem. Soc.* **1961,** *83,* 4657.
311. Allinger, N. L.; Miller, M. A.; Chow, L. W.; Ford, R. A.; Graham, J. C. *J. Am. Chem. Soc.* **1965,** *87,* 3430.
312. Kay, M. I.; Okaya, Y.; Cox, D. E. *Acta Crystallogr. Sect. B* **1971,** *27,* 26.
313. Dobler, M.; Dunitz, J. D. *Helv. Chim. Acta* **1965,** *48,* 1429.
314. Scott, L. T.; Brunsvold, W. R. *J. Am. Chem. Soc.* **1978,** *100,* 4320.
315. Bregman, J.; Hirshfeld, F. L.; Rabinovich, D.; Schmidt, G. M. J. *Acta Crystallogr.* **1965,** *19,* 227.
316. Van-Catledge, F. A.; Allinger, N. L. *J. Am. Chem. Soc.* **1969,** *91,* 2582.
317. Dewar, M. J. S.; Haddon, R. C.; Student, P. J. *Chem. Commun.* **1974,** 569.
318. Baumann, H. *J. Am. Chem. Soc.* **1978,** *100,* 7196.
319. Mislow, K.; Dougherty, D. A.; Schlegel, H. B. *Tetrahedron* **1978,** *34,* 1441.
320. Gleiter, R. *Tetrahedron Lett.* **1969,** 4453.
321. Van Alsenoy, C.; Scarsdale, J. N.; Schäfer, L. *J. Comput. Chem.* **1982,** *3,* 53.
322. Paquette, L. A.; Balogh, D. W.; Usha, R.; Kountz, D. K.; Christoph, G. G. *Science* **1981,** *211,* 575.
323. Cristoph, G. G.; Engel, P.; Usha, R.; Balogh, D. W.; Paquette, L. A. *J. Am. Chem. Soc.* **1982,** *104,* 784.
324. Jensen, J. L. *Prog. Phys. Org. Chem.* **1976,** *12,* 189.
325. Hagler, A. T.; Stern, P. S.; Lifson, S.; Ariel, S. *J. Am. Chem. Soc.,* **1979,** *101,* 813.
326. Ermer, O.; Bödecker, C.-D. *Chem. Ber.* **1981,** *114,* 652.
327. Hounshell, W. D.; Mislow, K. *Tetrahedron Lett.* **1979,** 1205.
328. Geneste, P.; Kamenka, J.-M.; Roques, R.; Declerq, J. P.; Germain, G. *Tetrahedron Lett.* **1981,** 949.
329. Bartell, L. S.; Kuchitsu, K.; deNeui, R. J. *J. Chem. Phys.* **1961,** *35,* 1211.
330. Martin, H.-D.; Kunze, M.; Beckhaus, H.-D.; Walsh, R.; Gleiter, R. *Tetrahedron Lett.* **1979,** 3069.
331. Anet, F. A. L.; Ghiaci, M. *J. Am. Chem. Soc.* **1980,** *102,* 2528.
332. Baas, J. M. A.; van de Graaf, B.; van Rantwijk, F.; van Veen, A. *Tetrahedron* **1979,** *35,* 421.
333. van de Graaf, B.; Baas, J. M. A.; Wepster, B. M. *Rec., J. Roy. Netherl. Soc.* **1978,** *97,* 268.
334. van de Graaf, B.; Baas, J. M. A.; Widya, H. A.; *Recl. Trav. Chim. Pays-Bas* **1981,** *100,* 59.
335. Ermer, O.; "Aspekte von Kraftfeldrechnungen"; Wolfgang Bauer Verlag: Munich, 1981.
336. Schmidbaur, H.; Blaschke, G.; Zimmer-Gasser, B.; Schubert, U. *Chem. Ber.* **1980,** *113,* 1612.
337. Tai, J. C.; Allinger, N. L. *J. Am. Chem. Soc.* **1977,** *99,* 4256.
338. Tai, J. C.; Allinger, N. L. *Tetrahedron* **1981,** *37,* 2755.
339. Liljefors, T.; Allinger, N. L. *J. Am. Chem. Soc.* **1976,** *98,* 2745.
340. Liljefors, T.; Allinger, N. L. *J. Am. Chem. Soc.* **1978,** *100,* 1068.
341. Lindner, H. J. *Tetrahedron* **1981,** *37,* 535.
342. Lindner, H. J. *Tetrahedron* **1974,** *30,* 1127.

343. Anet, F. A. L.; Rawdah, T. N. *J. Org. Chem.* **1980**, *45*, 5243.
344. Maier, W. F.; Schleyer, P. v. R. *J. Am. Chem. Soc.* **1981**, *103*, 1891.
345. Anet, F. A. L.; Krane, J. *Isr. J. Chem.* **1980**, *20*, 72.
346. Shirahama, H.; Osawa, E.; Chhabra, B. R.; Shimokawa, T.; Yokono, T.; Kanaiwa, T.; Amiya, T.; Matsumoto, T. *Tetrahedron Lett.* **1981**, *22*, 1527.
347. Lambert, J. B. *Accts. Chem. Res.* **1971**, *4*, 87.
348. Rüchardt, C.; Beckhaus, H.-D. *Angew. Chem.* **1980**, *92*, 417; *Angew. Chem. Int. Ed.* **1980**, *19*, 429.
349. Burkert, U.; Allinger, N. L. *J. Comput. Chem.* **1982**, *3*, 40.
350. Osawa, E. *J. Am. Chem. Soc.* **1979**, *101*, 5523.
351. Wipff, G. *Tetrahedron Lett.* **1980**, 4445.
352. Rondan, N. G.; Paddon–Row, M. N.; Caramella, P.; Houk, K. N. *J. Am. Chem. Soc.* **1981**, *103*, 2436.
353. Favini, G.; Simonetta, M.; Todeschini, R. *J. Comput. Chem.* **1981**, *2*, 149.
354. Casalone, G.; Pilati, T.; Simonetta, M. *Tetrahedron Lett.* **1980**, 2345.
355. Pilati, T.; Simonetta, M. *J. Chem. Soc. Perkin Trans. 2* **1977**, 1435.
356. Krebs, A.; Rüger, W. *Tetrahedron Lett.* **1979**, 1305.
357. Rüger, W. Dissertation, Universität Hamburg, **1981**.
358. Warner, P. M.; Peacock, S. *J. Comput. Chem.* **1982**, *3*, (in press).
359. Brunel, Y.; Coulombeau, C.; Rassat, A. *Nouveau J. de Chemie* **1980**, *4*, 662.
360. Baas, J. M. A.; van de Graaf, B.; Tavernier, D.; Vanhee, P. *J. Am. Chem. Soc.* **1981**, *103*, 5014.
361. Chiang, J. F.; Bauer, S. H., *J. Am. Chem. Soc.* **1966**, *88*, 420.
362. Avirah, T. K.; Cook, R. L.; Malloy, Jr., T. B. *J. Chem. Phys.* **1979**, *71*, 3478.
363. Van-Catledge, F. A. *J. Chem. Phys.* **1980**, *73*, 1476.
364. Anet, F. A. L.; Hartman, J. S. *J. Am. Chem. Soc.* **1963**, *85*, 1204.

5

Steric Energy, Heats of Formation, and Strain

Statistical Thermodynamics. Calculations of Vibrational Frequencies

As was pointed out in Chapter 1, molecular geometry can be defined in various ways. Which definition is the most useful depends on whether we are interested in the geometry of the molecule in the hypothetical motionless state (as when the geometry is optimized using ab initio methods to generate the molecule at the bottom of the potential well), or whether we are interested in an observable quantity like the geometry obtained by diffraction or spectroscopic methods. The different definitions, and the corresponding different structures, are a result of the molecular vibrations. A similar problem exists for the energy, and it is necessary to look into the meaning of the steric energies obtained by molecular mechanics calculations and their relation to observable energies.

In addition to the energy of the hypothetical motionless state, which is due to chemical bonding, molecules at 0 K contain zero-point vibrational energy. For a six-carbon alkane this zero-point energy is about 400 kJ/mol. At higher temperatures these molecules also contain translational, vibrational, and rotational energies, because of excitation of some of the molecules to higher quantum states. At room temperature the external rotational and translational degrees of freedom are fully excited, amounting to $3RT$ in the gas phase for all nonlinear molecules. However, vibrational modes are only partly excited; therefore, the vibrational energy must in principle be calculated for a particular molecule from the normal vibrational frequencies of that molecule. Vibrational energy (U_{vib}) is given in the harmonic approximation by the expression developed in statistical thermodynamics in Equation 5.1.

$$U_{vib} = \sum_{i=1}^{3n-6} h\nu_i \left(\frac{1}{2} + \frac{1}{e^{h\nu_i/kT} - 1} \right) \quad (5.1)$$

The internal energy (U) of the compound (in the gas phase) is the sum of the energy of the hypothetical motionless state, and the thermally averaged translational, rotational, and vibrational energies. The addition of RT converts this energy into the enthalpy (within the ideal gas approximation). When the enthalpy difference between two molecules that are isomers or conformers is calculated, the translational, rotational, and RT terms cancel to a very good approximation, leading to Equation 5.2, where ΔE_0 is the difference in the energies of the potential minima, and ΔU_{vib} is the difference in the vibrational energies.

$$\Delta H(A - B) = \Delta E_0(A - B) + \Delta U_{vib}(A - B) \tag{5.2}$$

With the information given by a molecular mechanics force field, it is possible in principle to calculate the vibrational energy contribution, because the normal vibrational frequencies can be calculated as outlined in pp. 17–22. This requires a knowledge of the second derivative matrix of the potential energy (the F matrix), Equation 5.3.

$$\mathscr{F}_x = \left(\frac{\partial^2 V}{\partial x_i \partial x_j}\right) \tag{5.3}$$

If the energy minimization has been carried out using the full-matrix Newton–Raphson (NR) method, this matrix is provided conveniently by the last cycle of the NR minimization, though in Cartesian, and not in mass-weighted Cartesian coordinates. The matrix conversion is easily carried out by matrix multiplication (Equation 5.4), where $\mathscr{F}_q$ is the force constant matrix in mass-weighted Cartesian coordinates, and M is a diagonal $3n \times 3n$ matrix, diag(M) = $m_1, m_1, m_1, m_2, m_2, m_2,$ ..., m_n, m_n, m_n, containing the atomic masses. The roots of the secular Equation 5.5 give the vibrational frequencies. (Equation 5.6).

$$\mathscr{F}_q = M^{-1/2} \mathscr{F}_x M^{-1/2} \tag{5.4}$$

$$|\mathscr{F}_q - E\lambda| = 0 \tag{5.5}$$

$$\nu = \frac{\sqrt{\lambda}}{2\pi} \tag{5.6}$$

The work of Schachtschneider and Snyder mentioned in pp. 20–22 was aimed directly at deriving a transferable valence or Urey–Bradley force field that could be used to predict the vibrational frequencies for a wide variety of hydrocarbons (1). Their force field, which works well only for strainless acyclic compounds, is somewhat different from a molecular mechanics force field, as discussed in Chapter 2. The force field as defined in the F matrix is a derived quantity in molecular mechanics, obtained from a large number of potential functions; it accounts for strained bond lengths and angles and nonbonded interactions automatically. The force constants contained in this F matrix are not really transferable, and the force field contains all off-diagonal force constants. The vibrational frequencies calculated with this type of force field are in principle superior to the ones calculated by Schachtschneider and Snyder's Urey–Bradley or valence force fields. Lifson and coworkers reported that the calculated vibrational frequencies for a number of compounds (cyclopentane, cyclohexane, ethane, butane, cyclodecane, ethylene, isobutene, cyclohexene, and *trans,trans,trans*-1,5,9-cyclododecatriene) were all within 20 cm^{-1} of the experimental values (2, 3, 4). A force field developed by Pickett and Strauss (5, 6) gave even better agreement (within 3.3 cm^{-1}), but their study was limited to a much smaller number of compounds and to only the low frequency vibrations that showed a strong conformational frequency dependence.

If we want to calculate not only good geometries and energies, but also vibrational frequencies, the difficulty of deriving the force field is much greater. In this case the minima must be located correctly, and the curvature of the potential surface near the energy minimum must be determined accurately. The anharmonicity of the potential surface affects the molecule's vibrational frequencies more than its geometry and energy. Although additional information becomes available for the parameterization through a large number of experimental vibrational frequencies (which is especially advantageous for the optimization of the force constants for stretching and bending), the additional data pose restrictions on the other parameters. In other words, a force field that must reproduce only a small number of properties can probably do a better job on these than can a more general force field parameterized on a larger number of different kinds of properties.

The calculated frequencies are useful for two different purposes. First, they can be compared with the experimental vibrational spectrum of a molecule, and they are helpful in the assignment of the observed bands to individual vibrational modes (the calculation also gives the normal modes of vibration in the eigenvectors associated with the frequency eigenvalues). It is mainly here that high precision of the calculated vibrational frequencies is desired. The second important use of these frequencies, as mentioned at the beginning of this chapter, is the calculation of the vibrational energy contribution to a molecule's thermodynamic properties. This calculation does not require highly accurate vibrational frequencies; when the vibrational frequencies are to be used only for this purpose, this poses only minimal restriction on the parameters used for the calculation of the geometry and energy. Also, vibrational frequencies and the related eigenvectors calculated by molecular mechanics force fields have been used recently to calculate the mean vibrational amplitudes and shrinkage corrections needed in electron diffraction work. A somewhat lower precision for the calculated frequencies is sufficient here also.

Only a few research groups have actually employed vibrational energy contributions explicitly in the calculation of conformational equilibria and heats of formation (2, 3, 6–10). Fortunately, the vibrational energy differences between conformations are small, usually less than 4 kJ/mol. The effects of zero-point vibrational energy and excitational energy work in opposite directions and partially cancel each other, because when the vibrational frequency (and, therefore, the zero-point vibrational energy) increases, the spacing between the vibrational energy levels widens, and the number of excited molecules decreases. The precision of the calculated energy differences sought in molecular mechanics calculations, ±0.4 kJ/mol, would seem, however, to require the evaluation of the vibrational energy. As a matter of fact, all of the early and most of the presently used force fields neglect any explicit account of this quantity. The success of these calculations can be explained only by assuming that the vibrational effects are implicitly included in a roundabout way by suitable parameter adjustments, which make the steric energies obtained from the molecular mechanics potential functions (with additive terms accounting for the bond energies) a good approximation to the thermally averaged energies. As a practical matter, it is difficult to parameterize a force field for the calculation of the energies at the minimum of the potential well because (a) experimental data are usually thermally averaged energies, and (b) the vibrational contribution can be determined only after the force field has already been parameterized. In most force fields the "steric energy" is, therefore, the thermally averaged energy relative to a hypothetical molecule with the same constitution, but with all bond lengths, and bond and torsion angles at their "strainless" values, and the atoms with van der Waals and electrostatic interactions corresponding to infinite separation of the atoms.

Molecular mechanics also allows thermodynamic functions other than the enthalpy to be calculated. With the calculated geometry, the mass, and the vibration frequencies, it is possible to calculate entropies, Gibbs energies, and equilibrium constants by means of statistical thermodynamics. For the evaluation of the entropy, the partition functions for translation (depending on the mass), rotation (depending on the moments of inertia, which are obtained from the geometry), and vibration (depending on the vibrational frequencies) can be evaluated by the standard methods of statistical thermodynamics. Such calculations have been reported by Boyd and coworkers (7–9), who found excellent agreement with experimental data for cyclohexane and benzene.

A somewhat different and complementary approach to the statistical thermodynamic methods discussed above has been introduced by McKenna and coworkers (67). They call it the **nonpotential energy effect** (**NPE**). The idea is to correctly reproduce the entropy-rich lower vibrational frequencies of molecules, especially the torsional frequencies. A simple valence force field is used, and useful results are obtained, although so far only for a small set of hydrocarbons and halides.

Heats of Formation

A fundamental quantity concerning the energy of a molecule is its heat of formation. Experimental values for heats of formation, together with attempts to predict these values, have contributed greatly to theories of structural chemistry. Usually the experimental method involves the determination of the heat of combustion of the compound and, for compounds that are not gases, the heat of vaporization (or sublimation). Sometimes heats of vaporization have been estimated rather than determined; this can introduce considerable uncertainty into the resulting experimental heat of formation.

Traditionally heats of formation have been calculated by increment addition (bond energy) methods (11–14). In molecular mechanics they are considered to consist of contributions from the formation of bonds (bond increments), the effects of strain represented by the steric energy, and contributions from statistical thermodynamics caused by population of vibrational levels, conformational mixing, and perhaps other terms. (These procedures differ somewhat in detail.)

In the preceding section it was shown how the (thermally averaged) enthalpy of one conformation of a molecule can be calculated relative to the energy at the bottom of the potential well by employing statistical thermodynamics. If more than one conformation is present, the enthalpy of the mixture (H_{mix}) must be calculated from the mole fractions (N_i) and heats (H_i) of the conformations present (Equation 5.7), using the mole fractions obtained from the Boltzmann distributions (Equation 5.8), where g_i is the statistical weight (the number of identical conformations) of conformer i, and ΔG_i^x is the Gibbs free energy excluding the entropy effects caused by the number of identical conformations.

$$H_{\text{mix}} = \sum_i N_i H_i \qquad (5.7)$$

$$N_i = \frac{g_i e^{-\Delta G_i^x/RT}}{\sum_i g_i e^{-\Delta G_i^x/RT}} \tag{5.8}$$

The foregoing describes the calculation of the enthalpy of a compound relative to the energy at the bottom of the potential well. If we want to know the heat of formation of the compound, which is the usual reference point, we must also know the energy at the bottom of the well. In principle, this can be found by summing bond energy terms (BE) and the steric energy (SE) of the molecule calculated from the force field. The bond energy terms must be evaluated empirically by fitting the heats of formation that are known for simple compounds. In theory, these can be calculated by ab initio methods, but in practice the results are rather poor.

While the above procedure is the full and proper method for the calculation of the heats of formation of compounds, some approximations have been made in practice. The most detailed studies of heats of formation so far carried out have always used some simplifications, the most usual one of which is to evaluate the bond energies at room temperature, and thereby to include the vibrational contribution directly in the BE term. It is not obvious that this can be done. This approximation has been made, and it seems to work remarkably well. In this case the formula used to evaluate heats of formation is Equation 5.9. The $4RT$ accounts for the translation, rotation, and PV term required to convert the energy of a nonlinear molecule to enthalpy. The steric energy (SE) that is added accounts for the strain, if any, in the molecule. The BE terms, summed over appropriate increments, allow for the bonding energy at room temperature.

$$\Delta H_f^\circ = 4RT + \text{BE} + \text{SE} \text{ (excluding linear molecules)} \tag{5.9}$$

The kinds of bond energy increments actually used are a matter of personal choice. Bond or group increments have been employed, and they are completely interconvertible (15). The method used by Schleyer et al., for example, employs the Franklin scheme (16) of four group increments (for CH_3, CH_2, CH, and quaternary carbon). An additional *gauche* methyl interaction increment, originally proposed by Schleyer et al. (17), is included in Boyd's force field (7–9). This term can account for various effects otherwise neglected, including low-order torsional terms (pp. 45–52) and spacing of vibrational energy levels as governed by the depth of the potential well (vide infra). Bond increments (CC and CH) and increments for structural features (primary, tertiary, and quaternary carbon) are used in the force fields developed by Allinger and coworkers (*18–22*). In each case these parameters are optimized by least squares fitting to experimental data, which is more straightforward than for most force field parameters because the heats of formation are a linear function of the parameters. The number of compounds utilized was limited mainly by the availability of experimental data.

There is a simple physical picture behind the calculation. The bond energy terms are just that, an energy change due to the formation of bonds. Structural terms are needed, in part, because of the van der Waals type of interactions between atoms bound to a common atom; these interactions are neglected in a valence force field. The geometry is corrected for these interactions by adjusting the values of the standard bond angles, and stretch–bend interactions, but the energies must be corrected separately as above. (Presumably force fields such as MUB-2, that include these interactions would require only small correction terms of this type, or none at all.) With this approach, the average deviation between calculated and experimental heats of formation for 32 hydrocarbons obtained in 1971 by Allinger et al. (*18*) was already close to the claimed average experimental error (1.92 kJ/1.63 kJ). With a different data set and without reoptimization of the parameters, Engler et al. (*15*) found for this force field (*18*) a standard deviation of 4.31 kJ from the experimental data. Their own force field gave a standard deviation of only 3.47 kJ, compared with a claimed average experimental error of 2.05 kJ (set of 49 compounds). These numbers are not quite what they seem, however, because different experimental values for the same quantities were sometimes used in parameterizing the two force fields. Consistent use of the experimental data would have brought the two standard deviations (4.31 kJ and 3.47 kJ) closer together.

An improved version of the Allinger force field (MM1, 1973) (*19*) gave better results than either of the earlier trials. Some statistical mechanical terms were also included in the 1973 calculation (*22*). First, the $3RT$ for translational and rotational energies, and the RT conversion factor for energies to enthalpies were added to the steric energies. Most vibrational energy terms were regarded as included in the increments—the bond stretching in the bond increments, and the bending terms in the structural feature increments. The torsional vibrations about bonds where internal rotation is relatively free needed additional attention, however. In an extensive study, Pitzer (*23*) had shown that thermal excitation of the rotation about a bond with a low rotation barrier (less than about 30 kJ) contributes about 1.25 kJ/mol to the energy of the molecule at 25°C. So this term was added to the above for each bond where appropriate, except for C–CH_3 bonds where it is included in the structure feature increment (*22*). (In ethane this term is already included twice, as two C–CH_3 rotational increments are included in two CH_3 group increments. The 1.25 kJ term must therefore be subtracted from the other energy terms.) These statistical mechanics terms do not require the evaluation of the vibrational frequencies and, therefore, are easy to calculate. A representative set of 38 compounds was examined by this method, and a standard deviation of 2.51 kJ/mol, compared with the average claimed experimental error of 2.13 kJ/mol (19), was obtained.

Major improvements in the potential functions further improved the fit between experimental and calculated heats of formation, as well as other properties of the compounds (20). The most important improvements were the introduction of low periodicity torsional energy terms and quite different van der Waals parameters. The same account of the statistical mechanical terms was used as previously, but the value of the torsional increment was changed from 1.25 kJ/mol to 1.51 kJ/mol. For a set of diverse hydrocarbons this treatment gave (MM2 force field) a standard deviation of 1.76 kJ/mol, while the reported average probable error in the experimental values is 1.67 kJ/mol. This is the best agreement between calculated and experimental heats of formation obtained to date with a method that avoided the calculation of vibrational frequencies. Detailed results are shown in Table 5.1 (20).

The more proper method for calculation of heats of formation, based on the full statistical mechanical treatment, was first used by Lifson and Warshel (2) and by Boyd (7–9). These authors calculated the heats of formation of only a small number of sample molecules, however; their work is not at all comparable with the lists of compounds examined in the work described above [see Table 5.1 (20)], so the abilities of these force fields in this respect cannot be judged. Extensive studies were carried out by Wertz (10), who developed a force field, in part by least-squares optimization of the parameters, that is capable of calculating geometries, vibrational frequencies, and heats of formation by the com-

Table 5.1

Heats of Formation; Gas; 25°C; kJ/mol; Alkanes (20)

Compound	Calculated	Experimental (63)	Calculated − Experimental
Methane	−75.19	−74.85	−0.33
Ethane	−81.13	−84.68	3.56
Propane	−103.64	−103.85	0.21
Butane	−124.98	−126.15	1.17
Pentane	−146.19	−146.44	0.25
Hexane	−167.49	−167.19	−0.29
Heptane	−188.74	−187.82	−0.92

Table 5.1 (Continued)

Compound	Calculated	Experimental (63)	Calculated − Experimental
Octane	−210.04	−208.45	−1.59
Isobutane	−134.60	−134.52	−0.08
Isopentane	−152.63	−154.47	1.84
2,3-Dimethylbutane	−177.74	−177.78	0.04
Neopentane	−169.66	−168.49	−1.17
Hexamethylethane	−224.51	−225.73	1.21
2,2,3-Trimethylbutane	−204.22	−204.81	0.59
Di-t-butylmethane	−247.07	−243.34	−3.72
3,3-Diethylpentane	−229.66	−233.34	3.68
Cyclobutane	26.40	26.69	−0.29
Cyclopentane	−76.44	−76.57	0.13
Cyclohexane	−123.55	−123.14	−0.42
Cycloheptane	−116.65	−118.07	1.42
Cyclooctane	−123.47	−124.39	0.92
Cyclodecane	−154.77	−154.31	−0.46
Methylcyclohexane	−154.56	−154.77	0.21
1,1-Dimethylcyclohexane	−181.46	−181.00	−0.46
trans-1,2-Dimethylcyclohexane	−181.21	−179.87	−1.34
cis-1,2-Dimethylcyclohexane	−174.43	−172.09	−2.34
Norbornane	−53.76	−51.88	−1.88
Cubane	622.75	622.16	0.59
Bicyclo[4.2.0]octane	−24.02	−26.74	2.72
Bicyclo[2.2.2]octane	−95.27	−99.37	4.10
cis-Bicyclo[3.3.0]octane	−95.27	−93.30	−1.97
trans-Bicyclo[3.3.0]octane	−65.81	−66.53	0.71
cis-Hydrindane	−127.44	−127.24	−0.21
trans-Hydrindane	−132.34	−131.59	−0.75
cis-Decalin	−171.67	−169.24	−2.43
trans-Decalin	−183.09	−182.17	−0.92
trans-syn-trans-Perhydroanthracene	−243.09	−243.17	0.08
trans-anti-trans-Perhydroanthracene	−217.65	−220.62	2.97
Protoadamantane	−86.73	−85.94	−0.79
Adamantane	−132.01	−132.88	0.88
1-Methyladamantane	−168.41	−169.74	1.34
Tetramethyladamantane	−278.07	−280.96	2.89
		Average probable error 1.67	Standard deviation, 1.76

plete statistical mechanical method, including a complete explicit evaluation of the vibrational energy. The force field used was simpler than the one in the MM2 calculation (20) discussed above, but the resulting calculated heats of formation agree with the experimental values to within the average quoted experimental error. When the effects of the vibrational energy treatment by the complete statistical mechanics method of Wertz were tested against various force fields for the same set of compounds (10), a standard deviation of 2.85 kJ/mol was obtained with the 1971 Allinger force field (18), 2.59 kJ/mol with the 1973 MM1 force field with the simple statistical mechanics treatment (22), and 1.55 kJ/mol with Wertz's force field (10). The average experimental error in ΔH_f° was 1.63 kJ/mol. The improvement is obvious, and the results are very good, even though the Wertz force field does not contain the low-order torsional terms now known to be desirable.

The heats of formation of adamantane and some of its derivatives have presented severe experimental problems. These have been discussed at some length (24, 25), and it suffices to say that not only are there several difficulties encountered in the determination of the heats of combustion in this particular class of compounds, but it is not certain that all of these difficulties are completely understood. In Table 5.2 are the experimental values for the heat of formation of adamantane that have been reported. Because none of these values has been withdrawn, all are considered valid. But the various values clearly do not agree to within the combined limits of the experimental errors assigned by the investigators.

We might also mention the methyl derivatives of adamantane and diamantane. Several values for heats of formation were determined experimentally for this group of compounds, and were compared with molecular mechanics calculations (24). The outstanding discrepancy between calculation and experiment is in the case of 2-methyladamantane, for which the values are -154.47 kJ/mol and -149.20 ± 2.59 kJ/mol respectively. The investigators who carried out the experimental work indicate (24) that the preferred value is the one from the molecular mechanics work. Finally, several of these compounds have been studied independently by two groups (24, 25). While the results essentially are

Table 5.2

Reported Experimental Values for the Heat of Formation of Adamantane (Gas, kJ/mol) (74)

-137.90 ± 0.79
-132.88 ± 1.34
-128.24 ± 4.10
-127.90 ± 3.77

internally consistent for each group, there is a systematic difference of about 8 kJ/mol between the two groups. The reason for this discrepancy is not known. The results of the McKervey group are in better agreement with the most recent molecular mechanics results (MM2), but the MM2 parameter set was derived including the McKervey results (with low weight), so the results are not completely independent. (The molecular mechanics results tend to agree more closely with the results of the McKervey group regardless of the weighting used).

The larger related molecule, diamantane, had a reported heat of formation of -136.40 ± 2.43 kJ/mol when the heat of formation was calculated (MM2) to be -143.47 kJ/mol. Subsequently, the heat of combustion was measured again by two independent groups, and the weighted mean obtained was -144.52 ± 1.72 kJ/mol (24), in good agreement with the molecular mechanics predictions.

These examples are outlined to illustrate an important point. In 1968 when heats of formation calculated by molecular mechanics disagreed seriously with the experimental values, the former were most likely in error. But in the intervening years the accuracy and reliability of molecular mechanics for this purpose have been improved greatly. Now such a disagreement (with hydrocarbons) strongly suggests that the experimental value is suspect.

Molecular mechanics is far superior to any other calculational method for obtaining heats of formation, and the standard deviations of the theoretical approaches have approximately reached the experimental errors, making the calculation a viable alternative to the experimental determination in applicable cases. It might be questioned whether further improvements in these calculations are possible. This can be expected. Because the molecular mechanics force field is based on a great many separate pieces of data, the average values for the parameters selected should permit the calculations to give an accuracy surpassing that of the individual measurements. In addition, because data from several laboratories are considered, the systematic errors resulting from data from any one source can also be reduced by averaging. Therefore, the calculational method should eventually be more accurate than the experimental method.

Heats of formation of alkenes have also been calculated by molecular mechanics (26, 27, 34), but here experimental data from heats of combustion are far fewer than for saturated hydrocarbons. On the other hand, while many heats of hydrogenation of alkenes in solution are known, solvation effects make these values difficult to utilize at present. Until recently, the solvent of choice for heats of hydrogenation was glacial acetic acid. Solvation effects are especially troublesome in this solvent (28). Only recently have heats of hydrogenation been measured in nonpolar solvents (29). These can be compared with the calculated heats of hydrogenation (calculated as the difference between the heats of formation of an alkene and the corresponding saturated hydrocarbon in the gas phase). This comparison is still not exact; it assumes that the alkene–alkane difference in ΔH_f° in an inert solvent (experiment) is the same as the difference in the gas phase (calculation). The error introduced is probably less than experimental error in most cases, and certainly this comparison is a big improvement over the acetic acid comparison.

In addition, the heats of solution of a number of alkenes and alkanes in acetic acid have now been measured (30, 31), and these permit correction of the heats of hydrogenation in acetic acid to the hydrocarbon phase. This is not quite what is desired, but again, it is an improvement over the raw acetic acid data.

The extensive heat of formation studies on the norbornane system are perhaps worthy of brief comment. These compounds have been important in many studies in physical organic chemistry over the years, and knowledge of their energy relationships would be helpful. Regrettably, the heat of formation data in this area have been in complete disarray until recently. Figure 5.1 shows calculated and experimental values (in parentheses) for the heats of formation of a group of these compounds, as obtained in recent studies (32). The calculated (MM2) values (33) are in satisfactory agreement with the experimental values determined by heats of hydrogenation, except for norbornene and norbornadiene. Regretably, good experimental values for these compounds were not available at the time of the force field work. It is now believed that the experimental values shown are correct, and the force field numbers are too negative. This seems to be largely a result of a bending deformation energy that is calculated to be too small in these compounds, and this error is expected to carry over to other similar compounds also. A variety of combustion data and other kinds of experimental values on these compounds are also available; the interested reader is referred to the literature for further discussion (32).

White (34) has parameterized his force field simultaneously for alkenes and alkanes. A comparison with other force fields is difficult because different compounds were studied, but White's work appears to give better structures, although poorer heats of formation, than MM1.

For molecules containing hetero atoms, the situation regarding heats of formation becomes less satisfactory. In the case of ethers, alcohols, and certain classes of compounds containing nitrogen, combustion gives well-defined products (CO_2, H_2O, N_2), and accurate thermochemical data can be obtained as with hydrocarbons. But often this is not the case with halides, compounds containing sulfur, or various other elements. Consequently, a molecular mechanics parameterization for heat of formation calculations on such compounds is inaccurate. Heats of formation for a number of functionally substituted compounds

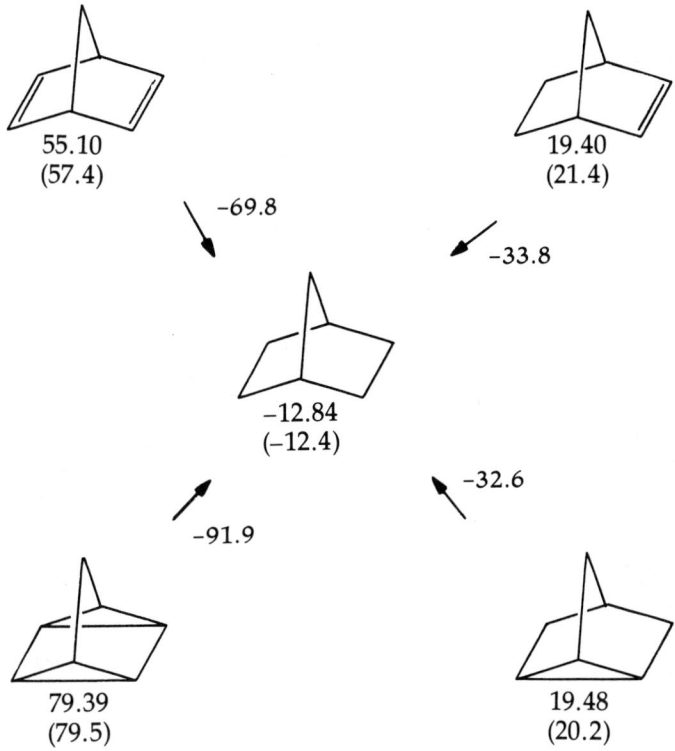

Figure 5.1. Calculated heats of formation (gas) and of hydrogenation for some compounds related to norbornane. The best experimental values are given in parentheses for comparison. The arrows indicate where heats of hydrogenation were actually measured.

have been calculated (22), and while the agreement with experimental data is not as good as that with hydrocarbons, it is reasonable. There are several problems here. First, there are necessarily more parameters, but there are usually fewer data with which to evaluate them. Second, functional groups contain dipoles that induce changes in the charge distribution in the hydrocarbon fragment to which the function is attached. This introduces further inaccuracies into the molecular mechanics model. Finally, many of the available data that one tries to fit in developing a force field concern equilibria in polar solvents. While the equilibria in hydrocarbons are not seriously affected by solvation, this is not the case for polar molecules. Hence, the effects of solvation, which are in general not well understood, further complicate the picture (See pp. 195 202).

Tables 5.1, 5.3, and 5.4 give some heats of formation calculated with the MM2 program and determined experimentally, which illustrate the

Table 5.3

Heats of Formation; Gas; 25°C; kJ/mol; Alcohols and Ethers (35, 36)

Compound	Calculated	Experimental (63)	Calculated − Experimental
Dimethyl ether	−183.38	−184.05	0.67
Methyl ethyl ether	−216.06	−216.40	0.33
Diethyl ether	−248.99	−252.13	3.14
Methyl propyl ether	−237.78	−237.73	−0.04
Methyl isopropyl ether	−251.37	−252.04	0.67
Methyl t-butyl ether	−285.93	−291.62	5.69
Dipropyl ether	−291.58	−292.25	0.67
Diisopropyl ether	−320.33	−318.82	−1.51
Isopropyl t-butyl ether	−352.17	−357.73	5.56
Dibutyl ether	−333.88	−333.97	0.08
Di-t-butyl ether	−366.73	−364.43	−2.30
Oxetane	−80.92	−80.54	−0.38
Tetrahydrofuran	−184.47	−184.18	−0.29
Tetrahydropyran	−224.68	−223.38	−1.30
Dimethoxymethane	−350.62	−348.40	−2.22
1,1-Dimethoxyethane	−388.44	−390.20	1.76
1,3-Dioxolane	−299.70	−301.67	1.97
1,3-Dioxane	−348.15	−350.24	2.09
1,4-Dioxane	−316.56	−315.93	−0.63
2-Methoxytetrahydropyran	−402.29	−399.57	−2.72
1,3-Dioxacycloheptane	−345.56	−346.44	0.88
Methanol	−200.16	−201.12	0.96
Ethanol	−232.46	−235.31	2.85
1-Propanol	−254.14	−255.94	1.80
2-Propanol	−272.25	−272.46	0.21
2-Methylpropanol	−283.76	−283.84	0.08
1-Butanol	−275.22	−275.27	0.04
2-Butanol	−292.75	−292.80	0.04
1,1-Dimethylethanol	−312.80	−312.63	−0.17
1-Pentanol	−296.44	−295.64	−0.79
1-Hexanol	−317.61	−316.52	−1.09
1-Heptanol	−338.78	−330.91	(−7.87)
1-Octanol	−359.99	−356.90	−3.10
Cyclopentanol	−243.34	−242.55	−0.79
Cyclohexanol	−291.71	−287.86	−3.85
Ethylene glycol	−383.30	−386.60	(3.31)
1,2-Propanediol	−431.91	−437.77	(5.86)
Standard deviation[a]			2.09

[a] Does not include three values in parentheses (36).

Table 5.4
Heats of Formation; Gas; 25°C; kJ/mol; Carbonyl Compounds (39,40)

Compound	Calculated	Experimental (63)	Calculated − Experimental
Acetaldehyde	−166.73	−166.23	−0.50
Propanal	−187.69	−190.16	2.47
2-Methylpropanal	−217.94	−218.61	0.67
Butanal	−208.87	−204.76	−4.10
Pentanal	−229.91	−227.82	−2.09
Hexanal	−251.08	−248.40	−2.68
2-Ethylhexanal	−295.56	−299.57	4.02
Acetone	−217.28	−217.15	−0.13
2-Butanone	−238.07	−238.57	0.50
2-Pentanone	−259.16	−259.07	−0.08
3-Pentanone	−257.53	−257.94	0.42
2-Hexanone	−280.41	−279.78	−0.63
3-Hexanone	−278.49	−278.28	−0.21
4-Heptanone	−299.45	−298.32	−1.13
5-Nonanone	−341.79	−344.93	3.14
6-Undecanone	−384.30	−387.23	2.93
4-Methyl-2-pentanone	−286.73	−291.21	4.48
4,4-Dimethyl-2-pentanone	−317.06	−320.49	3.43
2,6-Dimethyl-4-heptanone	−354.05	−357.69	3.64
Dineopentyl ketone	−415.18	−421.20	6.02
3-Methyl-2-butanone	−266.23	−262.59	−3.64
2-Methyl-3-pentanone	−286.60	−286.10	−0.50
3-Methyl-2-pentanone	−283.38	−284.09	0.71
2,4-Dimethyl-3-pentanone	−314.26	−311.29	−2.97
3,3-Dimethyl-2-butanone	−292.63	−290.66	−1.97
3,3-Dimethyl-2-pentanone	−304.39	−303.76	−0.63
3,3,4-Trimethyl-2-pentanone	−325.56	−328.44	2.89
3,3,4,4-Tetramethyl-2-pentanone	−343.97	−347.69	3.72
2,2-Dimethyl-3-pentanone	−313.55	−313.76	0.21
t-Butyl neopentyl ketone	−392.29	−393.92	1.63
Di-t-butyl ketone	−348.82	−345.77	−3.05
t-Butyl isopropyl ketone	−332.88	−338.23	5.36
Cyclopentanone	−194.10	−192.59	−1.51
Cyclohexanone	−230.87	−226.10	−4.77
cis-2-Hydrindanone	−248.19	−248.11	−0.08
trans-2-Hydrindanone	−250.96	−249.91	−1.05
8-Methyl-cis-hydrindan-2-one	−277.48	−286.86	9.37
8-Methyl-trans-hydrindan-2-one	−268.32	−275.06	6.74

Continued on next page.

Table 5.4 (continued)

Compound	Calculated	Experimental (63)	Calculated − Experimental
2-Norbornanone	−164.60	−168.20	3.60
Bicyclo[3.2.1]oct-2-one	−207.48	−216.73	9.25
Bicyclo[3.2.1]oct-3-one	−210.58	−221.33	10.75
Bicyclo[3.2.1]oct-8-one	−199.33	−193.30	−6.02
Bicyclo[3.3.1]nonan-9-one	−235.14	−239.74	4.60
Adamantanone	−239.62	−230.54	−9.08
Standard deviation			4.14

current state of the field. Alkanes (20), alkenes (26, 27), alcohols and ethers (35, 36), sulfides and mercaptans (37, 38), aldehydes and ketones (39, 40), alkynes (41, 42), halides (43, 44), amines (45), carboxylic acid and esters (46, 47), and silanes (48, 49) are the classes of compounds so far studied. Each new kind of bond requires a new set of parameters. Thus, a carbon–carbon double bond is different from a carbon–carbon single bond. It is also convenient to treat cyclopropane bonds as being different from alkane bonds in this context. Cyclopropane is too different from an alkane to treat it as such (which is also true of its chemistry). Cyclobutane needs no special treatment with respect to its heat of formation.

Strain Energy

Heats of formation are useful for a comparison of the relative energies of isomers. But to compare compounds that are not isomers, neither the steric energies nor the heats of formation are useful. Yet, chemists are often interested in quantitative comparisons in such cases. Such comparisons may be made through a quantity called the **strain energy**.

The earliest and most famous of the strain theories was that of von Baeyer (50), which dealt with the strain present in rings, and which was determined by qualitative considerations. Subsequent quantitative theories of strain energy have begun from a bond energy scheme. The basic idea is that simple strainless molecules exist, and large molecules are strainless if their heats of formation can be predicted as summations of the bond energies and other increments from the small strainless molecules. If the actual heat of formation is more positive than that calculated from these strainless increments, then the molecule is said to be strained. Apparently molecular mechanics is then an ideal way to calculate strain energies, because the energies corresponding to the deformations that occur in molecules are calculated in the course of the energy minimization.

There are many different definitions of strain energy in the literature, and while they are qualitatively similar, numerical values are meaningful only within the context of a particular definition. To find the strain energy of a molecule by molecular mechanics, we must relate the energy calculated for the actual molecule to that calculated for a hypothetical analog composed of the same number and kinds of strainless parts. At first thought, one might decide that the steric energy calculated by the force field would be the number desired. But we must recall that even strainless molecules do not have steric energies of zero, because not all of the potential functions are at their energy minima at the same time. So even simple molecules like ethane and propane have nonzero steric energies. Now if we wish to compare the strain energies for two compounds that are stereoisomers (configurational or conformational), the bond energy terms would be the same for each, and, therefore, the relative strain energies are given simply by the relative steric energies. But for compounds that are constitutional isomers, or that are not isomers at all, there is no simple way to interpret the steric energy in terms of strain.

As an illustration of the little significance of the absolute value of the steric energy, we may compare calculations of the heat of formation of tri-t-butylmethane as carried out by three different force fields (15). All three calculations give values for the heat of formation that are within 17 kJ/mol of one another, but the steric energies of these force fields differ by as much as 67 kJ. The nonbonded interactions amount to 86.9 kJ (Schleyer), 73.2 kJ (Boyd), and 7.91 kJ (Allinger, 1971). So why, then, are the calculated heats of formation from the three force fields so similar? The reason is that the "structural feature increments" appearing in the heat of formation calculations already contain a portion of the energy, which need not be calculated again with the potential functions, and thus does not appear in the steric energy. The steric energies and structural feature terms differ markedly from one force field to another, but their sums are (almost) force field independent.

Over the years many different definitions of strain energy have appeared in the literature. The reader must be aware of the fact that there is no generally accepted and unique definition. This situation arises because one must choose some kind of reference point to which strain energies will be referred. In practice it is convenient to have several different reference points, involving primary, secondary, tertiary, and quaternary carbons. These reference points may be determined by actual experimental measurements or by some kind of theoretical calculation. The choices are quite arbitrary, and no one scheme is fundamentally preferable to any other scheme. The present authors prefer a particular scheme and have used it since about 1970. We will explain this scheme in detail and show how other schemes differ from it.

First there is the question as to whether the strain energies for individual compounds should be determined experimentally or calculated. Strain energy, like the resonance energy of a pi system, is not directly measurable experimentally in any case; it must be determined by some kind of calculation based on reference compound information. So in our view, it is expedient to make the whole definition of strain energy a calculated quantity. If we want to calculate strain energies from experimental measurements on the individual compounds of interest, as in the Schleyer scheme (15, 17), we cannot determine strain energies for those compounds that have not been studied experimentally. While we emphasize that there is nothing fundamentally better about our approach, it does avoid experimental measurements on the compounds under consideration. Our approach uses calculated values, and thus the scheme is more widely applicable.

The next point concerns the strainless reference standards. We have chosen the normal alkanes from methane to hexane (the *anti* conformations only above propane) plus isobutane and neopentane as our strainless compounds. (Other compounds such as cyclohexane or higher normal alkanes have also been defined as strainless in the literature.) These compounds provide an accurately known set of data for reference purposes. Based on these compounds, we can assign (by least squares fitting) strainless increments to the C–C and C–H bonds, and to the primary, tertiary, and quaternary carbon units. (The secondary unit is redundant and has been omitted here). Any saturated hydrocarbon contains some combination of the above units, and the strainless heat of formation for a compound consisting of such units can be calculated by simply adding up the increments. The actual heat of formation can then be calculated by the standard molecular mechanics method, and the difference between these two values gives a calculated strain energy.

Actually, while the above is simple enough in principle, a few complications arise. We have found it convenient to define a quantity that we call **inherent strain;** this is a preliminary quantity that is obtained on the way to finding the total strain energy. The inherent strain is calculated by adding the strainless increments as mentioned and then calculating the heat of formation for the most stable conformation of the molecule in question with two assumptions. These are: (a) that the compound consists of a single conformation, and (b) that the molecule contains no open-chain C–C bonds requiring that special torsional increments be added. Propane, for example, meets these assumptions, and its inherent strain is calculated and needs no further change to yield the actual strain. However, butane meets neither of these assumptions.

Because of the mixture of *anti* and *gauche* conformations at room temperature, the actual compound will contain a "population increment" of about 1.25 kJ/mol, which is caused by the excess enthalpy of the *gauche* molecules. This excess enthalpy is calculated from a Boltzmann distribution, knowing the steric energy difference between the conformations. In addition, the central bond, belonging to an open chain and not to a methyl group, requires that a torsional enthalpy increment be added, as in calculation of the heat of formation (pp. 175–176). So the strain calculated for n-butane is the inherent strain analogous to that calculated for propane, plus these additional terms. For some purposes, the inherent strain is of interest, and for other purposes, total strain. By defining the two terms separately, we have all of the available information broken down as is usually most convenient.

According to these definitions of strain, while the small molecules previously mentioned are indeed calculated to be strainless, or at least to have negligible strains (because sometimes a revised heat of formation becomes available after the procedures have been set), molecules such as butane do contain strain because of the conformer population and torsional increments. Cyclohexane does not contain any strain from the latter causes, but in fact has a nonzero inherent strain. This can be viewed as a result of the *gauche* interactions within the ring itself. A representative selection of strain energies for hydrocarbons is shown in Table 5.5. Note that while the inherent strains of the normal hydrocarbons are zero (to within calculational error), the total strains are considerable, up to 20.21 kJ/mol in the case of decane. Hence we must be quite clear as to which kind of strain is being discussed. The isobutane and neopentane numbers are zero by definition, but the isopentane and 2,3-dimethylbutane numbers are roughly four and eight kJ/mol, corresponding to one and two *gauche* interactions, respectively. Proceeding down the table, we find that the more congested systems show much higher strain energies.

The cycloalkanes show strain energies that have been well understood for a long time. Cyclobutane shows mainly bending strain, whereas in cyclopentane the strain is mostly torsional. Cyclohexane shows a strain energy of 10.92 kJ/mol according to the present definition. The larger rings are strained, up to a maximum of 75.02 kJ/mol for the inherent strain in cyclodecane. For cyclododecane, the strain falls off. Cyclohexadecane has relatively little inherent strain, but the molecule is rather floppy, more like an open chain than like cyclododecane. Therefore, while the inherent strain is small, the total strain is much larger.

There has been considerable concern about the origin of strain in molecules, that is, just what kinds of interactions are responsible for the high energy contents of strained structures. In certain cases this is easy to discern. In cyclobutane most of the strain is from angle deformation,

Table 5.5

Strain Energies of Hydrocarbons (kJ/mol) (20)

Compound	Inherent Strain	Total Strain
Methane	0.0	0.0
Ethane	−0.13	−1.55
Propane	−0.04	−0.04
Butane	0.04	2.80
Pentane	0.0	5.73
Hexane	−0.04	8.66
Heptane	−0.13	11.55
Octane	−0.21	14.43
Decane	−0.38	20.21
Isobutane	0.0	0.0
Isopentane	4.31	6.11
2,3-Dimethylbutane	8.70	10.21
Neopentane	0.0	0.0
2,2,3-Trimethylbutane	20.50	22.01
Hexamethylethane	33.85	35.35
di-t-Butylmethane	35.27	38.28
3,3-Diethylpentane	36.36	42.38
Cyclobutane	111.25	112.76
Cyclopentane	33.97	33.97
Cyclohexane	10.92	10.92
Cycloheptane	38.37	42.13
Cyclooctane	59.20	59.20
Cyclononane	73.14	73.14
Cyclodecane	75.02	76.06
Cyclododecane	53.60	53.60
Cyclohexadecane	25.31	49.45
Norbornane	77.24	77.24
Cubane	695.05	695.05
Bicyclo[2.2.2]octane	59.79	59.79
Adamantane	43.64	43.64

while in cyclopentane there is some angle deformation and much torsional strain. However, in many other more complicated cases it is difficult to decide just what "causes" the strain. After the energy of the molecule is minimized, the strain is distributed in various degrees of freedom. However, it is not reasonable to say that these resulting deformations *caused* the strain, because to a large extent they may have occurred to *relieve* the strain. To some extent this partitioning of strain is also force-field dependent. This means that different models will partition the strain in different ways, even though they may predict the same final total strain. Which model gives a more "real" partitioning of strain depends on which is the better model. This brings us back to the question of which force field is better, often an unanswerable question at present.

One of the ideas behind the strain concept is the rationalization and prediction of the stability of molecules, but often the stability does not follow directly from the strain energy. Highly strained molecules are often very stable because of the high activation energies for reactions leading to less strained compounds. Examples of compounds of this class that have been studied by molecular mechanics are bridgehead olefins that violate Bredt's rule. For these, dimerization is often an important mechanism of stabilization. The stabilities of these olefins often do not parallel the total strain energies. However, the stabilities can be very well correlated with the strain energy decrease that accompanies the interconversion of the trigonal olefinic carbons to tetragonal carbons, the model for the latter being saturated hydrocarbons (68–70). An alternative approach to this problem is an estimation of the reactivity from the sum of the deformation energies (in- and out-of-plane bending and torsion) of the double bond, which does not consider the saturated product at all (71). Each approach thus attempts to separate the strain in the double bond from the strain in the rest of the molecule, such as that due to small rings or congestion. A similar approach has been used in studies of olefins that are sterically hindered at the double bond (72), but here the strain energy difference between the olefin and a saturated product is not very useful. Instead, the "strainless" reference in this case was chosen to be the two hypothetical species that are obtained when the two halves of the double bond are separated to infinity. The difference between the steric energies of the intact olefin and the two fragments has been referred to as the "fragmentation strain" and has been correlated successfully with the observed stability and also with the double bond vibration frequency (72, 73).

Resonance Energies. Heats of Formation of Conjugated Hydrocarbons

Since the early days of pi-electron theory, a continuing effort has been made to understand resonance energies. While these quantities may be calculated at any level of theory, relating them to experimental observables is not straightforward. The original definition of "aromatic" as applied to benzene and its relatives, after all, came from the fact that compounds of this class often had a pleasant fragrance. The meaning of the word "aromatic" has changed over the years, and it has never really been uniquely defined. The most general current usage is probably "benzene-like," but that means different things to different people. The earliest modern definition dealt with the heat of formation of benzene relative to that calculated for a hypothetical (unconjugated) cyclohexatriene (51). The thermochemical definition of aromaticity has never been completely satisfactory because the calculated and experimental quantities are never quite comparable. That is, the quantity that could be calculated could not be measured experimentally, and the quantity that could be measured could not be calculated. So other definitions of aromaticity are frequently used. A common one is the chemical shift of a proton attached to the system in question (52, 53). But this shift is dependent upon the "ring current," and this is not related in any simple way to the thermochemical definition of aromaticity. Thus, given molecules can be aromatic by one criterion but not by the other.

After many years of confusion, Breslow (54) and Dewar (55, 56) put forth a definition of aromaticity that is probably the most useful to date. It is clear and unambiguous, and deals only with calculated quantities. No reference to experiment is necessary. This removes one difficulty but does not completely solve the problem. It permits the (quantitative) calculation of whether or not a compound is aromatic (if certain boundary conditions are met), but it does not permit the calculated aromaticity to be related to anything that can be measured experimentally. According to their definition, we compare a Kékulé form of the system in question with a linear polyene containing an equal number of (formally) single and double bonds. The calculated energy of the system is either lower than, the same as, or higher than the reference, and the compound is said to be either aromatic, nonaromatic, or antiaromatic, accordingly.

Chemists are, on the whole, more interested in chemical reactivity than in thermochemical stability. The two do not go hand in hand. A simple example is butadiene. While simple MO theory says there is a "resonance energy" (although not by the definition of Dewar and Breslow, since they would take this as the nonaromatic reference point), experimentally it is found that the compound is much more reactive than a simple alkene. The rationale is simple enough. While the conjugation lowers the total energy, it raises the orbital energy of the highest occupied molecular orbital, thus making the compound more easily attacked by electrophilic reagents. It similarily lowers the energy of the lowest unoccupied molecular orbital, increasing susceptibility to attack by nucleophiles and radicals. Clearly, we must separate thermochemical stability from chemical reactivity in discussions of aromaticity.

The chemical reactivity of conjugated alkenes is certainly important, but thus far molecular mechanics has not contributed very much to our understanding of this problem. We shall discuss only thermochemical stability here.

As with saturated hydrocarbons, and other molecules, the most fundamental way to discuss relative stability is to look at heats of formation and strain energies. While MO theories have given information concerning resonance energies, it was impossible to translate these into accurate heat of formation calculations in any general way until the advent of molecular mechanics. Molecular orbital theory by itself, as previously applied to pi systems, does tell us rather well what the pi-electronic energy is under proper conditions, but it does not allow for the sigma system. Thus, the resonance energies calculated for *cis-* and *trans-*di-*t*-butylethylene would be the same according to these theories. Obviously the sigma system introduces a large difference between the isomers, and if this is not allowed for, we will not get useful results. Accordingly, molecular orbital methods that did not consider the sigma system could only deal well with comparisons where sigma strain was either absent or would cancel out. More modern calculations, such as with MNDO, which include all of the electrons (or at least all of the valence electrons), in principle will give heats of formation. But there are some practical limitations that have so far prevented such calculations from being very accurate. Molecular mechanics heats of formation are probably better than semi-empirical MO calculations by a factor of five to ten.

The MMP1 force field was extended to include heat of formation calculations in 1977 (57). This was some years after the force field had been developed for structure, and in some respects the heats of formation are better measures of the accuracy of some of the parameters. It was thus found that some revision of the original MMP1 (58) force field was necessary to give good heats of formation. This was a somewhat patchwork job, and ideally the structures and heats of formation should have been optimized simultaneously in development of the force field. Nonetheless, the modified force field did a creditable job in calculating heats of formation. A broad spectrum of conjugated hydrocarbons was selected, 65 compounds in all, which ranged from small ones including ethylene and butadiene, to the largest conjugated hydrocarbons for which heats of formation are known experimentally (C_{20} and C_{24}). The calculated heats of formation over this whole range of compounds had a mean deviation of 5.61 kJ/mol from the experimental values. Thus the calculation is seen to be general and of reasonable accuracy.

There are a few points that can be made with regard to this calculation. First, it is a vast improvement over any previously reported similar calculation. Comparisons are made in Table 5.6. The other calculational methods were restricted to compounds that are planar and contain little or no strain in the sigma system. Even so, the reported mean deviations between the calculated and experimental values were at best 9.71 kJ/mol, and were on the whole much poorer (Table 5.6). In the molecular mechanics calculation (57), even when nonplanar systems and highly strained systems were included, and a much larger test set of compounds was used, the mean deviation was only 5.61 kJ/mol. If the comparison is made with the sets of compounds actually examined by other workers, it can be seen that the deviations are always much smaller from the molecular mechanics work [the improvement is a factor of two relative to the set of compounds used by Lo and Whitehead (59, 60), and a much larger factor for all other sets]. Nonetheless, it seemed clear even at

Table 5.6

Comparison of Different Methods for Heat of Formation Calculations of Conjugated Hydrocarbons (kJ/mol; gas; 25°C) (57)

Method	Number of Compounds Examined	Reported Mean Deviations (kJ/mol)	MMP1 Mean Deviation on Same Set
Dewar/de Lano (64)	20	32.64	7.07
Dewar/Harget (65)	19	28.45	5.86
Lo/Whitehead (59, 60)	17	9.71	4.85
MINDO/3 (66)	10	36.19	2.26
MMP1 (57)	65	5.61	—

the time the molecular mechanics work was published that the results could be improved substantially by including the heats of formation directly in the original parameterization. Indeed, MMP2 level calculations, which are incomplete at this writing, indicate that the improvement will be substantial (61).

A few features of these calculations can be mentioned here. First, the heats of formation of a few compounds as calculated (57) with MMP1 showed very large deviations from the experimental values. For example, 2,2-paracyclophane was in error by 41.80 kJ/mol, while tetracene was in error by 33.35 kJ/mol. It is suspected that the latter value is an experimental problem because the compound oxidizes in air, and obtaining a reliable heat of combustion on such a compound is not easy. On the other hand, the paracyclophane contains an error that is largely calculational, because the two benzene rings come together in such a way that the pi-orbitals of the aromatic rings overlap with the sigma orbitals of the connecting sidechains, introducing an unusual kind of instability into the system. This effect has been discussed by Mislow (62). Compounds containing this peculiar feature should be recognized as being beyond the scope of the molecular mechanics method in its present form (page 155). If an occasional special case such as this is set aside, the molecular mechanics procedures as exemplified by MMP1 make it possible to calculate, in a general way, the same quantities that can be measured experimentally (heats of formation), and finally resonance energies can be related quantitatively to experiment in a rigorous way (57, 61).

Literature Cited

1. Schachtschneider, J. H.; Snyder, R. G. *Spectrochim. Acta* **1963**, *19*, 117.
2. Lifson, S.; Warshel, A. *J. Chem. Phys.* **1968**, *49*, 5116.
3. Ermer, O.; Lifson, S. *J. Am. Chem. Soc.* **1973**, *95*, 4121.

4. Warshel, A. In "Semiempirical Methods of Electronic Structure Calculation"; Segal, G. A., Ed.; *Modern Theor. Chem.*, Vol. 7 Part A; Plenum: New York, 1977; p. 133.
5. Pickett, H. M.; Strauss, H. L. *J. Chem. Phys.* **1970**, *53*, 376.
6. Pickett, H. M.; Strauss, H. L. *J. Am. Chem. Soc.* **1970**, *92*, 7281.
7. Chang, S.; McNally, D.; Shary-Tehrany, S.; Hickey, M. J.; Boyd, R. H. *J. Am. Chem. Soc.* **1970**, *92*, 3109.
8. Boyd, R. H. *J. Chem. Phys.* **1968**, *49*, 2574.
9. Boyd, R. H.; Sanwal, S. N.; Shary-Tehrany, S.; McNally, D. *J. Chem. Phys.* **1971**, *75*, 1264.
10. Wertz, D. H.; Allinger, N. L. *Tetrahedron* **1979**, *35*, 3.
11. Benson, S. W. "Thermochemical Kinetics"; J. Wiley and Sons: New York, 1968.
12. Gasteiger, J. *Comput. Chem.* **1978**, *2*, 85.
13. Gasteiger, J. *Tetrahedron* **1979**, *35*, 1419.
14. Gasteiger, J.; Jacob, P.; Strauss, U. *Tetrahedron* **1979**, *35*, 139.
15. Engler, E. M.; Andose, J. D.; Schleyer, P. v. R. *J. Am. Chem. Soc.* **1973**, *95*, 8005.
16. Franklin, J. L. *Ind. Eng. Chem.* **1949**, *41*, 1070.
17. Schleyer, P. v. R.; Williams, J. E.; Blanchard, K. R. *J. Am. Chem. Soc.* **1970**, *92*, 2377.
18. Allinger, N. L.; Miller, M. A.; Tribble, M. T.; Wertz, D. H. *J. Am. Chem. Soc.* **1971**, *93*, 1637.
19. Wertz, D. H.; Allinger, N. L. *Tetrahedron* **1974**, *30*, 1579.
20. Allinger, N. L. *J. Am. Chem. Soc.* **1977**, *99*, 8127.
21. Allinger, N. L.; Hirsch, J. A.; Miller, M. A.; Tyminski, I.; Van-Catledge, F. A. *J. Am. Chem. Soc.* **1968**, *90*, 1199.
22. Allinger, N. L. *Adv. Phys. Org. Chem.* **1976**, *13*, 1.
23. Pitzer, K. S.; Gwinn, W. D. *J. Chem. Phys.* **1942**, *10*, 428.
24. Clark, T.; Knox, T. McO.; McKervey, M. A.; Mackle H.; Rooney, J. J. *J. Am. Chem. Soc.* **1979**, *101*, 2404.
25. Steele, W. V.; Watt, I. *J. Chem. Thermodyn.* **1977**, *9*, 843.
26. Allinger, N. L.; Sprague, J. T. *J. Am. Chem. Soc.* **1972**, *94*, 5734.
27. Allinger, N. L.; Sprague, J. T., unpublished data.
28. Jensen, J. L. *Prog. Phys. Org. Chem.* **1976**, *12*, 189.
29. Rogers, D. W.; Papadimetriou, P. M.; Siddiqui, N. A. *Mikrochem. Acta* **1975**, 396.
30. Rogers, D. W.; Dagdagan, O. A.; Allinger, N. L. *J. Am. Chem. Soc.* **1979**, *101*, 671.
31. Rogers, D. W.; Dagdagan, O. A.; Allinger, N. L., unpublished data.
32. Rogers, D. W.; Choi, L. S.; Girellini, R. S.; Holmes, T. J.; Allinger, N. L. *J. Phys. Chem.* **1980**, *84*, 1810.
33. Sprague, J. T., unpublished data.
34. White, D. N. J.; Bovill, M. J. *J. Chem. Soc., Perkin Trans. 2* **1977**, 1610.
35. Allinger, N. L.; Chung, D. Y. *J. Am. Chem. Soc.* **1976**, *98*, 6798.
36. Allinger, N. L.; Chang, S. H.-M.; Glaser, D. H.; Hönig, H. *Isr. J. Chem.* **1980**, *20*, 51.
37. Allinger, N. L.; Hickey, M. J. *J. Am. Chem. Soc.* **1975**, *97*, 5167.
38. Allinger, N. L.; Hickey, M. J., unpublished data.

39. Allinger, N. L.; Tribble, M. T.; Miller, M. A. *Tetrahedron* **1972**, *28*, 1173.
40. Profeta, S., Jr., unpublished data.
41. Allinger, N. L.; Meyer, A. Y. *Tetrahedron* **1975**, *31*, 1807.
42. Rogers, D. W.; Dagdagan, O. A.; Allinger, N. L. *J. Am. Chem. Soc.* **1979**, *101*, 671.
43. Meyer, A. Y.; Allinger, N. L. *Tetrahedron* **1975**, *31*, 1971.
44. Meyer, A. Y.; Allinger, N. L.; Yuh, Y. *Isr. J. Chem.*, **1980**, *20*, 57.
45. Profeta, S., Jr., Ph.D. thesis, Univ. of Georgia, 1978.
46. Allinger, N. L.; Chang, S. H.-M. *Tetrahedron* **1977**, *33*, 1561.
47. Allinger, N. L.; Yuh, Y., unpublished data.
48. Tribble, M. T.; Allinger, N. L. *Tetrahedron* **1972**, *28*, 2147.
49. Zalkow, V.; Freierson, M. R., unpublished data.
50. von Baeyer, A. *Ber. Dtsch. Chem. Ges.* **1885**, *18*, 2269.
51. Pauling, L. "The Nature of the Chemical Bond", 3rd ed.; Cornell Univ. Press: Ithaca, N.Y., 1960.
52. Pople, J. A.; Untch, K. G. *J. Am. Chem. Soc.* **1966**, *88*, 4811.
53. Jackman, L.; Sondheimer, F.; Amiel, Y.; Ben-Efraim, D.; Gaoni, Y.; Wolovsky, R.; Bothner-By, A. *J. Am. Chem. Soc.* **1962**, *84*, 4307.
54. Breslow, R. *Acct. Chem. Res.* **1973**, *6*, 393.
55. Dewar, M. J. S. "The Molecular Orbital Theory of Organic Chemistry"; McGraw-Hill: New York, 1969.
56. Dewar, M. J. S. *Adv. Chem. Phys.* **1965**, *8*, 65.
57. Kao, J.; Allinger, N. L. *J. Am. Chem. Soc.* **1977**, *99*, 975.
58. Allinger, N. L.; Sprague, J. T. *J. Am. Chem. Soc.* **1973**, *95*, 3893.
59. Lo, D. H.; Whitehead, M. A. *Can. J. Chem.* **1968**, *46*, 2027.
60. Ibid., 2041.
61. Allinger, N. L.; Sprague, J. T.; Yuh, Y., unpublished data.
62. Dougherty, D.; Mislow, K. *Tetrahedron* **1978**, *34*, 1441.
63. Cox, J. D.; Pilcher, G. "Thermochemistry of Organic and Organometallic Compounds"; Academic: New York, 1970.
64. Dewar, M. J. S.; deLano, C. *J. Am. Chem. Soc.* **1969**, *91*, 789.
65. Dewar, M. J. S.; Harget, A. J. *Proc. R. Soc. London, Ser. A.* **1970**, *315*, 443.
66. Bingham, R. C.; Dewar, M. J. S.; Lo, D. H. *J. Am. Chem. Soc.* **1975**, *97*, 1294.
67. Chalk, C. D.; Hutley, B. G.; McKenna, J.; Sims, L. B.; Williams, I. H. *J. Am. Chem. Soc.* **1981**, *103*, 260.
68. Burkert, U. *Chem. Ber.* **1977**, *33*, 2237.
69. Martella, D. J.; Jones, M., Jr.; Maier, W. F.; Schleyer, P. v. R. *J. Am. Chem. Soc.* **1979**, *101*, 7634.
70. Maier, W. F.; Schleyer, P. v. R. *J. Am. Chem. Soc.* **1981**, *103*, 1891.
71. Ermer, O. *Z. Naturforsch., Teil B* **1977**, *32*, 837.
72. Burkert, U. *Tetrahedron* **1981**, *37*, 333.
73. Burkert, U., unpublished data.
74. Clark, T.; Knox, T. McO.; Mackle, H.; McKervey, A.; Rooney, J. J. *J. Am. Chem. Soc.* **1975**, *97*, 3835.

Heteroatoms

Electrostatic Interactions, Solvation

In most force fields for saturated and unsaturated hydrocarbons the nonbonded interactions are represented exclusively by the van der Waals potential. As outlined in pp. 22–36, this is justified as long as no permanent charges or dipoles exist in a molecule. In the latter case, dispersion forces are not the leading terms of the nonbonded interactions. The bond dipoles in hydrocarbons are quite small, approximately 0.3–0.55 Debyes for the C–H bond (1, 2), and hydrocarbons are usually regarded as nonpolar. Earlier studies (3) indicated that inclusion of charges in hydrocarbon force fields made only insignificant differences in structures and energies, although in the Lifson–Warshel force field it was claimed that inclusion of these charges was necessary to reproduce spectroscopic data (4). The later Ermer–Lifson force field, however, managed to reproduce the data well without such terms (5). Explicit account of (intermolecular) point charge interactions was found to be valuable, if not necessary, for the calculation of crystal packing of hydrocarbons by Williams, especially for correct crystal energies (6), but the consensus of opinion seems to be that charges are negligible or, at most, of borderline importance for the determination of intramolecular nonbonded interactions in hydrocarbons. It is in more polar compounds like ethers and halides that they have to be taken into account.

There are two approaches in current use for the treatment of atomic charges: the point charge model, in which the charges are treated as geometric points placed at the atomic nuclei (in principle they could be placed at other positions also, although this possibility has not been pursued); and the dipole model, in which point dipoles are placed in bonds, at the centers, or at the point of touching covalent radii. Neither of the above approaches has been studied in much detail, although a fair number of different calculations have indicated that the two approaches are nearly equivalent, and discrepancies between the predictions made by the two methods (7, 8) were later found to be caused by inadequacies in the older calculations (59, 66).

The procedure with the dipole method is to pick a set of bond moments together with moments for lone pairs, so as to reproduce roughly the observed dipole moments of common monofunctional molecules (9). For molecules that contain two or more dipoles sufficiently distant, overall dipole moments and interaction energies between dipoles can be calculated approximately using Jeans' equation (see Equation 2.11, p. 29). This approximation is good enough to give semiquantitative values for conformational equilibria in many molecular species, such as 2-halocyclohexanones (9) and 1,2-dihaloethanes (10). But it is fairly crude, and not up to the level of the rest of the molecular mechanics calculations.

To carry out a treatment by the point charge model, the values for the point charges must be determined. This can be done by a preliminary molecular orbital calculation to obtain the needed values from a Mulliken population analysis (see texts on quantum mechanics). The values, and often even the signs, of the net charges obtained from such calculations differ from one quantum mechanical method to another, but systematic differences between the magnitudes of the charges calculated by different methods present no serious problems within a force field, as long as the same method is applied throughout.

The simple semiempirical method of Del Re (11) has been used to calculate the charges for peptides (12), because the charges obtained from this method are known to reproduce experimental dipole moments. (The method was explicitly parameterized for this purpose.) The dipole moment depends, however, not only on the net atomic charges, but also on the polarization at the individual atoms (13, 14). This polarization is in part the result of mixing of s and p orbitals on the same atom, and has been referred to as a **hybridization term** (13). Thus, it is understandable that the dipoles obtained using only the net atomic charges calculated by general quantum mechanical methods are smaller than the experimental dipole moments or those from Del Re's method.

The effectiveness of electrostatic interactions is dependent on the dielectric constant of the medium between interacting charges or dipoles. To get a reasonable magnitude for these interactions in condensed phases, a dielectric constant of four has often been used in the Coulomb potential, together with the large charges from Del Re type of calculations (12). This is, however, equivalent to scaling the charges down by a factor of two and using a dielectric constant of one.

The determination of the point charges by a quantum mechanical method remains somewhat undesirable, because of the limitations to molecular size, and the computer time expense with large molecules. Alternatively, we might try to develop a set of standard charges, picked to reproduce known dipole moments. In most cases this is far too rough an approximation, but it can be improved considerably if we allow for induced dipole moments. When a dipole is placed near a polarizable system, an induced dipole, which is oriented so as to minimize the energy of the system, will be generated. Depending on the case, this orientation can be such as to increase, decrease, or leave unchanged the magnitude of the total dipole moment. The magnitude of the effect depends upon the polarizability of the system, its distance from the dipole, and its orientation with respect to the dipole. Smith and Eyring, in a series of papers beginning in 1951 (15–19), developed a method for treating these induced dipoles so as to be able to properly calculate dipole moments in more complicated molecules. In their treatment, a simplification was introduced to make the mathematics tractable by hand; this simplification was to neglect induction further than one atom away. In other words, if we have a series of 1-haloalkanes, the halogen would induce moments in the bonds attached to C_1, but what happened from C_2 onward was neglected. The application of this approach to the calculation of dipole moments of chloroethanes has been reported (20).

A better approximation is to include all of the bonds in the molecule, and this is not difficult with the advent of computers. Such a scheme was put forth by Allinger and Wuesthoff (21, 22). This method took into account the charge on each atom in the molecule interacting with each bond, and distorted the latter according to the longitudinal polarizabilities of the bonds. Interestingly, the final equations arrived at are equivalent to those arrived at by the Del Re method (11). This method gave good results for the dipole moments of reasonably complicated systems. A still better approximation is to also allow for the effect of induction on the transverse polarizabilities of the bonds. This procedure has now been developed in full (23). The dipole moments calculated are not much different from those obtained via the Wuesthoff scheme, which were already quite good, but the electrostatic energies are further improved.

Two examples here may suffice to show the usefulness of the method. $2\beta,3\alpha$-Dichlorocholestane might be thought, to a first approximation, to have a zero dipole moment, since the C–Cl bonds are diaxial and are oriented approximately 180° apart (21). However, the observed moment is 1.27 D, rather far from 0.0. Part of the problem comes from the fact that the dihedral angle between the chlorines is not 180°, but 157°, both by molecular mechanics and by x-ray crystallography. Using standard C–Cl moments, this angle accounts for about half of the observed moment. The other half is found to result from the moment induced in the remainder of the molecule by the polar C–Cl bonds (21).

Hence this example shows that if we really want to understand dipole moments to within an accuracy of a few tenths of a Debye unit, we are going to have to do two things: use geometries of molecular mechanics quality, and allow for induced dipole moments.

For a second example we might consider the following three compounds: dimethyl ether, diethyl ether, and tetrahydrofuran. The experimental dipole moments for these compounds are, respectively, 1.30, 1.17, and 1.63 D (*197*). The permanent bond moments are the same, and the angles at which they interact very nearly the same, so that a first approximation would indicate these numbers should all be equal. But obviously they are not. In one case the observed moment is less than in dimethyl ether, and in the other case it is greater. The induced moments satisfactorily explain the observed values (*85*). If we take dimethyl ether as our standard, in diethyl ether the induced moments in the principle conformation are oriented as shown:

Hence they oppose the permanent moment, and reduce the value as observed. With tetrahydrofuran, on the other hand, the induced moments are in the directions indicated. Thus they augment the principle moment, and lead to an increase in the observed value. These differences are well calculated by the methods of Wuesthoff and Dosen-Micovic (*21, 22, 23*).

Another approach to the nonquantum mechanical determination of point charges employs the concept of the equalization of electronegativities. Concerning bond formation between two atoms, the charge density between them is shifted towards the more electronegative atom; this changes the electronegativities of the atoms, which leads to a charge shift back towards the original situation. This scheme also allows for induction through bonds, including nonneighbors, but only along bonds (longitudinal polarizabilities); the smaller transverse polarizabilities are neglected. With a suitable damping factor for the electronegativity shifts, the charges, which follow the quantum-mechanically determined values, are arrived at. However, at present the scope of this method is unclear (*24*).

Another aspect of the electrostatic interactions in molecular mechanics is the evaluation of solvent effects. Molecular mechanics calculations are carried out for the isolated molecule, that is, the gas phase. For hydrocarbons and some other nonpolar molecules, the conformational and structural properties, and the energy relationships, are rather insensitive to phase.[†] Studies have shown that a *gauche* butane unit occupies a slightly smaller volume than does an *anti* unit, and molecules that contain more *gauche* butane units will pack more closely in the liquid phase than will their *anti* counterparts (25). This permits slightly greater van der Waals attractions between *gauche* molecules than between the *anti* forms, and lowers their energy, compared with the gas phase. But the differences are rather small (about 0.8 kJ/mol between axial and equatorial methylcyclohexane, for example). Any changes in bond length which occur when a gas is condensed to a liquid or solid are too small to detect. Angles may change slightly, as is shown in crystallographic studies where two molecules of the same substance in nonidentical environments often have some bond angles that differ by two or three degrees. The differences are usually small, but they are beyond experimental error. Torsion angles can often be deformed with the expenditure of only a small amount of energy, and these may differ considerably with phase.

The situation is more complicated for molecules that contain more than one bond dipole or more than one pair of appreciably charged atoms. While solvation of hydrocarbons by nonpolar solvents is of negligible importance in the present context, as explained above, solvation of charges and dipoles must be explicitly dealt with.

More recent theories separate solvation effects into macroscopic effects, due to the interactions of the solute and the bulk solvent treated as a continuum, and microscopic effects, which are due to interactions of the solute molecule with specific solvent molecules located at defined places in an association complex. Here we will discuss only the macroscopic effects, as the others appear to pose no special problems within the framework of molecular mechanics.

Solvent effects in molecular mechanics in most cases have been treated (25) by means of the solvent's dielectric constant, D. While bulk dielectric constants of solvents can be measured easily enough, the "effective" dielectric constant on a molecular scale may be substantially different. Thus, the bulk dielectric constant of benzene and hexane differ very little, but their effective dielectric constants, which measure the results they have on chemical equilibria between polar species, differ by a factor of approximately four. In the older method of dealing with solvation, effective dielectric constants are employed for the calculation of electrostatic interactions using Equations 2.10 and 2.11 on p. 29. This obviously has no effect on nonpolar hydrocarbons, but for polar mole-

---

[†] Of course, the total heat content of a substance changes very much with phase, according to the heat of sublimation, or heat of vaporization. These changes are not relevant in the present context.

cules, increasing the dielectric constant has the effect of decreasing the numerical values of the electrostatic interactions. In the conformational equilibrium of 1,2-dichloroethane, for example, going to a more polar solvent effects a decrease of the repulsion in the *anti* form by (intramolecular) dipole/dipole interaction. A larger effect of the same kind occurs for the *gauche* form, which contains a larger amount of electrostatic destabilization. The net effect is that the more polar *gauche* form tends to be increasingly favored in more polar solvents.

The simple methods discussed above are valid in principle when the dipoles or charges are separated by distances that are large compared with the charge distributions about the dipole or atom. This does not seem to be a very good approximation in many cases. Further, the values to use for the effective dielectric constants are not known, and, in fact, except over a very limited range of molecules, they may not be constant for a given solvent. As long as the space between interacting parts of the molecule are filled uniformly by solvent molecules, it is a reasonable model. But this is rarely the case.

Another way to approach the problem, the reaction field model, goes back to Onsager (26), and has been developed for molecular mechanics by Abraham (27). This model is based on the theory of interactions of a multipole solute molecule in a polarizable continuum. Interaction of a polar, but neutral, molecule with a polarizable solvent of dielectric constant D induces a "reaction field" in the solvent, and this is accompanied by an energy decrease, which to a first approximation is given by Equation 6.1 (a is the radius of the solvent cavity or the molecular radius, μ the dipole moment, and E^v and E^s the energies of the molecule in vapor and solvent, respectively). The larger the dielectric constant, the more the solvent/solute interaction will stabilize a molecule, especially if the molecule has a high dipole moment. This means that in this model, the equilibrium between a more and a less polar conformation will be shifted towards the more polar conformation with increasing dielectric constant of the solvent.

$$E^v - E^s = \frac{D-1}{2D+1} \frac{\mu^2}{a^3} \tag{6.1}$$

A better level of approximation is reached when, in addition to the polarization of the solvent by the solute dipole, the corresponding polarization is evaluated for the solute quadrupole; then, the polarization of the solute by the thus polarized solvent is considered (21–23, 27). The expression obtained for the solvent effect in this manner is given by Equation 6.2, where $k = \mu_A - \mu_B$, the difference in the dipole moments of the conformations to be compared, $x = (D-1)/(2D+1)$, and the quadrupole term, h, is given by Equation 6.3. The q_{ij}s are the elements of the quadrupole tensor, and α is the polarizability of the solute as determined from the refractive index. Both the dipole and quadrupole moments of the solute molecule are calculated from point charges determined as described before. While the dipole moment is determined uniquely, the quadrupole moment also depends on the origin chosen for the coordinate system. This difficulty in the application of this approach as well as in the determination of the molecular radius, also an ill-defined quantity, have been reasonably well overcome by suitable definitions for the charge-weighted center of gravity and radius.

$$\Delta E^s = \Delta E^v - kx\left(1 - \frac{2\alpha}{a^3}x\right) - 3hx/(5-x), \qquad (6.2)$$

$$h = \frac{3}{2a^5} \sum_{i,j=x,y,z}^{i \neq j} [4q_{ii}^2 + 3(q_{ij} + q_{ji})^2 - 4q_{ii}q_{jj}] \qquad (6.3)$$

Let us take as an example of this procedure the molecule 3,3-dimethylpentane-2,4-dione (28). This molecule has two conformations that are much more stable than the others in the gas phase, according to molecular mechanics. Of the conformations shown, D and F are the more stable. The latter is most stable by 0.4 kJ at an effective dielectric

constant of 1.5. The latter also has a larger dipole moment (4.3D vs. 0.2D), but a smaller dipole–dipole interaction energy (5.0 kJ/mol vs. 7.9 kJ/mol). According to the Jeans' approximation, the electrostatic energies of these conformations are reduced with increasing dielectric constant, so that for D = 8.0, the conformation with the larger electrostatic interaction (D) has its energy lowered further, and becomes more stable (the new electrostatic energies being, respectively, 0.92 kJ/mol and 1.46 kJ/mol). But, according to the Wuesthoff–Abraham method, the heats of solvation of the two conformations are, respectively, 3.8 kJ/mol and 5.4 kJ/mol, which tends to stabilize F rather than D. The latter is found to be correct experimentally (28).

The method of using different dielectric constants for different solvents in Equation 2.11 and the Abraham–Wuesthoff method usually give qualitatively the same result, and quantitatively similar results, but here is a case where they actually predict different things. The Abraham–Wuesthoff method is a better approximation, and perhaps not suprisingly, experiment agrees with the predictions arrived at in this way.

Silicon

Silicon is especially easy to add to a molecular mechanics hydrocarbon force field, because the fundamental tetrahedral unit of carbon is retained. The electrostatic interactions need not be treated explicitly, as no large electronegativity difference exists. The atoms lack lone pairs and may be treated as spherical. The only real difference compared with a hydrocarbon force field, then, is in the parameters, not in the potential functions. The most characteristic geometric features of silaalkanes are the extended bond lengths: C–Si (1.87 Å) as compared with C–C (1.54 Å), and Si–H (1.48 Å) compared with C–H (1.10 Å). As a consequence, most nonbonded interactions here are attractive, where the comparable quantities in hydrocarbons are repulsive.

Two parameterizations for silicon have been reported, one of the Allinger 1971 hydrocarbon force field (29, 30) and one of the 1968 version of this field (31). These force fields are rather similar in their feature of a hard hydrogen van der Waals potential, which differentiates them from most other force fields. It should not be taken as a strong confirmation of the validity of the calculated results when the two silane force fields agree in cases where experimental data are missing, since the same systematic error may have occurred in both calculations. Unfortunately, few experimental data are available in silaalkane conformational analysis, and because the results of molecular mechanics calculations can never be better than the empirical data used for the parameterization, there are limits to the credibility of the predictions.

The rotational barrier in methylsilane (6.99 kJ/mol) is much less than in ethane (12.5 kJ/mol), and according to the calculations, van der Waals repulsions between the hydrogens contribute nothing to the former. The experimental value must therefore be matched in the calculation exclusively by the torsional V_3 potential. 1-Silabutane and 2-silabutane are interesting analogs of n-butane, and are key compounds for the conformational analysis of the silaalkanes. For

1-silabutane a rotational potential function very close to that of n-butane was calculated. 2-Silabutane, on the other hand, was calculated (29–31) and subsequently found (32, 33) to have *gauche* and *anti* conformations with very similar energies. The reason is that the nonbonded interactions across the Si–C bond are over such long distances that they are quite weak. Thus, these interactions are similar in the two conformations, which therefore turn out to have very similar energies.

The conformational energies of silylcyclohexane and trimethylsilylcyclohexane, which both favor the equatorial conformation, are calculated to be similar with both force fields [6.15 (29, 30) and 5.27 kJ (31), and 16.24 (29, 30) and 14.27 kJ (31), respectively].

Silacyclopentane was calculated to be nonplanar, and to have a pseudorotation barrier much higher than cyclopentane, 16.44 kJ (29, 30), which corresponds well to the experimental estimate of 16.28 kJ (34, 35). Unlike cyclopentane, the high energy of the planar arrangement is not due almost entirely to Pitzer strain. Mainly because of the long C–Si bonds, bending energy is also an important contributor. Both bending and torsion are less favorable in the C_s conformation than in the C_2, which are the maximum and minimum points on the pseudorotational cycle, respectively.

According to calculations with one force field, the chair conformation of silacyclohexane is preferred by only 13.0 kJ/mol over the twist-boat (1,4-**T**), because torsional forces in the twist form are comparatively small (29, 30). The chair/twist energy difference for silacyclohexane was not reported with the other force field (31), but a detailed study of the 1,1-dimethyl derivative, for which the chair/chair ring inversion barrier is known, was carried out (36). The twist boat with C_2 symmetry (1,4-**T**) is also here calculated to be the most stable nonchair form with a relative energy of 18.3 kJ/mol, while an energy of 20.8 kJ/mol was obtained for the 2,5-**T** form. The surprising difference between the chair/twist energy

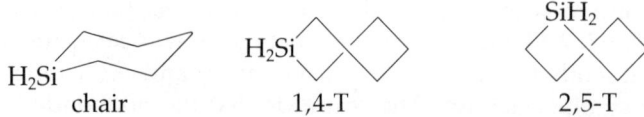

gap of the parent compound and the 1,1-dimethyl derivative may be due only to the differences in the force fields, as unfortunately neither of the two force fields was used to calculate the chair/twist energy difference for both compounds. The chair/twist ring inversion barrier was calculated in good agreement with experiment [23.0 kJ/mol (37)] by both force fields [22.8 kJ/mol (29, 30), and 21.2 kJ/mol (36)]. The transition state has C–Si–C–C approximately planar. The twist form is obviously in only a shallow well, with a calculated depth of 2.9 kJ/mol, in the chair/chair interconversion.

The conformational energies of methyl groups at various positions on the silacyclohexane ring were calculated (36), and the values for the 2-, 3-, and 4-positions (7.82 kJ, 6.95 kJ, and 5.31 kJ, respectively, with a preference for the equatorial isomer) are close to those in methylcyclohexane. The 1-methyl was calculated to have, however, a slight preference [0.17 kJ (29, 30), 0.84 kJ (36)] for the axial position. This value has recently been measured experimentally (38), and, indeed, the axial methyl has been found to be more stable (by 1.45 kJ/mol). The *t*-butyl group at the 1-position again prefers the equatorial conformation, but only by 5.36 kJ (36).

A further extension to cover polysilanes has been reported for one of the above force fields (39). The bond distances are even longer (Si–Si 2.36 Å) than in silaalkanes, and nonbonded interactions further lose their importance. The rotational barrier of disilane [5.10 kJ (40)] was also fit with a V_3 torsional potential. The rotational barrier in hexamethyldisilane was calculated to be lower (4.39 kJ) than in disilane, because of attractive van der Waals interactions between the methyl groups.

For tetrasilane, which is known to be a conformational mixture containing similar amounts of *gauche* and *anti* isomers (41), the *gauche* was calculated to be 0.38 kJ more stable than the *anti*. Delocalization of the sigma electrons, which has been shown to be more important in tetrasilane than in hydrocarbons (42) and which is expected to favor a planar arrangement of four silicon atoms, was neglected in the force field parameterization. For cyclopentasilane, electron diffraction and simple force field calculations (43) showed that a nonplanar structure is more stable than the planar D_{5d} structure. The more refined force field calculations (39) indicated that C_s and C_2 nonplanar structures are equally stable, and are in a free pseudorotation. This is expected as discussed for cyclopentane (pp. 89–91). The planar geometry is 7.61 kJ higher in energy. Cyclohexasilane, which was calculated to prefer an ideal chair conformation with 60° torsion angles, has not been studied experimentally, but the structure of the dodecamethyl derivative is known from x-ray crystallography, and it exhibits the chair conformation (44). The twist form of cyclohexasilane was calculated to have an energy of 8.16 kJ, with an inversion barrier (transition state with C_2 symmetry) of 17.32 kJ (39). An interesting derivative is hexadecamethylbicyclo[3.3.1]nonasilane, which, according to x-ray crystallography, adopts not the chair/chair form known in bicyclo[3.3.1]-nonanes, but a chair/half-chair conformation (45). The force field

calculation showed that neither a chair/chair, nor a chair/boat are energy minima, but that the form found in the crystal is, in fact, the only stable geometry (39).

Calculations on tetra(trimethylsilyl)methane were carried out, along with other highly hindered tetrahedroid compounds containing germanium, tin, or lead (39, 46) in place of silicon. These compounds do not possess T_d tetrahedral symmetry, but rather the trimethylmetal groups are all twisted, as was discussed in pp. 79–88 (46, 47). The parameters for higher members of Group IV were originally developed by Ouellette (48), who was able to calculate structures [for example of $C_2H_5GeH_3$ (49)] and rotation barriers of several derivatives quite successfully.

A study on calculations of triarylsilanes and their propeller conformations has been reported by Mislow et al. (50). There are few experimental data to compare the results with, except dipole moments.

Halogens

The halogens were among the first heteroatoms for which a systematic extension of a hydrocarbon force field was reported (9). Looking at the structures of alkyl halides, we might expect that calculations would be straightforward and that no complications would be encountered, since halides arise from hydrocarbons by simply replacing a hydrogen atom in the structure by a somewhat larger and more distant, but roughly spherical halogen atom. To calculate the structures and energies of polyhalides, we would expect that the electrostatic interactions between the charged atoms or between the carbon–halogen bond dipoles would have to be included (25). The first extensive studies on conformational stabilities and rotational barriers of halo-substituted ethanes were very encouraging, although in a number of cases systematic discrepancies seemed to exist. Some of these appeared to result from deficiencies of the electrostatic model used, and, therefore, from deficiencies of the force field, while others appeared to be due to unsatisfactory methods of energy minimization. So the larger rotational barriers of haloethanes calculated by Abraham and Parry are too high, due to the fact that they did not allow for relaxation of the molecules at the eclipsed geometry (bond lengths and angles) (51). The force field of Goursot–Leray and Bodot, which does not necessarily contain better parameters, but with which they minimized the energy with respect to all internal coordinates (52), gave much better results. Other faults of Abraham's force field in the calculation for fluorides were found to originate from an unsatisfactory choice of standard geometries at the energy minimum (53).

Both the point charge interaction model and the dipole/dipole interaction scheme have been applied to calculations on halides (10, 52–58); although no systematic comparison of the two approaches has been reported. A scheme for the determination of induced bond moments to be used in the framework of the dipole interaction method of electrostatic interactions in halides has been presented by Meyer (56–58), and leads to results superior to those of a related force field without account of induced moments (10). Induced moments are especially important for vicinal halogens (1,4-interactions), and most problems could be treated in a force field with a relatively simple scheme of electrostatic interactions, but where low periodicity torsional terms were employed for a careful handling of 1,4-interactions (59). The scheme that was employed contained only the carbon–halogen bond moment and the moment of the C–C bond(s) ending on the carbon carrying the halogen atom, as well as a mutual polarization term for neighboring (vicinal) carbon–halogen bonds.

The calculations of halides pertain at present to fundamental properties such as the conformational energies, rotational barriers, and the details of molecular geometries of relatively small mono- and polyhalides. The simplest conformational energies of the alkyl halides, and a chronically difficult case, are the *gauche–anti* energy differences of the n-propyl halides, especially n-propyl chloride. Because *gauche–anti* energy differences are closely related to the energy differences between equatorial and axial isomers of cyclohexane derivatives, both the

n-propyl chloride and chlorocyclohexane data must be fit with the same parameter set. It is, in fact, possible to correctly calculate the conformational energy (the A value) of chlorine in chlorocyclohexane (which pre-

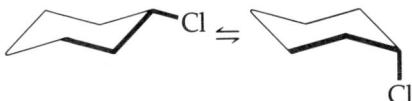

fers the equatorial position by 1.88 kJ/mol) with force field parameters that also give excellent results for a large number of other polymethyl and t-butylmonochloroethanes. But, with this and several earlier force fields the *anti* conformation of n-propyl chloride is calculated to have a lower enthalpy [by 1.00 kJ/mol (59)] than the *gauche* form, while experimentally the *gauche* form is reported to be favored over the *anti*. Unfortunately, it is not clear from the published data if the observed energy difference refers to an enthalpy or to a Gibbs free energy. Experimental work aimed at settling this point is currently underway (60).

While electrostatic interactions have significance in calculations of monohalides only in those force fields that take polarization of the molecule into detailed account by applying charges or CC and CH bond dipoles, these interactions are dominant for the conformations of vicinal dihalides with every type of electrostatic calculation. We also expect to be able to calculate the dipole moments of polyhalides by the addition of bond moments (reversing the usual method of conformational analysis of drawing conclusions about the conformation by measuring the dipole moment). If dipole moments of geminal or vicinal dihalides are calculated from the carbon–halogen bond moments derived from the dipole moment of a monohalide, the calculated dipole moments are too large. A simple remedy is to assume polarization of the hydrocarbon moiety and mutual polarization of carbon–halogen bonds, which decreases their bond moments. This approach has been used successfully by Meyer to calculate conformational properties of vicinal dichlorides (56), and geminal and vicinal polyfluorides (57), including not only the dipole moments, but also the relative energies. The approach can be further simplified by neglecting any polarization except that of the carbon–halogen bonds, and by using a nonpolarizable bond moment for the C–C bond as described above (59). Conformational energies and dipole moments of vicinal dihalides were calculated in good agreement with experimental data, and the results were best with full account of polarization.

A dichloride of interest is *trans*-1,4-dichlorocyclohexane, which by this latter approach is calculated to prefer the diequatorial conformation by 1.05 kJ/mol (56). Experimental values indicate a preference for the

diequatorial form of less than 0.71 kJ in the gas phase [electron diffraction (61, 62)], but in polar solvents the diaxial conformation is preferred. Extrapolation from solvents of low polarity to the gas phase indicates a preference for the diaxial form by 3.35 kJ/mol (63). If the extrapolated solution value is correct, the approach that assumes polarization only of the carbon–halogen bonds is not sufficiently detailed. Abraham and Rossetti (63) have pointed out that a preference for the diaxial conformation can be explained by a polarization of the C–H bonds to the axial hydrogens antiperiplanar to the chlorine atoms. However, Meyer's calculations including C–H polarization give a pref-

erence for the diequatorial conformation, although by a smaller margin than without C–H polarization. But, in his calculation the interaction still is a dipole/dipole interaction, and it is possible that a charge/charge interaction potential would give a larger stabilization of the diaxial conformation.

A related case has been encountered in 4-chlorocyclohexanone. The axial form is destabilized due to van der Waals interactions with the

ring, and yet the axial conformation is experimentally found to be preferred (64, 65). Collins and Kirk (7) have calculated the electrostatic contribution to the energy difference by both the charge and the dipole method, and found that dipole/dipole interactions would be insufficient to stabilize the axial over the equatorial form. The interaction of point charges chosen to fit the bond moments is much stronger, and this calculation yields a value in close agreement with the experimental value (7, 8, 64–66).

Solvent effects on the conformational equilibria of dihalides and haloketones have been studied by applying Abraham's reaction field method, and using point charges obtained by the modified Smith–Eyring method (21–23). The conformational energies in the gas phase were calculated in good agreement with the experimental values, and the solvent dependency could be reproduced fairly well (see pp. 195–202).

Oxygen

Oxygen occurs in organic molecules in two basically different bonding situations, which must be handled differently in molecular mechanics. Carbonyl oxygen (bound to only one other atom) was handled in a meaningful way much earlier than was ether oxygen, mainly because 1,4-nonbonded interactions, which can be described well by a superposition of van der Waals and torsional energies, are the only nonbonded interactions of conformational significance in carbonyl compounds. Ether oxygen needs a more detailed description of its nonbonded interaction potential. Both types of oxygen are found in carboxylic acids and esters, which will be discussed at the end of this section.

Electrostatic interactions are as important in oxa derivatives as in halides, and the same problems of dealing with small induced charges exist. In the carbonyl compound force fields reported to date, the dipole approximation could be applied in its simplest form, assigning bond moments only to the C=O bond and the C–C bonds ending on the carbonyl carbon. Calculations on haloketones, in which the point charge interaction method was employed for the calculation of electrostatic interactions, have been mentioned previously. For calculations on conformations of monoketones, and generally for compounds carrying no second polar group, the electrostatic interactions can obviously be neglected. If other polar groups exist in the molecule, it is not clear if the dipole approximation is sufficient, or if charge interactions improve the results. In ethers and alcohols, nonbonded interactions and electrostatic interactions are often quite important, and interactions of induced charges have been found to play an important role in the conformational energies of, for example, the 5-alkyl-1,3-dioxanes (67), and in the rotamers of alcohols (68).

While van der Waals interactions of hydrogens and halogens have C_∞ symmetry about their bond to carbon, the carbonyl and the ether oxygen only have local C_{2v} symmetry, and, therefore, possibly less symmetry in their nonbonded interactions than halogens. In more familiar terms, the question is whether the lone pairs of electrons on oxygen need to be explicitly included in the van der Waals potential, or whether a spherical oxygen potential is a sufficient approximation. In electron density difference maps from the final steps of x-ray diffraction

studies of oxygen compounds, the lone pairs show up in the carbonyl group plane at 120° angles to the C=O bond, and in ethers pointing to the missing corners of a tetrahedron. (Of course, the two sp^2 or sp^3 orbitals do not appear as two separate lobes as in the sketches usually shown, but rather as a fraction of a sphere of electron density.) Lone pairs have not been necessary or useful for calculations on carbonyl compounds, but in ether force fields, lone pairs were a welcome remedy for defects in the van der Waals potential of oxygen (67, 69, 70). Still this question has not really been answered. With a different hydrocarbon force field it may well be possible that lone pairs will be unnecessary. Lone pairs were introduced as though they were additional atoms attached to oxygen, and their positions were also optimized during energy minimization. Thus, in compounds containing many oxygen atoms, such as carbohydrates, the lone pairs add considerably to the time required for energy minimization, and, therefore, are not desirable.

Lone pairs of electrons not only have an effect on the van der Waals potential, but also on the electrostatic interactions. In some force fields (MM1, MM2) "bond" dipole moments have therefore been assigned to the lone pairs (69, 70). In others, where the monopole model (charge interactions) was used, a point charge was located along the bisector of the COC bond angle and adjusted to give good fits to experimental conformational energies (71–73).

A question not touched on in the preceding sections is whether or not the heteroatom has any effect on the van der Waals characteristics of a carbon to which it is attached. This would be especially important for carbonyl carbon, which is treated like an olefinic carbon with the same van der Waals potential, and with its peculiarities of the bending potential (in-plane and out-of-plane bending being calculated separately). Clearly the electron density at a carbonyl carbon is lower than at an olefinic carbon. There do not, however, appear to be experimental data available that will permit the choice of any better parameters at present. This problem of insufficient experimental data is frequently encountered in molecular mechanics parameterizations. It is important to know if parameters of related, well studied interactions have simply been transferred unaltered to apparently similar cases. Then if discrepancies between experimental and calculated properties arise, they can perhaps be removed by adjusting the assumed values of the parameters.

Aldehydes and Ketones. Van der Waals interactions alone were found to be insufficient to reproduce the two most characteristic ground state properties of simple carbonyl compounds, the eclipsing of the acetaldehyde methyl hydrogen by the carbonyl oxygen, and the eclipsing of the propionaldehyde methyl group by the carbonyl oxygen (74). Attractive interactions that go beyond van der Waals interactions are responsible, and torsional energy terms had to be adjusted as usual in such cases (9, 75, 76).

In the ground state of 2-butanone, the methyl group eclipses the carbonyl oxygen, but when the steric repulsion is much larger, as it is in 4,4-dimethyl-2-pentanone, the conformation with the t-butyl group eclipsing oxygen is less stable than the one with hydrogen in this posi-

tion (75, 76). When a secondary alkyl group is attached to a carbonyl group, as in cyclohexanecarboxaldehyde, the conformation with oxygen eclipsing one of the (ring) carbons is found to be most stable (75, 76).

For calculations on cyclic ketones, the conformations of the related hydrocarbon rings were used as starting geometries. Cyclopentanone was calculated to prefer the C_2 half-chair conformation, with the C_s envelope conformation being 13.5 kJ less stable (75, 76). Conformations with the carbonyl group at other positions of the cyclopentane ring do not correspond to energy minima. The contrast between cyclopentane and the corresponding ketone is noteworthy.

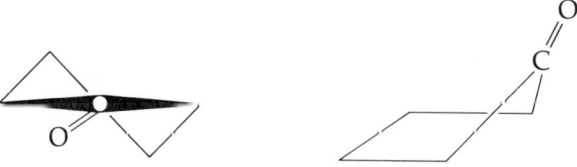

For cyclohexanone the most stable conformation is the chair, and a relative energy of only 11.38 kJ/mol is calculated for the C_2 twist-boat. Other high energy conformations are the C_1 twist-boat and the C_s boat at 15.78 kJ/mol and 22.30 kJ/mol, respectively. The ring inversion barrier was calculated (75, 76) to be 16.3 kJ, with C–C(O)–C–C coplanar, an experimental value being 16.7 kJ/mol (77).

C_2 C_1 C_s

The conformational energies of methyl groups at the 2- or 4-positions of cyclohexanone are similar to those found in cyclohexane, but in the 3-position the axial isomer has about 2.1 kJ/mol less energy than in the other isomers. This "3-ketone effect" is well reproduced by molecular mechanics calculations (9).

Methyl–methyl 1,3-synaxial interactions in the chair form, which are relieved in the twist-boat, might be expected to force 3,3,5,5-tetramethylcyclohexanone into one of the twist forms. However, the

chair was calculated to be 15.9 kJ/mol more stable than the boat, in agreement with the available experimental data (9).

Decalones and hydrindanones are important models for steroids, and were studied in some detail (9, 75, 76, 78, 80–82). 1-Decalone prefers the *trans* configuration. The two conformations of the *cis* form have calculated energies relative to the *trans* of 8.24 kJ/mol (steroid form) and 12.00 kJ/mol (nonsteroid form), while the experimental Gibbs free energies range from 4.6 kJ/mol to 13.0 kJ/mol for the *cis–trans* equilibrium (79). Similar results are found for the 10-methyl derivative, but for the 9-methyl derivative, the equilibrium lies on the side of the *cis* form. For 2-decalone, the *cis* form was calculated to be 9.2 kJ/mol less stable than the *trans* form (75, 76), while for the 10-methyl derivative, the calculated energy difference is only 0.8 kJ/mol (9, 75, 76). The nonsteroid form of 10-methyl-*cis*-2-decalone is more stable than the steroid form by 2.8 kJ/mol (9). (*See* opposite page.)

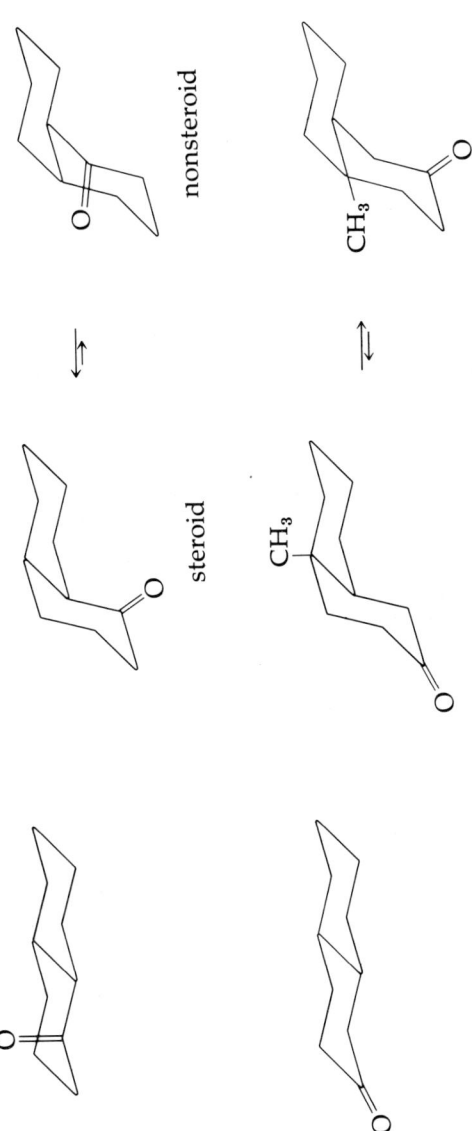

The chair–twist-boat conformational equilibria for several *trans*-2-decalones were studied by molecular mechanics. In all cases except the 1,1,5-trimethyl derivative, the chair–chair conformation was by far the most stable one (*80–82*). For the latter compound, depending on the force field applied, the conformation with the cyclohexanone ring in the twist form was calculated to be 2.5–20.5 kJ/mol less stable than the chair–chair form (*80–82*).

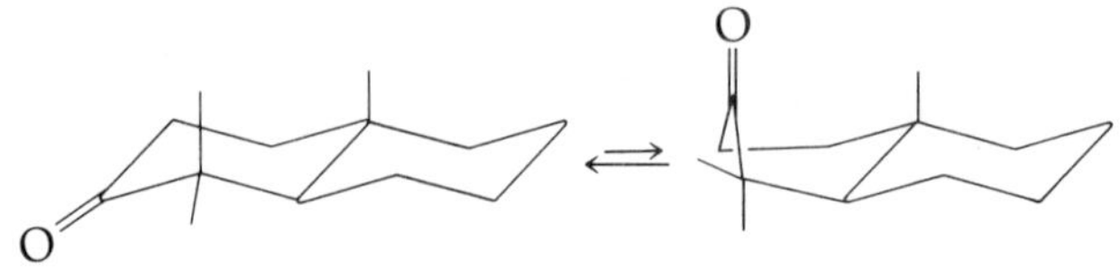

The same conformational situation of a cyclohexanone ring suffering 1,3-diaxial repulsions of methyl groups in the chair form exists in 4,4-dimethyl-5-α-androstan-3-ones (*83–85*), and this problem will be discussed in pp. 253–255.

For cycloheptanone, the chair pseudorotational family was calculated to be 8.4–16.7 kJ/mol more stable than the boat family. The C_2 twist-chair conformation is not the most stable form. The conformation with the carbonyl group in the 2-position is more stable by 1.05 kJ/mol (*75, 76*). Therefore, a mixture of these conformations is expected.

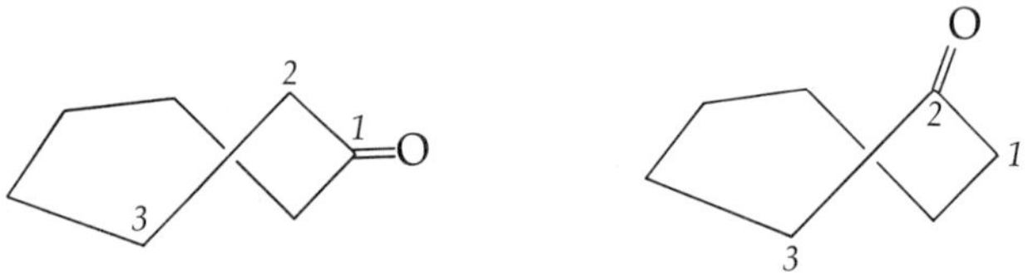

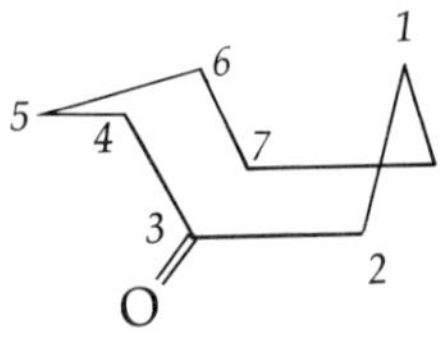

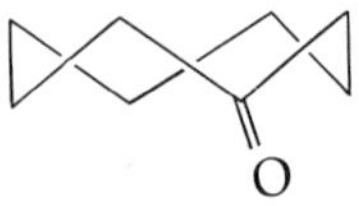

Cyclooctanone is most stable with the carbonyl group at the position numbered 3 in the diagram shown (*89*). The two C_s conformations of the boat–chair family had earlier been calculated to be the least stable of all boat–chairs, while a twisted crown and the other boat–chairs studied (and possibly other conformations not considered) were calculated to have intermediate stabilities (*75, 76*). In the parent cyclooctane the boat–chair conformation is preferred (pp. 100–104), and the transannular hydrogen–hydrogen van der Waals repulsions between hydrogens inside the ring at the positions numbered 3 and 7 in the above figure are particularly severe. Hence the fact that the calculations and experiment give this preferred conformation is not surprising (75, 76, 89).

For cyclononanone, the unsymmetrical conformation shown was

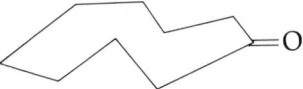

calculated to be the most preferred of the five considered. Geminal dimethyl substitution at C_4 and C_7 causes additional steric problems, and in the tetramethyl derivative, the C_2 conformation shown and other unsymmetrical conformations are the more stable ones (75, 76).

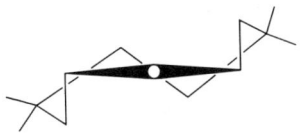

For cyclodecanone, the three possible boat–chair–boat forms were considered, and again the conformation shown was calculated to be the most stable (75, 76) for the same reasons outlined above for cyclooctanone.

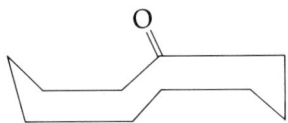

Several diketones have been studied, the most interesting being cyclohexane-1,4-dione. A number of experimental studies were inconsistent, and seemed to indicate an equilibrium between the boat, D_2 twist-boat, and C_1 twist-boat, with a preference for one or the other, depending on the method applied. All of the data can, however, be interpreted in terms of a flexible, rapidly pseudorotating D_2 twist-boat (86). The calculations indicate that the energy of the chair form is only 0.54 kJ/mol above that of the D_2 twist form. That the molecule appears to exist exclusively in the twist form would best be explained by the entropy favoring the flexible form. (See next page.)

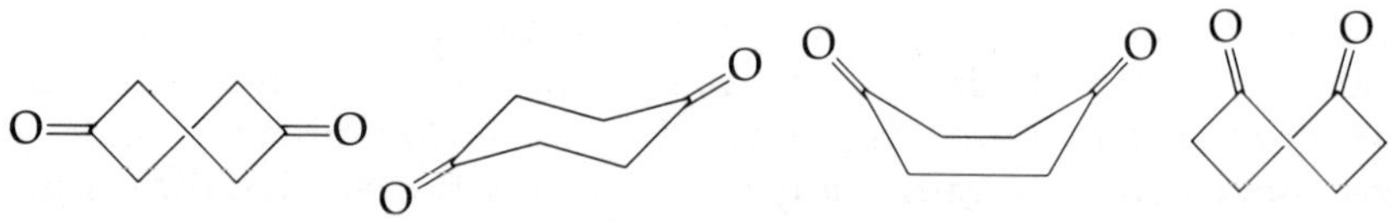

The triketo form of phloroglucin (cyclohexan-1,3,5-trione) was calculated to prefer the C_s boat form, with a small amount of the C_2 twist forms (2.26 kJ/mol), but no chair form present (75, 76).

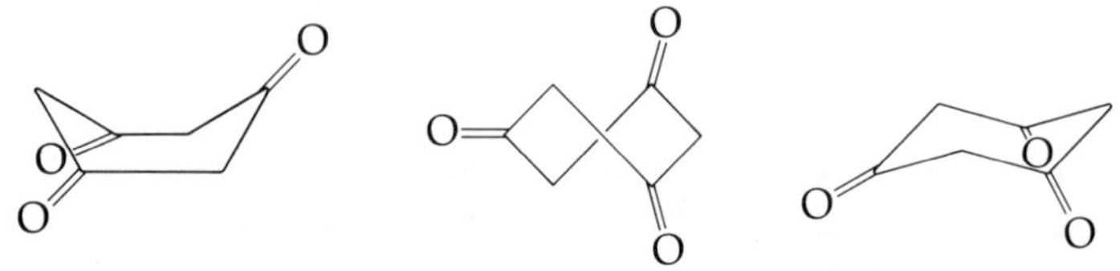

The conformation of cyclodecane-1,6-dione calculated to be most stable [and found in the crystal (87)] is the boat–chair–boat form **(1)**, in agreement with the results for the monoketone where the carbonyl group is found at the same position. Infrared spectral data indicate,

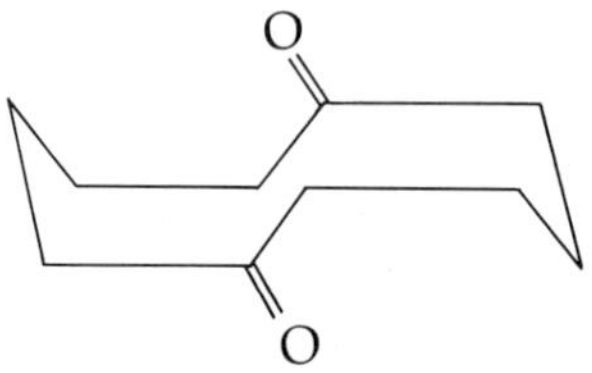

1

however, that in solution another conformation predominates (88), and NMR data seem to support the preference for a more symmetrical structure such as **(2)** (89). Cyclooctadecane-1,9-dione has a structure of this symmetry **(3)** in the crystal (90), and this structure was also found to be the best of those studied by molecular mechanics (MM2) (90).

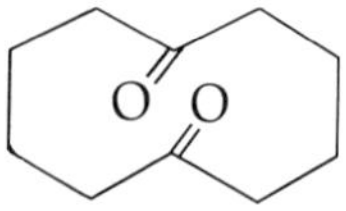

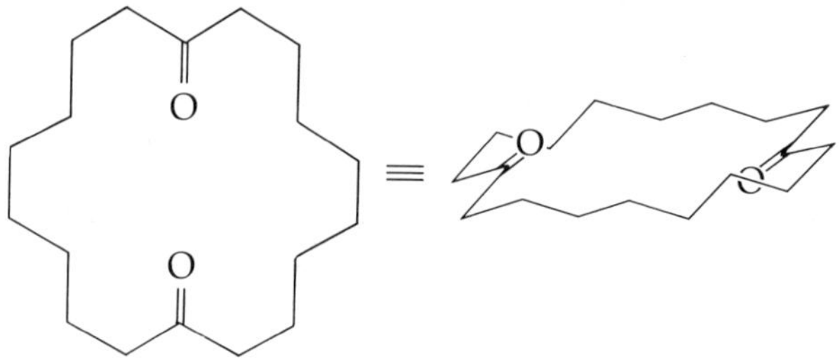

2   3

1,9-Cyclohexadecanedione and the corresponding mono- and diethyleneketals have also been examined by molecular mechanics (91). Generally a square conformation is preferred, with a ketal on a corner, or a ketone at other than a corner. The diketone is exceptional, preferring a rectangular (C_2) conformation.

Ethers, Acetals, Alcohols. Parameterization of a force field to cover the biligand oxygen found in ethers and alcohols is less simple than originally expected (9). To make the problems clearer, it is useful to discuss separately the different possible fragments that occur in ethers. Two of these, the CCCO and OCCO, have parallels in the calculations of halides, with similar problems being encountered. The fragments with a C–O bond as central unit (COCC and COCO) introduce additional problems.

Conformational energies of compounds with rotational isomerism in the CCCO fragment have been useful for the adjustment of the van der Waals properties of oxygen. The energies of the axial 5-alkyl-1,3-dioxanes are in part determined by van der Waals repulsions involving the ring oxygen atoms. The van der Waals radius of a spheri-

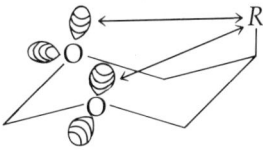

cal oxygen can be adjusted to fit the experimental energies, but to get sufficient repulsion from this source alone would require a very big oxygen. To keep the van der Waals radius of oxygen smaller than that of carbon, the nonbonded interactions can be made more repulsive in the direction of the 5-alkyl group, which is also the direction of the lone pairs on oxygen. This is conveniently done by adding the lone pairs as pseudoatoms with their own van der Waals potentials (67, 69, 70). There is a way of introducing the desired instability of the axial alkyl group without the lone pairs (see below), but an additional reason for including them stems from the geometry of the parent compound, dimethyl ether. The methyl groups here do not have threefold symmetry, and the presence of the lone pairs permits the reproduction of the experimental geometry of this compound, something that could not be done without the lone pairs.

One of the basic requirements of a hydrocarbon force field is the reproduction of the *gauche/anti* energy difference of *n*-butane, and the energy difference between equatorial and axial methylcyclohexane. In the same way, we expect the correct results for the *gauche/anti* energy difference of methyl ethyl ether, and of the conformations of 2-methyl-1,3 dioxane, which, according to conventional reasoning, are determined respectively by van der Waals repulsions between the terminal methyl groups or the transannular repulsions of the axial methyl group. However, it was not possible to calculate these quantities correctly on

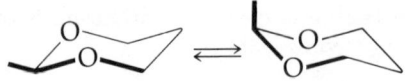

the basis of van der Waals and electrostatic interactions alone. The *gauche* energy comes out far too low. The apparent solution to the problem, as with the hydrocarbons, is the addition of a low periodicity torsional energy term to stabilize the CCOC *anti* form relative to the *gauche* (70, 92, 103). This may be interpreted as being a substitute for a more detailed treatment of the 1,4-interactions of the lone pairs (92). The *gauche* conformation is quite a lot higher in energy relative to the *anti* form in methyl ethyl ether than in butane (by 6.3 ± 0.8 kJ/mol vs. 3.8 kJ/mol, in the gas phase). The molecular mechanics value chosen to give the best results over a broad range of compounds settled on 7.2 kJ/mol for this energy. The energies of 2-methyl-1,3-dioxane and related compounds were then well calculated (70, 92). The heats of formation (70) were summarized in Chapter 5 (Table 5.3).

Several simple cyclic ethers have been studied by molecular mechanics (69, 70, 94, 95), and the results have been compared with experimental information. The agreement is usually good. To a first approximation, the ether is similar to the corresponding hydrocarbon, but there are some differences.

Oxetane is effectively planar, in contrast with the corresponding hydrocarbon, cyclobutane, which is strongly puckered and has a 6-kJ/mol barrier at the planar conformation (pp. 119–120). A small barrier at the planar conformation was detected from the microwave spectrum of oxetane, but the first vibrational level lies above this barrier, so the compound undergoes a wide-amplitude vibrational motion through the planar form (96). The molecular mechanics calculations also give a planar molecule. The calculated (70) C–O bond lengths are somewhat too short, however (1.430 Å compared with 1.449 Å experimentally) (96).

Tetrahydrofuran, like cyclopentane, is practically a free pseudorotor, both by calculation and by experiment. The geometry is unexceptional (69, 70, 95).

For tetrahydropyran, the calculations indicate that the chair conformation is very much more stable than other possibilities, as with cyclohexane (69, 70, 95). The bond lengths and angles are calculated to be similar to those in cyclohexane, with the C–O bond length being similar to those found with simple ethers. An electron diffraction study detected only the chair conformation (97), with a geometry in good agreement with that calculated.

Oxepane (oxacycloheptane) is calculated to have a stable twist-chair series of conformations (*98*). In this case, the oxygen can occupy any of four nonequivalent positions, so that the compound would be expected to exist as a mixture of four conformations. The electron diffraction data did not give sufficient information to determine the structure with certainty (*98*), but have been interpreted on the basis of two conformations, as proposed from the vibrational spectrum (*99*).

With oxacyclooctane, as with cyclooctane, the boat–chair is the stable conformation, and by placing the oxygen at the position shown, the serious transannular repulsion between hydrogens, which occurs in the corresponding hydrocarbon, can be relieved. Thus, calculations indicate that this particular conformation is the most stable one (*94, 100*).

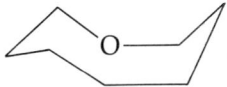

For calculations involving polyethers, three different cases must be distinguished: polyethers where the oxygen atoms are separated by three or more carbon atoms; the glycol ethers; and the acetals. Heats of combustion (*101*) show that the geminal oxygens (acetal) lead to a strong stabilization of the molecule, while the vicinal oxygens (glycol) lead to a strong destabilization. If the oxygens are separated by three or more carbons, they are essentially independent. These facts are readily understood from elementary principles. The acetal shows the strong anomeric type of resonance, which leads to stabilization. The glycol structure places together the positive ends of strong dipoles, which is electrostatically disfavored. But there is more to this latter case, which is sometimes referred to as a *gauche* effect (*102*), meaning the *gauche* conformation is more stable relative to the *anti* than van der Waals and electrostatic interactions would suggest.

The problems posed by the glycol ethers, and their solutions, are already known from studies with vicinal dihalides, where, in addition to van der Waals and electrostatic interactions, low periodicity torsional potentials were employed to fit the experimental data (*56–59, 103*). Such torsional terms could account for through-bond effects in glycol ethers if necessary (*103*). It was found that the experimental data for the 5-alkoxy-1,3-dioxanes can be reproduced either with (*70*) or without (*92*) employing such terms. However, explicit account of a *gauche* effect is mandatory in acetals. It has long been known in carbohydrate chemistry that an anomeric oxygen prefers an axial position in which the COCO fragment is in a *gauche* conformation. This anomeric effect is not a consequence of van der Waals or electrostatic interactions, or at least not of these alone, but mainly results from through-bond hyperconjugation, which stabilizes a torsion angle of 90°. In this geometry, electron donation from the HOMO on oxygen (the p_z lone pair orbital) into the

C–O antibonding sigma orbital involving the other oxygen is most effective (104, 105). Such an effect cannot, of course, be expected to be reproduced by a molecular mechanics force field, unless it is specifically introduced. A low periodicity (V_2) torsional potential is an ideal way to account for the rotational part of the effect (69, 70, 92). This procedure does not, however, deal correctly with the C–O bond lengths, which also change with the degree of conjugation. One way of handling the problem is to make r_0 a function of environment (106).

With a parameterization that takes the aforementioned details into account, the properties of ethers could be well calculated, including *gauche/anti* energy differences of acyclic compounds, and the equatorial/axial energy differences of 2-, 4- and 5-methyl-1,3-dioxanes (67, 70, 92). The chair form of 1,3-dioxane was calculated to be only 16.7–20.1 kJ/mol more stable than the most stable twist form, the 1,4-twist (69, 70, 107, 108). Experimental data were known for such equilibria only for polymethyl derivatives, and have been interpreted as indicating much higher energies for the nonchair forms (109). Molecular mechanics calculations indicate that 1,3-dioxanes usually prefer the chair conformations, except in a few exceptional cases, where a 1,3-synaxial methyl/methyl interaction occurs in the chair conformation, and where the resulting conformation is the 1,4-twist with no pseudoaxial methyl group (107). Otherwise, the chair form is more stable than any twist form.

Calculations were reported for 2-oxabicyclo[3.3.1]nonane, 3-oxabicyclo[3.3.1]nonane, and several substituted 2,4-dioxabicyclo[3.3.1]-

1,4-twist   2,5-twist

nonanes (*110, 199*). For 2,4-dioxabicyclo[3.3.1]nonane the conformation with a chair in the cyclohexane ring and a boat in the 1,3-dioxane ring was found to be 8.28 kJ more stable than a double chair form (*110*), in agreement with experiment. The MM2 force field also gives this conformation as the preferred one (*95*).

Recent ab initio calculations (*111*) suggest that the largest barrier in methyl ethyl ether (where the two methyls go past each other) may be significantly higher than previously recognized. The same could well be true for the COCO fragment. Because electron correlation has not been included yet in the ab initio calculations, their reliability is still uncertain. But if these barriers are indeed higher than realized previously, the molecular mechanics conformational energies for the twist-boat forms would be too low.

The conformational energies calculated (and found experimentally) for methyl groups at C_5 of 1,3-dioxanes with other methyl groups at C_4 and/or C_6 show once more the superiority of molecular mechanics over more classical methods of prediction of conformational energies by assuming the additivity of A-values. In 4,4,5-trimethyl-1,3-dioxane and in *ref*-4, *cis*-5, *trans*-6-trimethyl-1,3-dioxane, two *gauche* butane interactions of 3.56 kJ/mol are counted in the equatorial isomers (**1** and **3**), but only one such interaction in the axial forms (**2** and **4**).

1	**2**	**3**	**4**

In addition, in the axial form there is a transannular interaction of a 5-methyl group in a 5-methyl-1,3-dioxane (3.56 kJ/mol). So the simple approach would predict identical energies for the axial and equatorial conformations. But molecular mechanics calculations and experimental data both show that the equatorial isomer is more stable than the axial form by up to 5.65 kJ/mol (*92*).

1,3,5,7-Tetroxocane is experimentally a rather well-studied ether, and its properties may be contrasted with those of cyclooctane itself. Replacing one or a few methylene groups in cyclooctane by oxygens (or nitrogens) makes rather little difference in the overall conformational equilibrium, as experiments and calculations on the oxacyclooctanes have shown (*94*). But in tetroxocane there are four conformations that might be considered as of possible importance. In the crystal, the crown conformation is found (*112*), apparently deformed from the symmetrical (D_{4d}) conformation observed in solution (*113–115*) by crystal packing forces. Calculations indicate (*70, 94*) that one of the boat–chair conformations (**1**) is most stable in the gas phase, followed in turn by the boat–boat (D_{2d}), the crown (D_{4d}), and finally the other boat–chair (2). The dipole moment of the D_{4d} conformation is much greater than the

| chair–chair (crown) | boat–boat | boat–chair (1) | boat–chair (2) |
| D_{4d} | D_{2d} | C_s | C_s |

moments of the other conformations, and, hence, this conformation is stabilized by solvation in solution, or in the crystal where continuous stacks of molecules are aligned (*112*). The other three conformations have energies that are calculated to remain about the same, but the crown conformation is increasingly stabilized by increasing dielectric constant, and becomes the most stable conformation by the time the dielectric constant reaches five. Experimentally, the crown conformation predominates in solution (*113–115*), and the boat–chair (1) is also present, so the agreement with experiment here is good. While earlier work indicated the crown was a minor conformation in the gas phase (*113*), a recent electron diffraction study finds D_{4d}/C_s in the ratio 32:68 (*196*). Because of the unfavorable entropy of the D_{4d} (symmetry number 4), these values give a lower enthalpy for the crown of 1.7 kJ/mol.

Calculations on ortho esters have been carried out using the force fields (*69, 70, 92*) developed for acetals. The structures and conformational energies calculated for trimethoxymethane (*116*) and 1,1,1-trimethoxyethane (*117*) are in agreement with results from electron diffraction. The conformational energy of 2-methoxy-1,3-dioxane was also correctly reproduced, confirming the additivity of *gauche* interaction terms here (*92*).

Calculations on the conformations of alcohols (the rotameric isomers about the C–O bond) suffer somewhat from the scarcity of experimental data. The rotational barrier of methanol is easily reproduced by adjusting the V_3 torsional constant of HCOH torsion, but other rotational barriers are not available. The vibrational and NMR spectra of alcohols show some dependency on the rotameric equilibrium, and molecular mechanics calculations account very well for the HCOH coupling constants in the NMR spectra (*68*).

Hydrogen Bonding. In the MM2 force field (and also in MM1), hydrogen bonding occurs as a result of the electrostatic and van der Waals interactions. No special account of it is taken. It is found with carboxylic acids (pp. 224–225) that the most stable arrangement is the dimer, and both the bond energy and geometry calculated agree reasonably well with experiment. Methanol molecules used as a test case dimerized with reasonable geometries and hydrogen-bond energies (118).

The *gauche* conformation of ethylene glycol is calculated (with MM2) (70) to be more stable than the *trans* by 3.35 kJ/mol at a dielectric constant of 1.5, in agreement with experiment (119). This agreement with MM2 is partly due to the V_2 term for OCCO, which was chosen to be -3.35 kJ/mol. This value also leads to a satisfactory axial–equatorial equilibrium for 5-methoxy-1,3-dioxane. The corresponding hydroxyl compound was studied in some detail (70, 92). The staggered conformations for the hydroxyl (axial and equatorial) with the hydrogen over the ring or out away from the ring all correspond to energy minima in the calculations. The relative energies were determined. In cyclohexanol, the most stable conformation has the hydroxyl equatorial in the C_I conformation (68, 70). In the dioxane, the equatorial hydroxyl also has a C_I conformation, which is more stable than the C_s by the rather large amount of 6.78 kJ/mol (70). The C_I conformation for the axial hydroxyl has an energy of 1.00 kJ/mol above that of the most stable equatorial conformation. However, the C_s axial conformation is best of all, being 5.77 kJ/mol more stable than the most stable equatorial conformation. This stability in the main results from an electrostatic effect (hydrogen bonding). The hydrogen bond is therefore calculated to be of the bifurcated type, and it does not point towards one of the oxygens (70, 92). This is because the electrostatic energy does not vary rapidly with torsion angle while the hydrogen is over the ring, but the torsional barrier does. Hence, the position of the hydrogen is governed by the latter. This system has been studied experimentally, and it has been concluded that the strongly predominant observable molecular species (by coupling constants in the proton NMR) (120) has the bifurcated hydrogen bond in the axial conformation. Thus, the calculation is in good agreement with experiment on this point.

From these limited examples, it may be concluded that most if not all of the data concerning hydrogen bonding can be accounted for automatically as a result of electrostatic and van der Waals interactions. In other force fields and in more quantitative work, it was found that the potential functions used for the van der Waals interaction had to be modified to reproduce hydrogen bond geometries and energies, either by neglecting the oxygen–hydrogen van der Waals repulsion at small distances (121), or even by using an attractive oxygen–hydrogen interaction (93, 122). To reproduce the hydrogen bonding potential obtained earlier from ab initio calculations, the MM1 force field has been modified by making O . . . H van der Waals interactions angle dependent, and by adding a Morse potential for this interaction (195).

Force fields for alkanes, ethers, and alcohols have been reported that do not take lone pairs into account, and that use transferable point charges for electrostatic interactions. One of these (123) was designed to calculate structures, relative conformational energies, and vibrational frequencies on the basis of the philosophy of the consistent force fields (see pp. 36–38). Only a small fraction of the conformational problems outlined above were considered in this work, however, and the important *gauche/anti* equilibria in ethers (which required the introduction of lone pairs and low-order torsional terms in other force fields) were not studied with this force field. The second force field (192), based on White's hydrocarbon field, was designed to calculate geometries of macrocyclic polyethers, but it also gives geometries, relative energies, and heats of formation of simple ethers (excluding acetals). The force field was employed to calculate possible conformations of several crown ethers (9-crown-3, 12-crown-4, and 18-crown-6) as well as other macrocyclic polyethers. The calculated structures were compared with those found in the crystals of free and complexed ethers (192). Some macrocyclic polyethers were discussed on page 108.

Calculations have been reported for carbohydrates, but until quite recently only with very simple force fields, and usually only with partial geometry optimization. For these reasons, all of the calculations on carbohydrates are discussed in Chapter 7, together with those on other molecules that have been studied only by more approximate methods.

Carboxylic Acids and Esters. Force fields for calculations on carboxylic acids and esters are desirable because of the frequent occurrence of these structural elements in natural products. One detailed extension of a molecular mechanics hydrocarbon force field to this class of compounds has been reported (124). An important property of carboxylic acids to be reproduced by the force field is the preference of the hydroxyl proton for the position *cis* to the carbonyl group, which is due to dipole/dipole interactions. Numerical values for the *cis/trans* energy difference and the rotational barrier in formic acid have been deduced from ab initio calculations, and the results obtained from the parameters determined in this way for the same quantities in other carboxylic acids show good agreement with experiment. The van der Waals properties of the acidic proton were chosen to allow for hydrogen bonding between two carboxylic acids to give the observed cyclic dimer geometry and hydrogen-bond energy without employing an additional hydrogen-bond potential besides van der Waals and dipole/dipole interaction. (*See* discussion on hydrogen bonding in previous section.)

For isobutyric acid the conformation with the methyl group nearly eclipsing the carbonyl oxygen was calculated to be 0.88 kJ/mol more stable than the one with hydrogen eclipsing this oxygen (124). A similar preference for the unsymmetrical conformation was found for cycloalkane carboxylic acids.

Succinic acid was calculated to have a lower enthalpy (0.54 kJ/mol) in the *anti* isomer if a dielectric constant of 1.0 was employed, but this conformation was less stable if a higher dielectric constant was chosen (124). The *gauche* form is a *dl* pair, so the equilibrium contains more *gauche* than *anti* isomer in any case, but more so in solvents of high dielectric constant. The limiting Gibbs free energy difference is 2.93 kJ/mol at very high dielectric constants.

Parameters for esters were deduced from small model compounds, as usual. Calculations on lactones are the most interesting area of application for an ester force field, and such calculations have been carried out with three slightly different force fields (124–126, 201). A typical feature of esters is the tendency of the CC(=O)OC fragment to be coplanar, although with a smaller barrier to rotation than in an olefin. For this reason, the conformations of lactones are related more closely to the conformations of the corresponding cyclic olefins than to those of the cycloalkanes. γ-Butyrolactone prefers the conformation with four planar ring atoms, and the β carbon out of the plane of the remainder of the five-membered ring. This permits the ester fragment to be planar while the alkane part is puckered, leading to a very stable structure. δ-Valerolactone was at first calculated (MM1) to be somewhat more stable in the boat conformation than in the half-chair, which is the preferred form in cyclohexene. Later, more refined calculations (MM2) put the half-chair at slightly greater stability, which has been confirmed experimentally from the microwave spectrum (127).

E (kJ/mol)   2.5   0.0

Medium and large ring lactones have also been studied in some detail with the MM2 force field (201). The seven-membered lactone was calculated to contain three conformations in equilibrium, but two of them had negligible concentrations. The most stable conformation is the chair, with the energies for the boat and for the *trans* conformations being higher as indicated.

	chair	boat	trans
E (kJ)	0.0	11.38	22.2

With the eight-membered ring, the situation becomes even more complicated. Here there are predicted to be four conformations, all of which contribute reasonable populations. Three of these are *trans* about the ester linkage, and one is *cis*. One of the *trans* conformations is preferred in the gas phase, the energies being as indicated. But if the mixture is put into a solvent of high dielectric constant, the *cis* conformations, having a much higher dipole moment than any of the *trans* conformation, is appreciably stabilized, and it becomes the most stable conformation. It is known from dipole moment measurements (in benzene solution) that seven-membered or smaller lactone rings are exclusively *cis*, and those containing nine or more members are exclusively *trans* (128). The eight-membered lactone is a mixture, with the *cis* predominating somewhat in benzene solution (128), and the calculations are consistent with these facts.

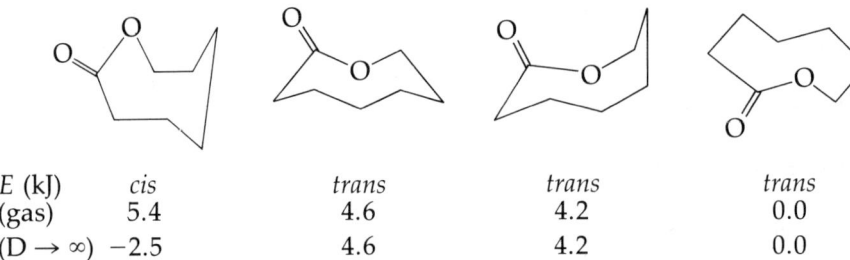

E (kJ)	cis	trans	trans	trans
(gas)	5.4	4.6	4.2	0.0
(D → ∞)	−2.5	4.6	4.2	0.0

The larger lactone rings have been given a cursory examination. Much of the strain in the cyclodecane ring comes from two pairs of hydrogens that are internal to the ring. By placing a lactone linkage at the positions shown, one hydrogen of each pair is removed, and the strain for this compound is greatly reduced relative to the hydrocarbon. (See also the conformations of cyclooctanone and cyclodecanone for the same effect.) This seems to be the likely structure for the ten-membered lactone.

or, in another view,

Similarly, for the twelve-membered lactone, much of the strain results from repulsions between hydrogens within the ring. Placing the lactone linkage at the position shown relieves half of these repulsions, and this is the probable conformation for this compound.

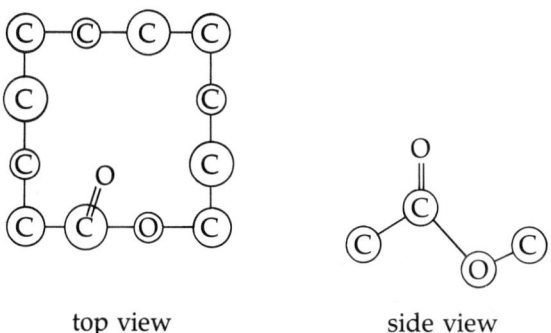

top view   side view

Conformational studies of lactones using the MM1 force field and lanthanide shift reagents have been described (129).

A lactone force field has been applied to the calculation of the stability of several stereoisomeric γ-lactones in the field of terpenoids. The relative stabilities of α-santonin (1) and β-santonin (2) and their 6-epimers (3 and 4), as well as isomeric germacranolides (5 and 6) were obtained in excellent agreement with experimental data (125, 126). (See next page.)

Nitrogen

Nitrogen appears in various functional groups in organic molecules. The nitrile group poses no special problems (9), and is comparable to an alkyne function. The amide nitrogen has been dealt with at various levels for application in peptide force fields. For a discussion of more approximate amide force fields (with rigid bond lengths and angles) and references to more precise ones, see Chapter 7. No force field has yet been reported for the imine nitrogen. Only recently has a detailed amine force field been worked out. There is no doubt but that the amine nitrogen has been the most troublesome atom yet included in molecular mechanics force fields. Many of the problems arising in amine force fields also occur in ether force fields, and only the inclusion of low-order torsional energy terms has yielded satisfactory results in these cases (130). Azo compounds have been studied in some detail by two different groups (131, 132). Kao and Huang used ab initio calculations to deduce some of the necessary force constants required. Finally, calculations on ammonium salts have been reported, apparently without problems (133).

Whether or not the lone pair on nitrogen is essential is unclear, as is also true in the ether case. In nitrile and amide force fields, lone pairs have not been used. The effects of the lone pair for spectroscopic and other properties of amines are, however, so ubiquitous, and the position of the lone pair is easy to define, so that in the most elaborate amine force field to date, a lone pair was employed. The van der Waals properties of the lone pairs were regarded as very important in earlier days of conformational analysis, when, for example, the question of whether the hydrogen on piperidine occupies the equatorial or the axial position was related to the relative steric requirements of hydrogen and lone pair (134, 135). However, the calculations show that the transannular interactions between hydrogens and the lone pair are all attractive, not repulsive (130, 136), and that the van der Waals characteristics of the lone pair cannot be uniquely defined by the piperidine equilibrium. The van der Waals characteristics of the lone pair are, on the other hand, helpful when bond angles of small amines are to be fit by the force field. Also, explicit inclusion of the lone pair can better mimic the known electron distribution in an amine.

Observed trends in the C–N bond lengths in simple amines have not been reproduced by the molecular mechanics model so far. While in hydrocarbons steric crowding leads to C–C bond stretching, the C–N bond lengths are known to decrease in the series methyl-, dimethyl-, trimethylamine (137–139). On the other hand, good agreement between calculated and experimental bond angles was obtained, and the rotational barriers of N-methyl groups were successfully reproduced (130).

A considerable problem arises in situations where a CCNH *gauche/anti* equilibrium exists. The compound studied most extensively in this regard is piperidine, which has this torsion angle *anti* in its equatorial NH form. A host of experimental data exists concerning the equilibrium, and the most recent consensus favors the equatorial form slightly over the axial. Open chain analogs containing this fragment are ethylamine and isopropylamine. While in the latter compound, the form with C_s symmetry and the minimum number of *gauche* CH_3/H interactions seems to be favored by 0.50 kJ/mol (140), the ethylamine equilibrium shows conflicting experimental data. Most recently a preference for the form with the maximum number of *gauche* CH_3/H interactions was deduced (*141, 142*). Unfortunately, it was impossible to simultaneously fit all of these conflicting data by one force field. Because of the apparently reliable piperidine equilibrium position, the parameters were fit to this equilibrium (*130*). These parameters led to an assignment of the conformational equilibrium in ethylamine that is the reverse of that most recently reported in the literature [of which, however, the authors did not appear to be very confident (*130, 141*)].

Other important fragments in amines are the CCCN fragment, which was studied in *n*-propylamine and *sec*-butylamine, and the CCNC fragment of methylethylamine. Low periodicity torsional energy terms were mandatory here, similar to the situation found in ethers.

With the force field fit to the previous data (electrostatic interactions were calculated by the dipole interaction model, including a N–lone pair moment), calculations were carried out on a large number of amines, mainly cyclic ones (*130*). Triethylamine was calculated to prefer the propeller-like conformation with C_3 symmetry, with a second conformation (C_1), however, only 0.88 kJ/mol less stable.

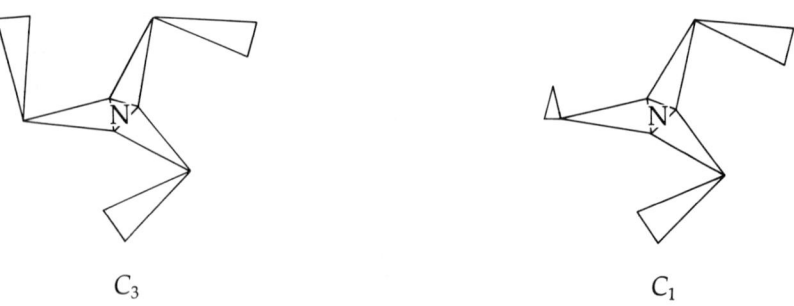

Azetidine was treated as a system different from other amines for reasons outlined in connection with cyclobutane (pp. 23–24) and 116–121). With these parameters, the ring pucker, the equatorial preference of the NH proton, and the barrier to ring inversion all fit well (*130*). For pyrrolidine, for which few experimental data are available, a

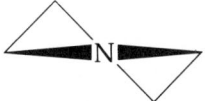

preference for the half-chair form shown was calculated. Other participants of the pseudorotational cycle are on the average 1.26 kJ/mol less stable, and the planar geometry has a relative energy of 18.28 kJ/mol.

The piperidine N–H problem has been alluded to above, and the preference for the equatorial form is solely a consequence of torsional energy terms with a still unresolved physical origin. The N-methyl group in N-methylpiperidine clearly prefers the equatorial position, and after the torsional terms were adjusted for methylethylamine, the energy of the axial methyl group was calculated to be 10.46 kJ/mol (*130*), midway between the available experimental values (*134, 135*).

A series of polymethylated C-methylpiperidines was studied and showed the expected trends. The unusually small calculated preference for the equatorial position of the 2-methyl group in 1,2,3,3-tetramethylpiperidine (2.5 kJ/mol) appears to have no experimentally known counterpart (*130*).

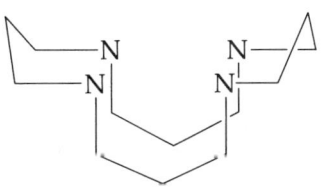

The geometry of the only large monocycle studied, 1,5,9,13-tetraazacyclohexadecane (**1**), was calculated in close agreement with the one found in the crystal (*143*).

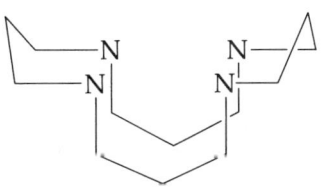

1

Many bicyclic systems with the nitrogen at the bridgehead and at other positions were studied (130). In the former, the molecule is in an equilibrium of *cis*- and *trans*-fused rings, which differs from the hydrocarbon analog in that there is a conformational and not a configurational relationship here. Pyrrolizidine was calculated to prefer two isoenergetic *cis* conformations (crown **2a**, and boat–chair, **2b**) over the *trans* (**2c**) by 11.3 kJ/mol. This is considerably less than the *cis/trans* energy difference in the hydrocarbon (26.6 kJ/mol), but this is anticipated, as the *cis* form has CCNC *gauche* (or near *gauche*) torsion angles, which are *anti* in the *trans* isomer. The *gauche/anti* energy difference is, however, much higher in the CCNC than in the CCCC unit (144).

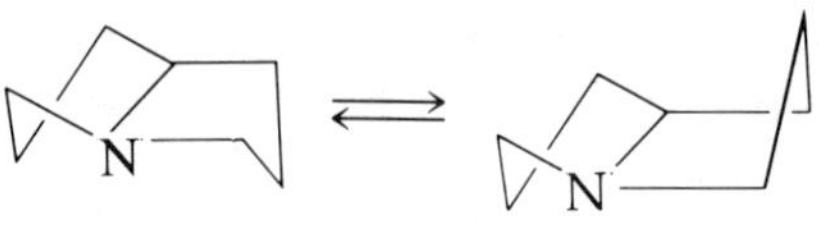

2a   **2b**

Indolizidine (**3**), being a piperidine derivative with a chair conformation, prefers the *trans* conformation, [calculated 10.92 kJ/mol (130), experimental 10.0 kJ/mol (145, 146)].

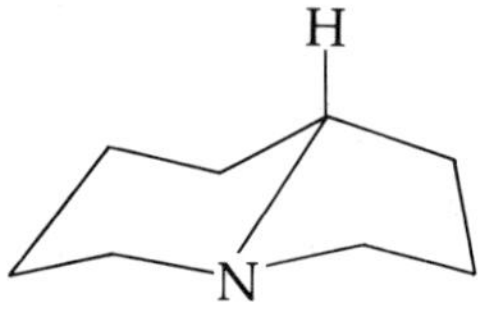

3

Quinolizidine (**4**) strongly prefers the *trans* form, with the calculated *cis/trans* energy difference (14.23 kJ mol) halfway between the available experimental values of 10.9 kJ/mol and 18.4 kJ/mol (145, 147).

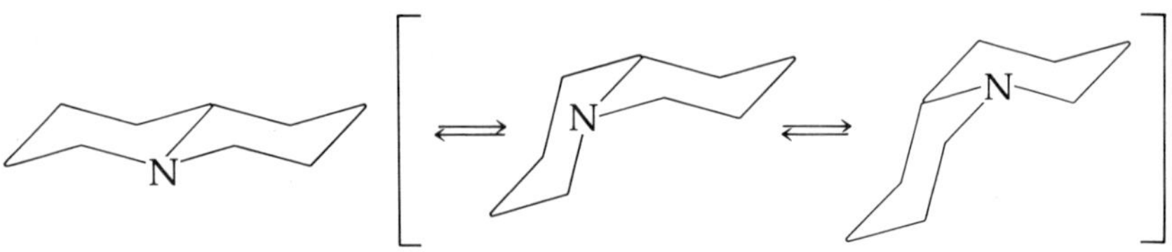

4

Quinuclidine (**5**) was calculated to prefer a C_3 (twisted) structure, analogous to the corresponding hydrocarbons (page 112). 1-

Azabicyclo[3.3.3]undecane (manxine, 6,) is highly strained, and the strong bond angle deformations at the nitrogen (CNC 115.6°) and in the alpha position (CCN 119°) account for the unusual spectroscopic properties and the low pK_a value (147–149, 200). The CCN angle (130) is large enough to be sp^2 in character, rather than sp^3.

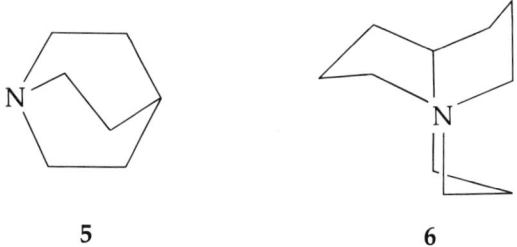

5   6

The bridged piperidine derivatives nortropane (7, R = H) and 9-azabicyclo[3.3.1]nonane (8) prefer conformations with the piperidine rings in the chair form, in the latter case by 8.54 kJ/mol, over a chair–boat. [The corresponding preference is calculated to be 9.67 kJ/mol in the hydrocarbon (150).] The NH proton of nortropane shows a 0.46 kJ/mol preference for the axial position, due to van der Waals repulsion by the bridge, but the N-methyl group in the same frame remains equatorial by 3.35 kJ/mol.

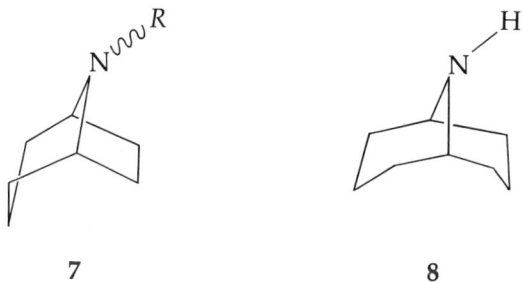

7   8

Much experimental information is available about perhydroquinolines and -isoquinolines (azabicyclo[4.4.0]decanes), which are easily studied by molecular mechanics. These compounds, and especially the *cis* isomers, which exist as conformational equilibria of two chair–chair conformations (151), have been helpful in checking the calculations for amine conformational equilibria. The calculations show, however, that the energy differences in the relaxed molecules appear mainly as angle bending and not as van der Waals terms, and, therefore, the results cannot be safely transferred to molecules with different backbones and different probable angle bending requirements. The calculated equilibrium positions agree well in most cases with the experimental values, with a few notable exceptions, namely the 1-, 6-, and 10-methyl derivatives of *cis*-2-azabicyclo[4.4.0]decane (9), where the differences between the experimental and calculated values are approximately 4 kJ/mol in each case. While some of the discrepancy may be due to experimental error, it seems likely at present that these are cases in which a number of small errors happen to add, instead of cancel, resulting in the discrepancies mentioned. These errors will perhaps be reduced when more accurate experimental data regarding simple systems become available.

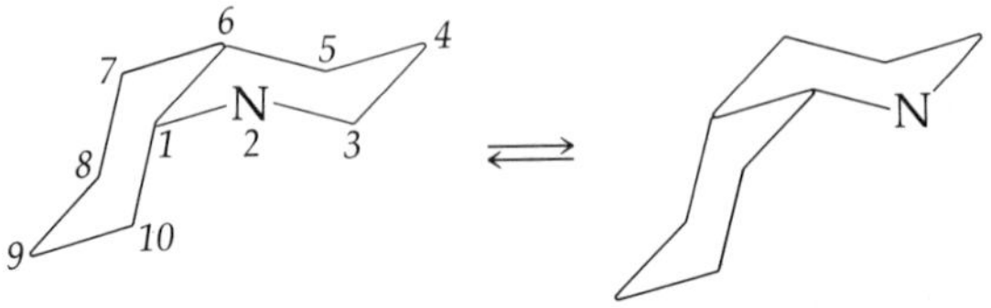

9

Electrostatic interactions are important in piperidones, and the axial orientation of the N–H proton in 4-piperidone is calculated to be destabilized by dipole interactions. An interesting result is also found in

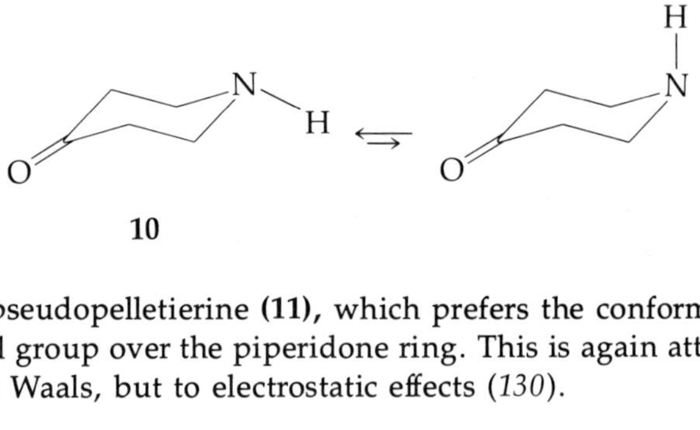

10

N-*t*-butylpseudopelletierine (11), which prefers the conformation with the *t*-butyl group over the piperidone ring. This is again attributed not to van der Waals, but to electrostatic effects (130).

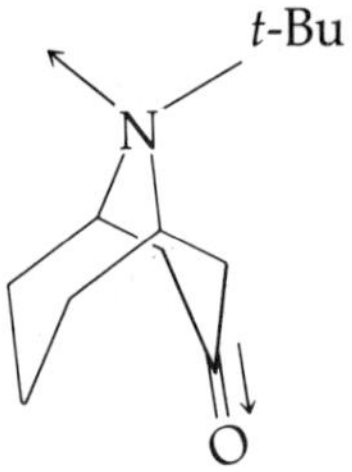

11

Sulfur

The conformational analysis of sulfur derivatives is in many regards similar to that of the sila compounds. The major difference relative to the hydrocarbons is again the extended bond length, leading to a decreased importance of repulsive van der Waals interactions. As with the silanes, bond stretching is easier and angle bending is less easy than in hydrocarbons.

In the extension of the 1973 Allinger hydrocarbon force field to monothiols and monosulfides, no electrostatic interactions were found necessary to fit the experimental data for structures and energies (152). Considerably more experimental data are available for these compounds than for silaalkanes, which permits a more reliable parameterization. Thus, it was possible to include four-membered rings in the calculations, employing stretch–bend and torsion–bend cross terms as in the cyclobutanes. The rotational barrier of methanethiol served to fit the appropriate torsional potential function. As in methylsilane, the net van der Waals interactions were about equal in the eclipsed and staggered conformations.

Calculations on a number of thiols were carried out using this force field (152). One result obtained was the preference of ethanethiol (1) for the *gauche* conformation, which amounts to 0.88 kJ (calculated), in good agreement with an estimate from calorimetric data (153, 154). This result is not from attraction between the SH-hydrogen and the methyl group, but [in this force field (152)] from repulsion between the SH-hydrogen and the vicinal hydrogens on the CH$_2$ group in the *anti* form. The situation here may be contrasted with that in ethanol where the *anti* conformation is of lower enthalpy (68, 70).

1

A large list of thioethers was studied in depth with this force field (152), but only a few examples will be discussed here. In 2-thiabutane (2), the *anti* form was calculated to be favored by 1.21 kJ. The experimental data indicate similar energies for the two conformers. 3-Methyl-2-thiabutane (3) also prefers by 1.00 kJ the conformation with the minimum number of *gauche* methyl interactions. The other conformation does not assume a fully C_s symmetrical structure, but rather has twisted methyl groups.

The structure calculated for thiacyclobutane was different from the experimental one available at that time, but later an improved experimental structure was published, which was then in full agreement with the force field structure. The ring is puckered, as are most saturated four-membered rings (155).

Thiacyclopentane has a high pseudorotational barrier of 11.7 kJ/mol (156, 157), similar to that in silacyclopentane. The calculation showed the half-chair (C_2) form to be more stable than the envelope (C_s) by 8.74 kJ, in agreement with experiment. It is uncertain whether the C_s form is an energy minimum or a maximum.

Thiacyclohexane (4) has only one twist-boat energy minimum, the TB-1, which is calculated to be 16.86 kJ less stable than the chair, in excellent agreement with an experimental estimate (158).

Other molecules for which calculations were carried out included 2- and 3-thiadecalin, 3-thiabicyclo[3.3.0]decane, 5-thiabicyclo[2.1.1]hexane, and 7-thiabicyclo[2.2.1]heptane, whose structural and energy data all agreed well with experimental values (*152*).

Molecules with more than one sulfur atom, but separated by carbon atoms, show no pronounced conformational peculiarities, but disulfides and polysulfides have some surprising features. The rotational potential about a sulfur–sulfur bond is quite different from that in ethane, and more like that in hydrogen peroxide, which has its minima at a torsion angle of 90°. Fourier decomposition of the complete rotational potential into cosine terms requires at least a V_2 and a V_3 term (pp. 32–36), which also must be included in the torsional potential function of the molecular mechanics force field (*159*). A molecule will then adopt a conformation in which the most favorable balance is found between the conformational requirements of the hydrocarbon part (minimum nonbonded strain and minimum bond eclipsing) and the preference of the disulfide part for the 90° torsional angle.

A molecule that demonstrates the latter point nicely is 5*H*,8*H*-dibenzo(*d*, *f*)(1,2)dithiocin (**5**), which has two conformations available, a chair (**5a**) and a tub (**5b**). The balance of all forces favors the chair, as is found in the crystal of this compound, although the CSSC torsion angle is more favorable in the tub form (*159*).

5a   **5b**

1,2-Dithiane is calculated to have a chair conformation, which is preferred over the most stable twist form by 16.32 kJ/mol. The situation is similar, but the preference much less, in the case of 1,2,4,5-tetrathiane (**6**) (4.6 kJ). The latter compound shows some striking conformational differences, relative to cyclohexane (Figure 6.1). There are two stable twist conformations (**6b** and **6b′**), with C_2 symmetry (mirror images). But instead of the pseudorotational itinerary where there would be unsymmetrical twist forms and classical boats, these latter conformations are simply points on the slope; only the two twist forms of C_2 symmetry correspond to energy minima. Furthermore, these two forms are not interconverted by a pseudorotational type of motion, as the barrier for that is extremely high. Rather, they interconvert by going through the chair conformation. The C_{2v} conformation is especially high in energy,

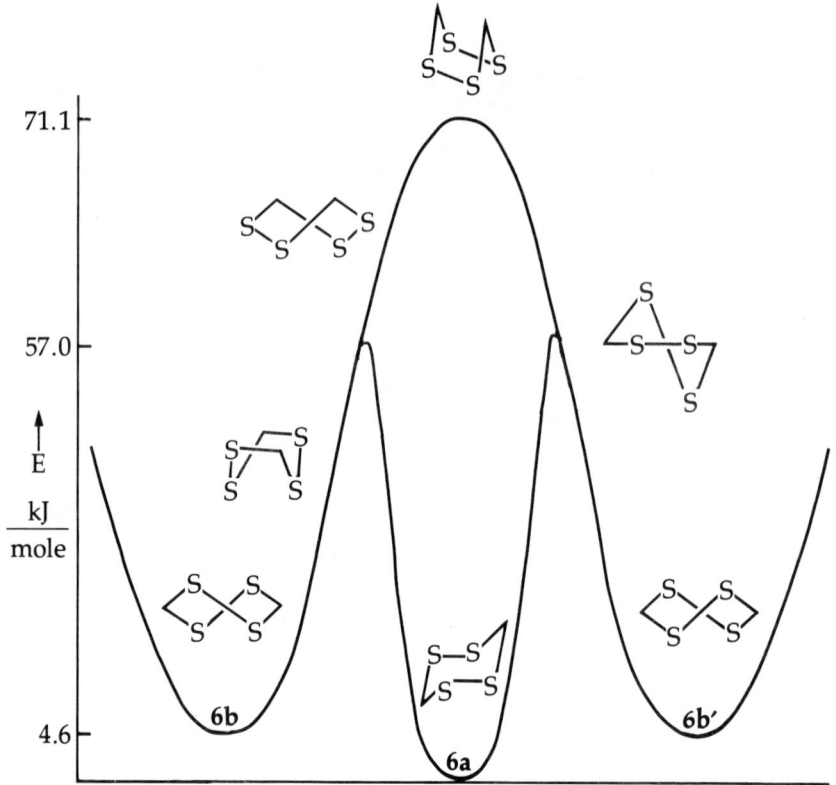

Figure 6.1. The conformational potential function for 1,2,4,5-tetrathiane.

and corresponds to the maximum. It has the very unfavorable eclipsed orientation of both disulfide units.

Replacement of the four hydrogens in 1,2,4,5-tetrathiane by methyl groups further stabilizes the twist form [as in cyclohexane, where the chair/twist energy difference was calculated to drop from 22.2 kJ in the parent compound to 13.0 kJ in the 1,1,4,4-tetramethyl derivative (*159*)]. The twist form of 3,3,6,6-tetramethyl-1,2,4,5-tetrathiane is calculated to be 2.93 kJ more stable than the chair. This has been found to be correct experimentally from the NMR spectra (*160*) and the crystal structure (*198*) of this compound. The twist-chair interconversion barrier was calculated to be 65.3 kJ, while the experimental value is 66.9 kJ.

For 1,2,3-trithiane, the only nonchair energy minimum was calculated to be the symmetrical boat, not the twist form. The calculated inversion barrier (51.5 kJ) fits the experimental value (55.2 kJ). Additional sulfur atoms in the ring, as in 1,2,3,4-tetrathian, 1,2,3,4,5-pentathian, or in cyclohexasulfur, increase the chair/nonchair energy difference, and pseudorotation of nonchair conformations was calculated to be easier with a larger number of sulfur atoms (159).

All the simple symmetrical and unsymmetrical disulfides would be expected to possess three conformations with CSSC near 90°, and the CCSS angle near 60° or 180°. CNDO/2 calculations indicated, however, that the CCSS torsion angle should be 0°, not 180°, in the most stable form of methyl ethyl disulfide (161, 162). Indirect evidence has been interpreted as indicative of still another energy minimum, with this torsional angle about 30°. This interpretation rests primarily on the fact that a structure of this kind occurs rather frequently in protein crystals (161, 162). Molecular mechanics calculations on methyl ethyl disulfide found the three potential minima as qualitatively predicted, and contrary to the CNDO/2 results; ab initio calculations, using the optimized geometries of the molecular mechanics force field, agreed well with the molecular mechanics results (163). It seems clear that the CNDO/2 prediction of a most stable CCSS eclipsed conformation is wrong. Raman spectral studies were interpreted as indicating three conformers, two with similar energies, and the third 1.2–3.3 kJ less stable (164), in agreement with the prediction of the molecular mechanics calculation. Severe steric hindrance as in the elusive bis-tri-t-butylmethyldisulfide will force the disulfide linkage into the *trans* (180°) conformation (193, 194).

Dihydrogentrisulfide and dimethyltrisulfide were studied by molecular mechanics, and electrostatic interactions are found to be responsible for the preference of the *trans* over the *cis* conformers (with a SSSC torsion angle of 80.9° in the latter molecule). This study also includes the interconversion barriers, which are predicted to be in the same range as those in disulfides (165). The same type of calculations have even been attempted for the tetrasulfides (166). (The torsional potential function obtained in Ref. 166 does not show the dominating effect of the large V_2 term as expected, and cannot result from the force field quoted.)

A polysulfide force field should also, in principle, be capable of handling elemental sulfur, and, with some modifications, the above force field could be used for the calculation of structures and energies of polysulfur cycles (167). The results for the small cycles, cyclotetrasulfur and cyclopentasulfur, are only of limited significance, because alternative structures, which lack normal covalent bonds (biradicals), may exist and cannot be handled by the method. Both molecules are calculated to be highly strained. Cyclohexasulfur is calculated to prefer the chair form by 63.2 kJ, with a chair → twist inversion barrier of 97.5 kJ. Its structure agrees very well with an experimental one. Cycloheptasulfur is most stable in the twist-chair conformation, which pseudorotates with a very low barrier of 1.25 kJ (chair), similar to the corresponding hydrocarbon. Cyclooctasulfur, the most stable sulfur allotrope, does not prefer the chair–boat as does cyclooctane, but prefers the D_{4d} crown conformation (by 43.1 kJ), which is also calculated to be quite flexible. Again, excellent

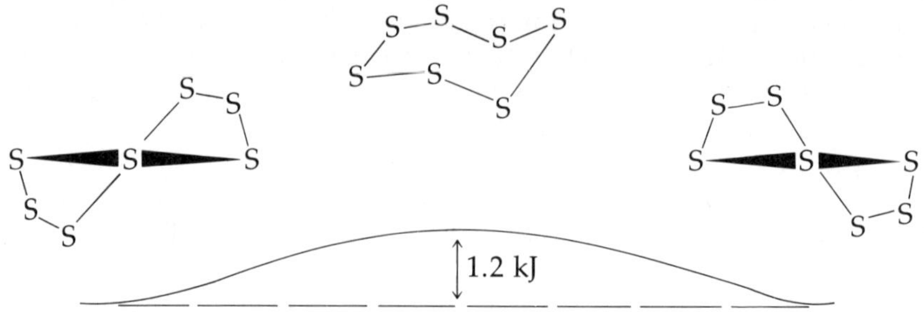

S_7

S_8: D_{4d}, Chair–boat

agreement between calculated and experimental structures is found (*167*). The contrast between the preferred conformations of this molecule and cyclooctane stems from the dihedral angle of the minimum energy conformations, near 90° and 60°, respectively. Cyclononasulfur is predicted to adopt an unsymmetrical geometry that is 16.74 kJ more stable than the next most stable conformation examined. The D_{5d} crown conformation of cyclodecasulfur was calculated to be more stable than the C_1 or C_{2h} conformations by over 29.3 kJ. It was subsequently found

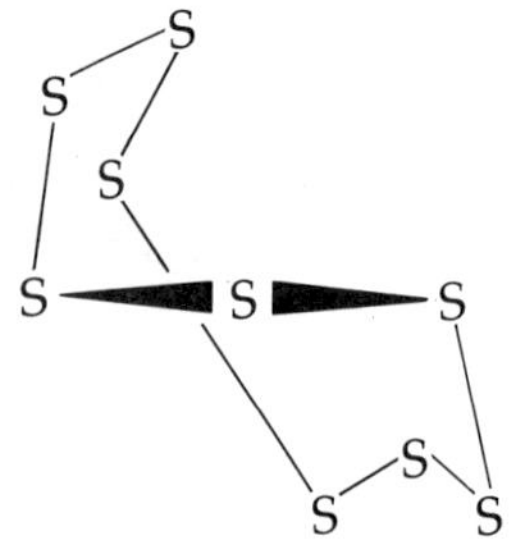

S_9 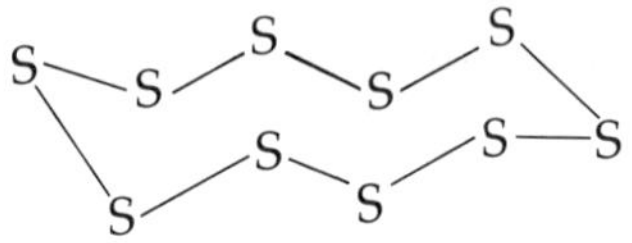 S_{10}

(*168*, *169*) that the D_2 conformation (which was not studied) exists in the crystal.

Of the still larger sulfur rings, only the well known cyclododecasulfur needs to be mentioned. It adopts a regular D_{3d} structure with alternating *cis–trans* bonds, according to both calculation and experiment.

It needs to be pointed out that the studies on large sulfur rings were cursory only (*167*), and conformations not considered may prove to be more stable than the ones examined. This has already happened in the case of cyclodecasulfur mentioned above.

Heats of formation, and, therefore, also the strain energies, of the sulfur homologues can be calculated quite accurately, and the calculations provide a valuable predictive method not only for the structures of the compounds, but also for the stabilities of yet unknown allotropes.

Force field calculations have also been carried out on a variety of sulfoxides (*170*). Electrostatic interactions are important in these compounds, and these, together with the usual kinds of interactions, were sufficient to reproduce the experimental data. Oxygen was treated as a van der Waals sphere in the force field, like the carbonyl oxygen. The calculated preference of thiane-1-oxides for the axial conformation was found with the force field used to be due to a repulsion of the equatorial oxygen by its vicinal hydrogens—not to an attraction of the axial oxygen

atom by its synaxial hydrogens as earlier supposed. The equatorial conformation of the sulfoxide oxygen becomes more favorable in 3,3-dimethylthian-1-oxide because of van der Waals repulsion from the syn-axial methyl group. In 1,3-dithiane-1-oxide, the same preference for the equatorial form is found experimentally and in the calculations, but here it is due to the electrostatic interactions of the ring thioether sulfur and the S–O bond dipole. Because of these two effects, the equilibrium in 5,5-dimethyl-1,3-dithiane-1-oxide is completely on the side of the equatorial conformation. The detailed study of nonchair conformations of several six-membered sulfoxide rings showed a preference for the chair form in all cases, but also a pseudorotational behavior of the nonchair forms that was quite different from that found in cyclohexane.

Other Heteroatoms

Another third-row atom whose incorporation into molecular mechanics force fields has been pursued in some depth is phosphorus. Even more than for sila derivatives, parameterization is hampered because of the scarcity of experimental data on geometries and energies of organic phosphorus derivatives. One force field for phosphorinanes which gives a reasonable fit with the available experimental data has been reported (*171*). For higher oxidation states of phosphorus, the difficulties in obtaining good fits to the experimental data are considerably more severe, and no force field for tetraligand derivatives (phosphoric acid derivatives with a doubly bound oxygen, for example) has yet been published. A force field approach to the stabilities and conformational interconversions of pentaligand phosphorus that demonstrates what topological problems are to be overcome for molecules with a valency higher than four has been reported (*172*). Pentavalent phosphorus compounds usually adopt a geometry with the ligands pointing to the corners of a trigonal bipyramid (TP), where the equatorial and axial PX

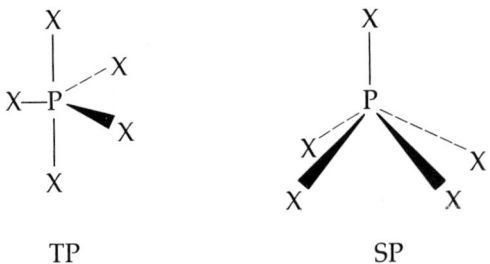

TP   SP

bond lengths and XPX bond angles are different. When ring strain is sufficiently unfavorable in the TP, a square pyramid (SP) geometry may be preferred, as in some catechol derivatives (*173*). (*See* opposite page.) This geometry is also discussed as an intermediate or transition state for the Berry pseudorotational ligand interchange (*174, 175*).

Two factors usually determine the preference of a ligand for one or the other of the positions in the TP, the electronegativity of the ligand (the more electronegative one prefers the axial position), and strain. The application of molecular mechanics to predict the ligand positions in the TP and its geometry, the TP ⇌ SP equilibrium, and the study of the Berry pseudorotation mechanism have been explored by Holmes et al (*172*).

In calculations on acyclic pentavalent phosphorus compounds, the only interactions that influence the calculated energies in the usual molecular mechanics model are the van der Waals interactions. For the molecular mechanics approach to be useful at all, it was found mandatory to include the 1,3-interactions to account for both the steric and electronegativity effects. But the 1,3-interactions are numerically much larger than the rest of the nonbonded interactions, and to avoid their completely overriding the others, Holmes et al. used a scaled 1,3-interaction potential with a weight of 10–20% of the ordinary van der Waals interactions. This idea was introduced earlier by Altona and Sundaralingam (*176*). (A Urey–Bradley force field would presumably give a similar result.) To include the electronegativity effect, the van der Waals interactions were further modified following VSEPR theory. Similar to the way the van der Waals center of hydrogen is shifted in hydrocarbon force fields, the van der Waals centers of the ligands were relocated along the P–X bonds, depending on the electronegativity of the ligand X. A first estimate of the relocation factor for every element possible as a ligand was calculated from the electronegativity difference, but the values finally used were optimized to fit experimental data of phosphoranes.

The number of stretching and bending parameters in phosphoranes is much larger than that in hydrocarbons, because one set of parameters applicable to both the TP and the SP geometries could not be found. It would seem that the topologically different positions in the molecule must have some different parameters, like a 120° strainless bond angle between equatorial ligands, and a 90° angle between equatorial and axial ligands. The PF_5 molecule, for example, will always adopt the geometry of the strainless parameters in a valence force field, no matter if this is the TP or the (high energy) SP geometry. For catechol derivatives, the molecule adopted a geometry close to SP even when the TP parameters were used, but with unsatisfactory precision. The degree to which the SP is approached can be guided to some extent by the 1,3-interactions.

It may still be possible that by a suitable choice of parameters, especially for the nonbonded interactions, a single set can be found that will work for both the TP and SP. The pseudorotation process that exchanges the ligands by the Berry mechanism has been simulated by varying the standard bond length and angle parameters continuously from the values for the TP to the SP. The latter geometry was calculated to be an energy minimum (an intermediate). This may, however, have been predetermined by the parameterization process.

A force field has also been developed for halophosphazenes (*177*), and the agreement obtained between calculated and experimental geometries, energies, and vibrational spectra of perhalophosphazene cyclic oligomers was taken as a sign that the bond properties here are independent of molecular size.

$$\{\underset{X}{\overset{X}{P}}=N\}_n$$

The structures of several heterotrypticenes containing Group IV or Group V elements (phosphorus, tin) have been calculated and used for the prediction of ^{13}C and ^{31}P NMR chemical shifts (*178*).

Finally, force fields have even been developed for octahedral metal complexes, and have been applied to calculate preferred conformational and configurational equilibria of chelate complexes, mostly of cobalt(III). The energy depends mainly on interactions within the chelating organic ligand, and, therefore, the force fields can again follow the general lines of hydrocarbon force fields. The complexing ligand atoms have either been nitrogen or oxygen. Rings containing five or six atoms (including the metal) have been studied, and the same conformations as in cyclohexane and cyclopentane were found to be possible (*179–190*).

Metal chelate complexes of 1,3-diaminopropane and related diamines with cobalt form six-membered rings whose chair and boat conformations can be combined to give sixteen different conformations. The most stable conformation of the *tris*-1,3-diaminopropane complex is

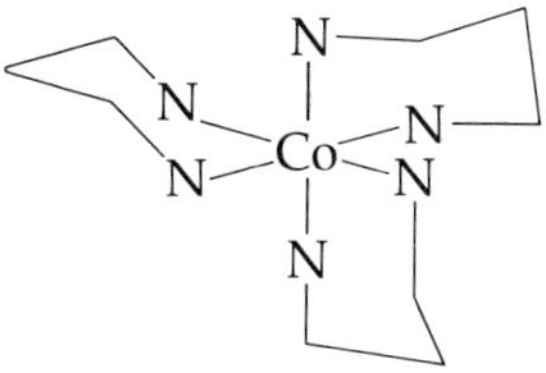

1

the *tris*-chair (**1**), and the same is doubtlessly true for the complexes that have equatorial methyl groups in the chair (*179–182*). Early calculations indicated that the complex formed by R,R-2,4-diaminopentane and Co^{3+} is more stable in the twist conformation with pseudoequatorial methyls (*179, 180, 181*), but later calculations gave a preference for the *tris*-chair form (*182*).

The complexes of diaminoethylene ligands were the subject of early force field calculations by Corey and Bailar (*183*). More recently, refined calculations with energy minimization have been performed on such complexes (*184–191*). Especially interesting are the complexes formed by multidentate ligands like triethylenetetramine. The complexes studied usually contained an additional, unsymmetrical ligand, and diastereomers were possible as in **2** and **3**. The good agreement found between the calculated and experimental geometries and energies indicates that the method is applicable in the field of metal chelates.

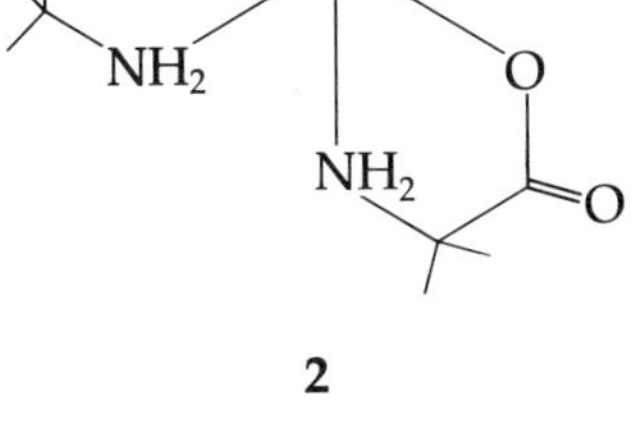

2 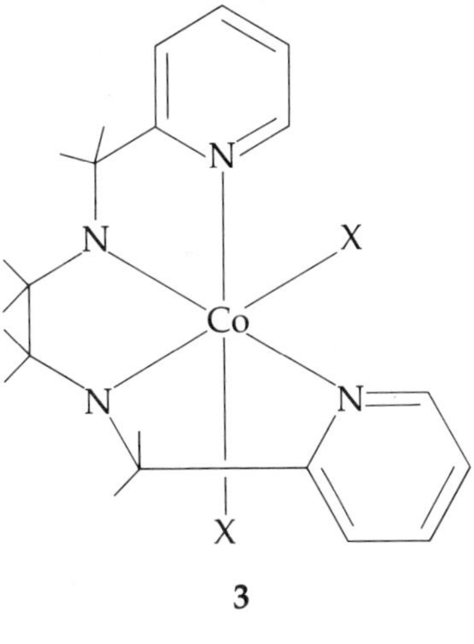   **3**

Some calculations on molecules containing the heavy Group IV atoms germanium, tin, and lead have been mentioned earlier on pages 85 and 205.

Literature Cited

1. Hiller, R. E., Jr.; Straley, J. W. *J. Mol. Spectrosc.* **1960**, *5*, 24.
2. Wiberg, K. B.; Wendoloski, J. J. *J. Am. Chem. Soc.* **1976**, *98*, 5465.
3. Miller, M. A., unpublished data.
4. Lifson, S.; Warshel, A. *J. Chem. Phys.* **1968**, *49*, 5116.
5. Ermer, O.; Lifson, S. *J. Am. Chem. Soc.* **1973**, *95*, 4121.
6. Williams, D. E. *Acta Crystallogr., Sec. A* **1974**, *30*, 71.
7. Collins, L. J.; Kirk, D. N. *Tetrahedron Lett.* **1970**, 1547.
8. Stolow, R. D. in "Conformational Analysis"; Chiurdoglu, G., Ed.; Academic Press: New York, 1971; p. 251.
9. Allinger, N. L.; Hirsch, J. A.; Miller, M. A.; Tyminski, I. J. *J. Am. Chem. Soc.* **1969**, *91*, 337.
10. Meyer, A. Y.; Allinger, N. L. *Tetrahedron* **1975**, *31*, 1971.
11. Del Re, G. *J. Chem. Soc.* **1958**, 4031.
12. Scheraga, H. A. *Adv. Phys. Org. Chem.* **1968**, *6*, 103.
13. Pople, J. A.; Beveridge, D. L. "Approximate Molecular Orbital Theory"; McGraw Hill: New York, 1970.
14. Dewar, M. J. S. "The Molecular Orbital Theory of Organic Chemistry"; McGraw Hill: New York, 1969.
15. Smith, R. P.; Ree, T.; Magee, J. L.; Eyring, H. *J. Am. Chem. Soc.* **1951**, *73*, 2263.
16. Smith, R. P.; Eyring, H. *J. Am. Chem. Soc.*, **1952**, *74*, 229.
17. Ibid., **1953**, *75*, 5183.
18. Smith, R. P.; Mortensen, E. M. *J. Am. Chem. Soc.* **1956**, *78*, 3932.
19. Smith, R. P.; Rasmussen, J. J. *J. Am. Chem. Soc.* **1961**, *83*, 3785.
20. Mark, J. E.; Sutton, C. *J. Am. Chem. Soc.* **1972**, *94*, 1083.
21. Allinger, N. L.; Wuesthoff, M. T. *Tetrahedron* **1977**, *33*, 3.
22. Dosen-Micovic, L.; Allinger, N. L. *Tetrahedron* **1978**, *34*, 3385.
23. Dosen-Micovic, L., Ph.D. Dissertation, Univ. of Belgrade, Yugoslavia, 1979.
24. Gasteiger, J.; Marsili, M. *Tetrahedron Lett.* **1978**, 3181.
25. Eliel, E. L.; Allinger, N. L.; Angyal, S. J.; Morrison, G. A. "Conformational Analysis"; Wiley-Interscience: New York, 1965.
26. Onsager, L. *J. Am. Chem. Soc.* **1936**, *58*, 1486.
27. Abraham, R. J.; Bretschneider, E. in "Internal Rotation in Molecules"; Orville-Thomas, W. F., Ed.; Wiley-Interscience: New York, 1974; p. 481.
28. Allinger, N. L.; Dosen-Micovic, L.; Viskosil, J. F., Jr.; Tribble, M. T. *Tetrahedron* **1978**, *34*, 3395.
29. Tribble, M. T.; Allinger, N. L. *Tetrahedron* **1972**, *28*, 2147.
30. Zalkow, V.; Frierson, M. R., unpublished data.
31. Ouellette, R. J.; Baron, D.; Stolfo, J.; Rosenblum, A.; Weber, P. *Tetrahedron* **1972**, *28*, 2163.
32. Matsuura, H.; Ohno, K.; Sato, T.; Murata, H. *J. Mol. Struct.* **1979**, *52*, 1.
33. Ibid., **1979**, *52*, 13.
34. Durig, J. R.; Natter, W. J.; Kalasinsky, V. F. *J. Chem. Phys.* **1977**, *67*, 4756.
35. Shen, Q.; Hilderbrandt, R. L.; Mastryukov, V. S. *J. Mol. Struct.* **1979**, *54*, 121.

36. Ouellette, R. J. *J. Am. Chem. Soc.* **1974**, *96*, 2421.
37. Jensen, F. R.; Bushweller, C. H. *Tetrahedron Lett.* **1968**, 2825.
38. Carleer, R.; Anteunis, M. J. O. *Org. Magn. Reson.* **1979**, *12*, 673.
39. Hummel, J. P.; Stackhouse, J.; Mislow, K. *Tetrahedron* **1977**, *33*, 1925.
40. Pfeiffer, M.; Spangenberg, H. J. Z. *Physik. Chem. (Leipzig)* **1966**, *232*, 47.
41. Ensslin, W.; Bergmann, H.; Elbel, S. *J. Chem. Soc., Faraday Trans. 2* **1975**, *71*, 913.
42. Bock, H.; Ensslin, W. *Angew. Chem.* **1971**, *83*, 435; *Angew. Chem. Int. Ed. Engl.* **1971**, *10*, 404.
43. Smith, Z.; Seip, H. M.; Hengge, E.; Bauer, G. *Acta Chem. Scand., Ser. A* **1976**, *30*, 697.
44. Carrell, H. L.; Donohue, J. *Acta Crystallogr., Sect. B* **1972**, *28*, 1566.
45. Stallings, W.; Donohue, J. *Inorg. Chem.* **1976**, *15*, 524.
46. Iroff, L. D.; Mislow, K. *J. Am. Chem. Soc.* **1978**, *100*, 2121.
47. Bartell, L. S.; Clippard, F. B., Jr.; Boates, T. L. *Inorg. Chem.* **1970**, *9*, 2436.
48. Ouellette, R. J. *J. Am. Chem. Soc.* **1972**, *94*, 7674.
49. Durig, J. R.; Lopata, A. D.; Groner, P. *J. Chem. Phys.* **1977**, *66*, 1888.
50. Hummel, J. P.; Zurbach, E. P.; DiCarlo, E. N.; Mislow, K. *J. Am. Chem. Soc.* **1976**, *98*, 7480.
51. Abraham, R. J.; Parry, K. *J. Chem. Soc. B* **1970**, 539.
52. Goursot-Leray, A.; Bodot, H. *Tetrahedron* **1971**, *27*, 2133.
53. Abraham, R. J.; Loftus, P. *Chem. Commun.* **1974**, 180.
54. Heublein, G.; Rühmstedt, R.; Kadura, P.; Dawczynski, H. *Tetrahedron* **1970**, *26*, 81.
55. Ibid., **1970**, *26*, 91.
56. Meyer, A. Y. *J. Mol. Struct.* **1977**, *40*, 127.
57. Ibid., **1978**, *49*, 383.
58. Meyer, A. Y. *J. Comput. Chem.* **1980**, *1*, 111.
59. Meyer, A. Y.; Allinger, N. L.; Yuh, Y. *Isr. J. Chem.* **1980**, *20*, 57.
60. Kuchitsu, K., personal communication.
61. Atkinson, V. A.; Hassel, O. *Acta Chem. Scand.* **1959**, *13*, 1737.
62. Ellestad, O. H.; Klaboe, P. *J. Mol. Struct.* **1975**, *26*, 25.
63. Abraham, R. J.; Rossetti, Z. L. *J. Chem. Soc., Perkin Trans. 2* **1973**, 582.
64. Lousalot, F.; Loudet, M.; Gromb, S.; Metras, F.; Petrissans, J. *Tetrahedron Lett.* **1970**, 4195.
65. Kirk, D. N. *Tetrahedron Lett.* **1969**, 1727.
66. Meyer, A.; Burkert, U. *J. Mol. Struct.*, in press.
67. Burkert, U. *Tetrahedron* **1977**, *33*, 2237.
68. Ibid., **1979**, *35*, 209.
69. Allinger, N. L.; Chung, D. Y. *J. Am. Chem. Soc.* **1976**, *98*, 6798.
70. Allinger, N. L.; Chang, S. H.-M.; Glaser, D. H.; Hönig, H. *Isr. J. Chem.* **1980**, *20*, 51.
71. Plyamovatyi, A. K.; Dashevski, V. G.; Kabachnik, M. I. *Dokl. Akad. Nauk SSSR* **1977**, *234*, 1100.
72. Tvaroska, I.; Bleha, T. *Coll. Czech. Chem. Comm.* **1978**, *43*, 922.
73. Tvaroska, I.; Bleha, T. *Biopolymers* **1979**, *18*, 2537.
74. Allinger, N. L.; Hickey, M. J. *J. Mol. Struct.* **1973**, *17*, 233.
75. Allinger, N. L.; Tribble, M. T.; Miller, M. A. *Tetrahedron* **1972**, *28*, 1173.

76. Profeta, S., Jr., unpublished data.
77. Anet, F. A. L.; Chmurny, G. N.; Krane, J. *J. Am. Chem. Soc.* **1973**, *95*, 4423.
78. Allinger, N. L.; Tribble, M. T. *Tetrahedron* **1972**, *28*, 1191.
79. Allinger, N. L.; Lane, G. A.; Wang, G. L. *J. Org. Chem.* **1974**, *39*, 704.
80. Schubert, W.; Schäfer, L.; Pauli, G. H. *Chem. Commun.* **1973**, 949.
81. Askari, M.; Ostlund, N. S.; Schäfer, L. *J. Am. Chem. Soc.* **1977**, *99*, 5246.
82. Schäfer, L.; Chiu, N. S.; Askari, M. *J. Mol. Struct.* **1978**, *48*, 445.
83. Allinger, N. L.; Burkert, U.; De Camp, W. H. *Tetrahedron* **1977**, *33*, 1891.
84. Midgley, J. M.; Parkin, J. E.; Whalley, W. B. *J. Chem. Soc., Perkin Trans. 1* **1977**, 834.
85. Burkert, U.; Allinger, N. L. *Tetrahedron* **1978**, *34*, 807.
86. Allinger, N. L.; Wertz, D. H. *Rev. Latinamer. de Quim.* **1973**, *4*, 127.
87. Dunitz, J. D., in "Perspectives in Structural Chemistry"; Dunitz, J. D.; Ibers, J. A., Eds.; John Wiley & Sons: New York, 1968; Vol. II, pp. 1–70.
88. Alvik, T.; Borgen, G.; Dale, J. *Acta Chem. Scand.* **1972**, *26*, 1805.
89. Anet, F. A. L.; St. Jacques, M.; Henrichs, P. M.; Cheng, A. K.; Krane, J.; Wong, L. *Tetrahedron* **1974**, *30*, 1629.
90. Allinger, N. L.; Gorden, B. J.; Newton, M. G.; Norskov-Lauritsen, L.; Profeta, S., Jr., *Tetrahedron* **1982**, in press.
91. Allinger, N. L.; Gorden, B.; Profeta, S., Jr. *Tetrahedron* **1980**, *36*, 859.
92. Burkert, U. *Tetrahedron* **1979**, *35*, 1945.
93. Burkert, U., unpublished results.
94. Burkert, U. *Z. Naturforsch. Teil B* **1980**, *35*, 1479.
95. Allinger, N. L.; Yuh, Y., unpublished data.
96. Creswell, R. A. *Mol. Phys.* **1975**, *30*, 217.
97. Breed, H. E.; Gundersen, G.; Seip, R. *Acta Chem. Scand. Ser. A* **1979**, *33*, 225.
98. Dillen, J.; Geise, H. J. *J. Mol. Struct.* **1980**, *64*, 239.
99. Bocian, D. F.; Strauss, H. L. *J. Am. Chem. Soc.* **1977**, *99*, 2866.
100. Anet, F. A. L.; Degen, P. J.; Krane, J. *J. Am. Chem. Soc.* **1976**, *98*, 2059.
101. Mansson, M. *Acta Chem. Scand.* **1972**, *26*, 1707.
102. Wolfe, S. *Acc. Chem. Res.* **1972**, *5*, 102.
103. Allinger, N. L.; Hindman, D.; Hönig, H. *J. Am. Chem. Soc.* **1977**, *99*, 3282.
104. Jeffrey, G. A.; Pople, J. A.; Radom, L. *Carbohydr. Res.* **1974**, *38*, 81.
105. David, S.; Eisenstein, O.; Hehre, W. J.; Salem, L.; Hoffman, R. *J. Am. Chem. Soc.* **1973**, *95*, 3806.
106. Jeffrey, G. A.; Taylor, R. *J. Comp. Chem.* **1980**, *1*, 99.
107. Burkert, U. *Tetrahedron* **1979**, *35*, 691.
108. Pickett, H. M.; Strauss, H. L. *J. Am. Chem. Soc.* **1970**, *92*, 7281.
109. Kellie, G. M.; Riddell, F. G. *Top. Stereochem.* **1974**, *8*, 225.
110. Peters, J. A.; Bovée, W. M. M. J.; Peters-van Cranenburgh, P. E. J.; van Bekkum, H. *Tetrahedron Lett.* **1979**, 2553.
111. Burkert, U. *J. Comput. Chem.* **1980**, *1*, 285.
112. Chatani, Y.; Yamauchi, T.; Miyake, Y. *Bull. Chem. Soc. Jpn.* **1974**, *47*, 583.
113. Kobayashi, M.; Kawabata, S. *Spectrochim. Acta, Part A* **1977**, *33*, 549.
114. Dale, J.; Ekeland, T.; Krane, J. *J. Am. Chem. Soc.* **1972**, *94*, 1389.
115. Anet, F. A. L.; Degen, P. J. *J. Am. Chem. Soc.* **1972**, *94*, 1390.
116. Spelbos, A.; Mijlhoff, F. C.; Faber, D. H. *J. Mol. Struct.* **1977**, *41*, 47.

117. Spelbos, A.; Mijlhoff, F. C.; Renes, G. H. *J. Mol. Struct.* **1978**, *44*, 73.
118. Hönig, H., unpublished data.
119. Podo, F.; Nemethy, G.; Indovina, P. L.; Radics, L.; Viti, V. *Mol. Phys.* **1974**, *27*, 521.
120. Jochims, J. C.; Kobayashi, Y. *Tetrahedron Lett.* **1976**, 2065.
121. Hagler, A. T.; Huler, E.; Lifson, S. *J. Am. Chem. Soc.* **1974**, *96*, 5319.
122. McGuire, R. F.; Momany, F. A.; Scheraga, H. A. *J. Phys. Chem.* **1972**, *76*, 375.
123. Melberg, S.; Rasmussen, K. *J. Mol. Struct.* **1979**, *57*, 215.
124. Allinger, N. L.; Chang, S. H.-M. *Tetrahedron* **1977**, *33*, 1561.
125. White, D. N. J.; Sim, G. A. *Tetrahedron* **1973**, *29*, 3933.
126. Guy, M. H. P.; Sim, G. A.; White, D. N. J. *J. Chem. Soc., Perkin Trans. 2* **1976**, 1917.
127. Philip, T.; Cook, R. L.; Malloy, T. B., Jr.; Allinger, N. L.; Chang, S.; Yuh, Y. *J. Am. Chem. Soc.* **1981**, *103*, 2151.
128. Huisgen, R.; Ott, H. *Angew. Chem.* **1958**, *70*, 312.
129. Kojima, Y.; Kalo, N.; Terada, Y. *Tetrahedron Lett.* **1979**, 4667.
130. Profeta, S., Jr., Ph.D. thesis, Univ. of Georgia, 1978.
131. Crans, D. C.; Snyder, J. P. *Chem. Ber.* **1980**, *113*, 1201.
132. Kao, J.; Huang, T.-N. *J. Am. Chem. Soc.* **1979**, *101*, 5546.
133. Terui, Y. *J. Chem. Soc., Perkin Trans. 2* **1975**, 118.
134. Blackburne, I. D.; Katritzky, A. R.; Takeuchi, Y. *Acc. Chem. Res.* **1975**, *8*, 300.
135. Lambert, J. B.; Featherman, S. I. *Chem. Rev.* **1975**, *75*, 611.
136. Allinger, N. L.; Hirsch, J. A.; Miller, M. A. *Tetrahedron Lett.* **1967**, 3729.
137. Lide, D. R., Jr. *J. Chem. Phys.* **1957**, *27*, 343.
138. McKean, D. C. *Chem. Commun.* **1971**, 1373.
139. Wollrab, J. E.; Laurie, V. W. *J. Chem. Phys.* **1969**, *51*, 1580.
140. Krueger, P. J.; Jan, J. *Can. J. Chem.* **1970**, *48*, 3229.
141. Manocha, A. S.; Tuazon, E. C.; Fateley, W. G. *J. Phys. Chem.* **1974**, *78*, 803.
142. Durig, J. R.; Li, Y. S. *J. Chem. Phys.* **1975**, *63*, 4110.
143. Smith, W. L.; Ekstrand, J. D.; Raymond, K. N. *J. Am. Chem. Soc.* **1978**, *100*, 3539.
144. Allinger, N. L.; Burkert, U.; Profeta, S., Jr. *J. Comp. Chem.* **1980**, *1*, 281.
145. Aaron, H. S.; Ferguson, C. P. *J. Org. Chem.* **1975**, *40*, 3214.
146. Johnson, C. D.; Jones, R. A. Y.; Katritzky, A. R.; Palmer, C. R.; Schofield, K.; Wells, R. J. *J. Chem. Soc.* **1965**, 6797.
147. Leonard, N. J.; Coll, J. C.; Wang, A. H.-J.; Missavage, R. J.; Paul, I. C. *J. Am. Chem. Soc.* **1971**, *93*, 4628.
148. Coll, J. C.; Crist, D. R.; Barrio, M. D. C. G.; Leonard, N. J. *J. Am. Chem. Soc.* **1972**, *94*, 7092.
149. Wang, A. H.-J.; Missavage, R. J.; Byrn, S. R.; Paul, I. C. *J. Am. Chem. Soc.* **1972**, *94*, 7100.
150. Mastryukov, V. S.; Popik, M. V.; Dorofeeva, O. V.; Golubinskii, A. V.; Vilkov, L. V.; Belikova, N. A.; Allinger, N. L. *Tetrahedron Lett.* **1979**, 4339.
151. Vierhapper, F. W.; Eliel, E. L. *J. Org. Chem.* **1977**, *42*, 51.
152. Allinger, N. L.; Hickey, M. J. *J. Am. Chem. Soc.* **1975**, *97*, 5167.
153. Smith, D.; Devlin, J. P.; Scott, D. W. *J. Mol. Spectrosc.* **1968**, *25*, 174.

154. Barnes, A. J.; Hallam, H. E.; Howells, J. D. R. *J. Chem. Soc., Faraday Trans. Part 2* **1972**, *68*, 737.
155. Karakida, K.; Kuchitsu, K.; Bohn, R. K. *Chem. Lett.* **1974**, 159.
156. Hubbard, W. N.; Finke, H. L.; Scott, D. W.; McCullough, J. P.; Katz, C.; Gross, M. E.; Messerly, J. F.; Pennington, R. E.; Waddington, G. *J. Am. Chem. Soc.* **1952**, *74*, 6025.
157. Crowder, G. A.; Scott, D. W. *J. Mol. Spectrosc.* **1965**, *16*, 122.
158. McCullough, J. P.; Finke, H. L.; Hubbard, W. N.; Good, W. D.; Pennington, R. E.; Messerly, J. F.; Waddington, G. *J. Am. Chem. Soc.* **1954**, *76*, 2661.
159. Allinger, N. L.; Hickey, M. J.; Kao, J. *J. Am. Chem. Soc.* **1976**, *98*, 2741.
160. Bushweller, C. H.; Bhat, G.; Letendre, L. J.; Brunelle, J. A.; Bilofsky, H. S.; Ruben, H.; Templeton, D. H.; Zalkin, A. *J. Am. Chem. Soc.* **1975**, *97*, 65.
161. VanWart, H. E.; Shipman, L. L.; Scheraga, H. A. *J. Phys. Chem.* **1975**, *79*, 1428.
162. Ibid., **1975**, *79*, 1436.
163. Allinger, N. L.; Kao, J.; Chang, H.-M.; Boyd, D. B. *Tetrahedron* **1976**, *32*, 2867.
164. VanWart, H. E.; Cadinaux, F.; Scheraga, H. A. *J. Phys. Chem.* **1976**, *80*, 625.
165. Snyder, J. P.; Harpp, D. N. *Tetrahedron Lett.* **1978**, 197.
166. Askari, M. *Tetrahedron Lett.* **1979**, 3173.
167. Kao, J.; Allinger, N. L. *Inorg. Chem.* **1977**, *16*, 35.
168. Reinhardt, R.; Steudel, R.; Schuster, F. *Angew. Chem.* **1978**, *90*, 55.
169. Reinhardt, R.; Steudel, R.; Schuster, F. *Angew. Chem. Int. Ed. Eng.* **1978**, *17*, 57.
170. Allinger, N. L.; Kao, J. *Tetrahedron* **1976**, *32*, 529.
171. Allinger, N. L.; von Voithenberg, H. *Tetrahedron* **1978**, *34*, 627.
172. Deiters, J. A.; Galucci, J. C.; Clark, T. E.; Holmes, R. R. *J. Am. Chem. Soc.* **1977**, *99*, 5461.
173. Brown, R. K.; Holmes, R. R. *J. Am. Chem. Soc.* **1977**, *99*, 3326.
174. Berry, R. S. *J. Chem. Phys.* **1960**, *32*, 933.
175. Holmes, R. R.; Deiters, J. A. *J. Am. Chem. Soc.* **1977**, *99*, 3318.
176. Altona, C.; Sundaralingam, M. *J. Am. Chem. Soc.* **1970**, *92*, 1995.
177. Boyd, R. H.; Kesner, L. *J. Am. Chem. Soc.* **1977**, *99*, 4248.
178. Hellwinkel, D.; Schenk, W.; Blaicher, W. *Chem. Ber.* **1978**, *111*, 1798.
179. Niketic, S. R.; Woldbye, F. *Acta Chem. Scand.* **1973**, *27*, 621.
180. Ibid., **1973**, *27*, 3811.
181. Ibid., **1974**, *28*, 248.
182. Niketic, S. R.; Rasmussen, K.; Woldbye, F.; Lifson, S. *Acta Chem. Scand., Ser. A* **1976**, *30*, 485.
183. Corey, E. J.; Bailar, J. C. *J. Am. Chem. Soc.* **1959**, *81*, 2620.
184. Buckingham, D. A.; Sargeson, A. M. *Top. Stereochem.* **1971**, *6*, 219.
185. Buckingham, D. A.; Creswell, P. J.; Dellaca, R. J.; Dwyer, M.; Gainsford, G. J.; Marzilli, L. G.; Maxwell, I. E.; Robinson, W. T., Sargeson, A. M.; Turnbull, K. R. *J. Am. Chem. Soc.* **1974**, *96*, 1713.
186. Buckingham, D. A.; Dwyer, M.; Gainsford, G. J.; Janson Ho, V.; Marzilli, L. G.; Robinson, W. T.; Sargeson, A. M.; Turnbull, K. R. *Inorg. Chem.* **1975**, *14*, 1739.

187. Brubaker, G. R.; Euler, R. A. *Inorg. Chem.* **1972**, *11*, 2357.
188. Favas, M. C.; Kepert, D. L. *J. Chem. Soc., Dalton Trans.* **1978**, 793.
189. Niketic, S. R.; Rasmussen, K. *Acta Chem. Scand., Ser. A* **1978**, *32*, 391.
190. Hald, N. C. P.; Rasmussen, K. *Acta Chem. Scand., Ser. A* **1978**, *32*, 753.
191. Jeffrey, G. A.; Taylor, R. *J. Comput. Chem.* **1980**, *1*, 99.
192. Bovill, M. J.; Chadwick, D. J.; Sutherland, I. O.; Watkin, D. *J. Chem. Soc., Perkin Trans. 2* **1980**, 1529.
193. Jorgensen, F. S.; Snyder, J. P. *J. Org. Chem.* **1980**, *45*, 1015.
194. Boyd, D. B.; Lipkowitz, K. B. *J. Comput. Chem.* **1981**, *2*, 324.
195. Taylor, R. *J. Mol. Struct.* **1981**, *71*, 311.
196. Astrup, E. E. *Acta Chem. Scand., Ser. A.* **1980**, *34*, 85.
197. McClellan, A. L. "Tables of Experimental Dipole Moments"; W. H. Freeman & Co.: San Francisco, 1963.
198. Korp, J. D.; Bernal, I.; Watkins, S. F.; Fronczek, F. R. *Tetrahedron Lett.* **1981**, 4767.
199. Peters-van Cranenburgh, P. E. J. i.; Peters, J. A.; Baas, J. M. A.; van de Graaf, B.; de Jong, G. *Rec. Trav. Chim. Pays-Bas* **1981**, *100*, 165.
200. Alder, R. W.; Arrowsmith, R. J.; Casson, A.; Sessions, R. B.; Heilbronner, E.; Kovač, B.; Huber, H.; Taagepera, M. *J. Am. Chem. Soc.* **1981**, *103*, 6137.
201. Allinger, N. L.; Chang, S. H.-M., unpublished data.

7

Large Molecules

Introduction, Steroids

Although molecular mechanics can deal with much larger molecules than quantum mechanical calculations can handle, the application of the methods described in Chapters 2 and 3 to real macromolecules is limited. Computer programs have been written that will allow complete relaxation of all bond lengths, bond angles, and torsion angles, and they can deal with molecules containing up to a few hundred atoms. In connection with a block-diagonal, Newton–Raphson energy minimization routine, the geometry optimization of a steroid with roughly 70 atoms takes about 10 minutes with present large computers. Several molecular mechanics calculations on steroids have indeed been published in the literature, and the geometries obtained usually show good agreement with x-ray structural data (*1–7, 9*).

The molecular mechanics geometry (MM1) of androsterone (**1**) has been compared with the x-ray crystal structure and with the structure determined from a Dreiding model (*1*). The usual convex nature of the steroid toward the beta face is seen both in the crystal structure and in the calculation. It is much less evident in the Dreiding model, and, because of the long lever arm involved in the length of the steroid molecule, if the C/D part of the molecule is positioned correctly in the Dreiding model, the oxygen at C_3 is too high by about 0.9 Å. The MM1 calculation puts it in the correct position. Thus a Dreiding model gives a geometry that contains larger errors than is widely recognized at present. However, in another study (*2*), the amount of energy required to bend the steroid significantly was found to be small compared with the typical binding energies of steroids to their receptors. The importance of exact geometries in steroid–receptor interactions clearly seems to be worth further study.

Molecular mechanics calculations have served to interpret circular dichroism curves (*8*), ^{13}C NMR shifts in steroids (*6*), and the dipole moments of 3,17-diketo steroids (2) (9, 11). Conformational studies of steroids are numerous. Examples include the conformational study of C/D cis- and trans-fused Δ^8-11-keto steroids (3); their configurational equilibrium was reported recently (12).

Other calculations were carried out on the conformational equilibrium of the A ring in 4,4-dimethylandrostan-3-ones (4) and -androst-5-ene-3-ones (5) (9, 11, 13). (These calculations were mentioned also on page 214.) Earlier calculations (MM1) had indicated that the ring A chair conformation in the former compound (4a) was more stable than the twist-boat (4b), but only by 0.8 kJ/mol. The crystal structures of derivatives show a chair (9, 10), and the spectroscopic properties of the compound in solution are consistent with a chair (11). The MM2 force field gives slightly higher torsional barriers than did MM1, and it indicates that the chair form is more stable than the boat by 3.0 kJ/mol, which is consistent with experiment. On the other hand, the Δ^5 compound (5) was calculated to be predominately in the twist-boat conformation (5b), as found experimentally (11).

An α-hydroxyl group in the 6-position of 4 leads to an interaction of the *syn*-diaxial type with the α-methyl at C_4, which must amount to about 8 kJ/mol (14). The A-ring would therefore be expected to adopt a boat conformation to relieve this repulsion. The crystal structure (15) shows ring A to be in a boat conformation.

In other studies the energies of the A-ring conformers of $\Delta^{5(10)}$ steroids (16) and of steroids with B/C *cis* ring fusions (7) were calculated. Substituents may introduce enough strain into such compounds that the all-chair conformation (CCC) is less favored than the one with the B and C rings in boat conformations (CBB). The conformations were predicted correctly in cases in which the conformational energies differed by more than 8.4 kJ/mol, but in cases in which CCC and CBB forms have almost the same energy, the calculations (MM1) appear to favor the boat by about 4 kJ/mol too much (7), similar to the case of compound 4 discussed previously.

In principle the androstanes can exist as a total of 64 stereoisomers (6 asymmetric centers). Of the 16 that we might call the "more natural ones," namely those with the 18- and 19-methyl groups β, most would be of quite high energy. In fact, only four of them are known (17). The stereochemistry most commonly found in nature is the 5α, 8β, 9α, 14α or *trans-anti-trans-anti-trans* (6). This stereochemistry is a kinetic product resulting from the cyclization of squalene, but it is not the most thermodynamically stable isomer of the parent ring system. Rather, the 14β isomer (7) is the most stable, although the presence of a C_{17} side chain tends to stabilize the 14α isomer (6) (18). Equilibration experiments (17) gave relative energies of 7.5 kJ/mol, 0.0 kJ/mol, 11.3 kJ/mol, and 6.3 kJ/mol for the 5α 14α (6), 5α 14β (7), 5β 14α (8), and 5β 14β (9), in line with what we might expect from studies on the model methyldecalin and methylhydrindane ring systems (pp. 109–110) (19). (*See* next page for 6–9.) Calculations with a rather old force field reproduced this information semiquantitatively (17). (The 8α and 9β isomers were not known experimentally and have not been studied by molecular mechanics.)

The progesterone molecule contains an acetyl group at C17. The orientation of this side chain was studied many years ago by rotatory dispersion (157), and dipole moment measurements (158). Schmit and Rousseau (159) discussed the side chain conformations for this system,

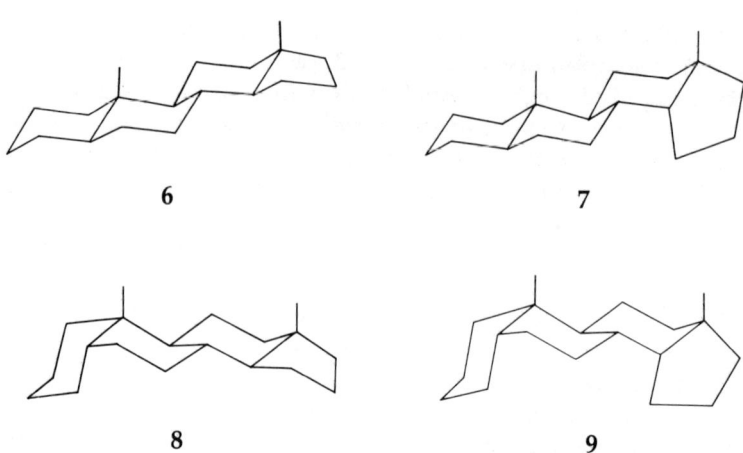

6

7

8

9

and made proposals on the basis of molecular mechanics calculations using the GEMO program (160). More recently Duax (161) examined crystallographic data on eighty-five compounds of this class, and concluded that the crystallographic data were in disagreement with those proposals. Still more recently, Profeta (162) has repeated these molecular mechanics calculations using the MM2 program, and found that there is no conflict between the x-ray data, the rotatory dispersion data and dipole moments, and his force field calculations. The only conflict is between the calculations of Schmit and Rousseau and all other data. Profeta concluded that Schmit and Rousseau had apparently used a force field that was inadequate for the task. The importance of using a well-developed force field to obtain reliable numbers cannot be overemphasized.

Another study dealt with the rotational isomerism in the side chain of the 20-epimers of 5α-pregnan-3β,20-diol (20). The side chain minimized nonbonded strain with the steroid backbone mainly by angle bending, not by torsion.

It was found experimentally (21) that the 11-oxo-9α-estradiol ring system (10a) is thermodynamically unstable with respect to the unnatural 9β isomer (10b), and treatment with base furnishes an equilibrium mixture (ΔG 6.15 kJ/mol). Molecular mechanics (MM1) showed (21) that the instability of the 9α isomer (10a) is a result of the unfavorable interaction between the carbonyl oxygen and the hydrogen at C1. In the 9β isomer (10b) the 9-hydrogen is equatorial and the bond from C9 to C10 is axial with respect to the C-ring. (See opposite page for 10a and 10b.) This puts the aromatic A-ring down almost perpendicular to the C-ring, and it completely eliminates the unfavorable repulsion.

                    10a   ⇌   10b

As the above examples illustrate, molecular mechanics can be considered an established method for the study of conformational problems in steroids, despite their size. However, with regard to the computer times required for geometry optimization, steroids are not typical of large molecules, because large torsion-angle adjustments are usually not necessary during energy minimization. More typical are cyclic peptides, or large hydrocarbon rings. These usually require up to 30 minutes of CPU time for geometry optimization from a poor starting geometry at this molecular size. For economical calculations with still larger molecules, simplifications have been introduced. In the following sections, the calculations discussed are mainly those in which bond lengths and angles, the so-called hard variables, have been kept at fixed values, and only torsion angles have been varied to optimize nonbonded and torsional energies. A common relationship in the three classes of compounds discussed here—carbohydrates, nucleosides/nucleotides, and peptides—is that calculations on dimers were used to predict the conformations of the corresponding polymers. Applications of the molecular mechanics potential functions to simple homopolymers like polyethylene are not considered in this book because the problems of polymer conformational analysis are quite different from those of the conformational analysis of small molecules or polypeptides. An excellent book is available on this topic (22). For more recent calculations on homopolymers see Reference 23.

Carbohydrates

Only a few of the calculations on conformational energies and anomeric equilibria in carbohydrates that have been undertaken since Reeve's pioneering computational studies (24) fall in the field of molecular mechanics. Most early calculations did not consider the precise molecular geometry, but only checked for the occurrence of *gauche* or *anti* interactions, like the well-known calculations of Angyal, which consist of the addition of interaction increments (25, 26). The first application of molecular mechanics potential functions and energy minimization to the conformational analysis of carbohydrates goes back to work from Ramachandran's group (27). The geometry of the carbohydrate ring, and all exocyclic bond lengths and angles, were kept at the fixed values usually found in x-ray crystal structures. Exocyclic torsion angles were optimized roughly, not employing a torsional potential function (27). The $^4C_1$ conformation of the monosaccharide α-D-glucopyranoside (11) was found in this first study to be at least 17 kJ/mol more stable than any other conformation.

This study immediately encountered the main problem in calculations of carbohydrate conformations, namely the very large number of

$^1C_4$ $^4C_1$

11

conformations that depend upon the position of the hydroxyl protons. The number of calculations necessary for a complete study (3^n, where n is the number of bonds about which three different conformations are possible) was regarded as unreasonable. In the course of time, various ways out of this dilemma were devised. In this first study (27), the authors determined what they regarded as the minimum energy hydroxyl conformation by the following procedure. First, the preferred position of the proton on the anomeric hydroxyl group was determined; then the proton on O2 was added, and the energy was minimized by incremental rotation of this proton about the C2–O2 bond. The same procedure was repeated in turn for each hydroxyl group around the ring. Then, again starting with the anomeric hydroxyl group, the position of the proton was readjusted, and after four to five iterations around the ring, the optimization converged (27). The same procedure was used for the optimization of the exocyclic conformation of the pyranose acetates. The minimum energy geometry obtained was independent of the initial geometry (28).

Rao and coworkers have refined the method in the course of calculations of conformational and configurational equilibria for a large number of pyranoses (29–35). The first calculations correctly predicted the preference for the $^4C_1$ conformation for all hexopyranoses. They also correctly predicted the direction of the $^4C_1 \leftrightarrows {}^1C_4$ equilibrium for pentopyranoses (29). The results were improved dramatically by allowing bond angle relaxation for the hydroxyl oxygen atoms, while in all of these calculations the pyranose ring was kept rigid. The axial substituents were found to tilt up to 4.5° (the CH_2OH group) out from the ring to minimize nonbonded strain (*30*). A further major improvement was found in an explicit estimation of the entropy for the conversion of the calculated enthalpies to the observed Gibbs free energies (*31*). The simple approach taken was to assume that the entropy differences occur only because of the restriction of the conformations of the hydroxyl groups, and that vibrational, translational, and external rotational entropy effects are identical in the different conformations. As a rule of thumb, only two conformations were allowed for axial hydroxyl groups (with the hydroxyl proton pointing out from the ring), but three were allowed for equatorial hydroxyl groups. The entropy term was obtained as $S = 2.3R \log P$, where R is the gas constant, and P is the number of possible (allowed) conformations (*31*). Although this procedure overestimates the energy difference between the three conformations for the axial hydroxyl groups, it gives a simple counting scheme for an estima-

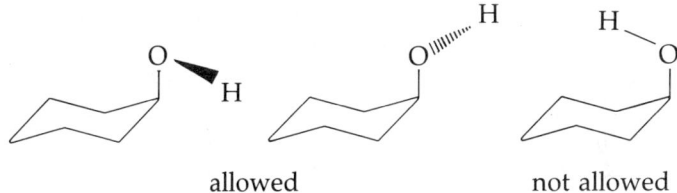

allowed   not allowed

tion of the entropy difference; it results in a significant improvement in the calculated equilibrium positions. Still, the calculations are not quite consistent because the enthalpy term used is not the weighted enthalpy over all conformations but just that of the minimum energy conformation.

The Gibbs energies obtained by this procedure were still at variance with the experimental equilibria and with the values calculated by Angyal's scheme (*25, 26*), but they could be brought into good agreement when a constant term was added to account for the axial preference of the anomeric hydroxyl (the anomeric effect). Earlier claims by the same group that the anomeric effect is accounted for automatically by the sum of the van der Waals and electrostatic interactions (*30*) are not correct (*31*).

The same method was employed for calculations on aldopyranose pentaacetates (*32*) and tetraacetates (*33*), as well as on methyl-D-aldopyranosides (*34*). The calculations on the acetates showed a less pronounced bending of the axial substituents out from the ring, away from the ideal geometry. This was interpreted to mean that the strain between synaxial acetate groups is much less than that between hydroxyl groups (*32, 34*). The addition of a term accounting for the anomeric effect was found to be necessary in all cases, but the results were not quite as close to the experimental values as those for the free sugars. The same approach was also used in a study of the isomeric cyclitols (35).

The entropy correction in calculations that allow for only two conformations of the axial hydroxyl groups, and likewise for the acetates, has been criticized by Rees and Smith (36), who point out that all conformations should be included in the calculation, and that the conformational and anomeric equilibria should be obtained from a Boltzmann distribution. However, the prohibitively large number of possible conformations induced these authors to employ statistical methods to calculate average energies for the conformations differing only in their sidechain conformations. The force field and energy minimization method used were very similar to that of Rao and co-workers and kept the ring fixed, but no extra term for the anomeric effect was employed. The conformations selected were chosen by a random sampling (Monte Carlo) technique, and the procedure is claimed to give a much better description of the entropy difference between the conformations and anomers. Good agreement between the calculated and the experimental conformational Gibbs free energies for most pentopyranoses lacking synaxial hydroxyl/hydroxyl interactions was found. An exception was D-xylose (12), which has three and four equatorial hydroxyl groups in the α and β forms, respectively, and exhibits a greater stability than calculated for the all-equatorial conformation. The error in the calculation here was attributed to solvation by water. The xylose molecules are thought to fit especially well into the solvent structure of water, which affords extra stabilization. The disagreement with experiment found in the case of pentopyranoses exhibiting 1,3-synaxial

α β

12

strain is thought to be caused by the rigid bond angles, which were not optimized in this study (36).

These calculations were extended by Dunfield and Whittington (37) in a study of hexopyranoses, optimizing not only torsion angles but also bond angles, keeping only the bond lengths fixed. This yielded geometries in remarkable agreement with those found in the crystal, with ring flattening for the sterically more crowded examples, and bond angle variations ascribed to the anomeric effect. The calculated conformational energies generally agreed very well with the experimental data, and where this was not the case, solvent effects were invoked as in the xylose case previously described.

None of these calculations of monosaccharide conformations is entirely satisfactory from the point of view of the energy minimization procedures, because none of them relaxed all of the internal coordinates. More general energy minimization methods have been applied only in studies aiming at the calculation of the molecular geometry. Lugovskoi and Dashevskii (38) calculated the geometries of several carbohydrates employing a Newton–Raphson method for geometry optimization as early as 1970. Kildeby et al. calculated the structures and energies of the chair, twist, and boat conformations of α- and β-D-glucopyranoside, also employing a Newton–Raphson method for energy minimization, but with ad hoc parameters in their force field (39). Side chain conformations were not optimized for all 729 conformations, but only for a set that was regarded as important by qualitative arguments, which numbered 21 for the α anomer and 8 for the β anomer. The geometry obtained was regarded as the global minimum for the gas phase, and possibly for nonpolar solvents as well. In the crystal, other exocyclic torsion angles are found, which is not surprising because in the crystal the exocyclic conformation is largely determined by intermolecular hydrogen bonds. The calculations include no terms accounting explicitly for the anomeric effect, which is known, for example, to cause an increased bond length for the exocyclic axial C1–O bond of α-D-glucopyranoside. This bond lengthening is not expected to be reproduced by the force field employed, because of the hyperconjugative nature of the anomeric effect, which is not allowed for in the force field.

The MM1 force field with its parameterization for ethers (pp. 217–222) has been applied to the calculation of the structures of eight 1,6-anhydropyranoses, and generally good agreement between the calculated and experimental (x-ray) structures was found (40). The application of the same force field to pyranosides (41) also gave satisfactory results, with the exception of the geometry at the anomeric carbon. The calculated C–O bond lengths are all very similar, unless different parameters are employed for the standard bond lengths of endocyclic C–O bonds and of C–O bonds equatorially and axially attached to the ring (41). The experimentally observed bond length differences are a consequence of the anomeric effect not accounted for in the original MM1 force field.

Conformational and anomeric energies of 2-deoxy-2-aminohexopyranoses (of glucose, galactose, and mannose) were calculated by an approximate treatment without energy minimization, and employing Rao's entropy scheme (42).

Recently, molecular mechanics calculations have been applied to determine geometries and energies of cyclic acetals formed by alditols (43) and by pentofuranoses (44). The alditol calculations gave generally good agreement between calculated and experimental conformational equilibrium positions, but disagreed with the relative energies obtained in equilibration experiments. However, the latter disagreement seems to be due mainly to solvent effects (the experiments were carried out in concentrated sulfuric acid, while the calculations refer to the gas state) (43). The second study (44) deals with the conformational, configurational, and constitutional equilibria of 2,3- and 3,5-O-alkylidene derivatives of pentofuranoses with a cis-2,3-diol (ribose and lyxose, 13–16). The high energy of trans-fused 3,5-O-alkylideneribofuranoses (14) was also found in the computations, but the lyxo derivatives are, according to the calculations, relatively stable in the cis-fused 3,5-form (16). The relative energies of the anomers were in all cases in full agreement with experiment, and the same was found for the relative energies of ribo- and lyxo-derivatives (44). Disagreements with experimental data exist for some C-glycosides, but these are probably caused either by

structural differences between the compounds studied experimentally and in the computations (X = carboxymethyl or cyanomethyl groups were approximated in the calculations by methyl groups), or by inadequate treatment of solvent effects (44).

Some studies of the conformations of ribose will be discussed in the following section (pp. 265–274) because of their close relationship to nucleoside and nucleotide conformations.

In the conformational analysis of di- and polysaccharides, the molecular mechanics method has been employed to study the conformations about the two C–O bonds linking the two monosaccharide residues, Φ and Ψ. In the earliest application of molecular mechanics to carbohydrate conformations, rotation about these bonds of a glucose molecule in an optimized geometry linked to a second such molecule was examined (27).

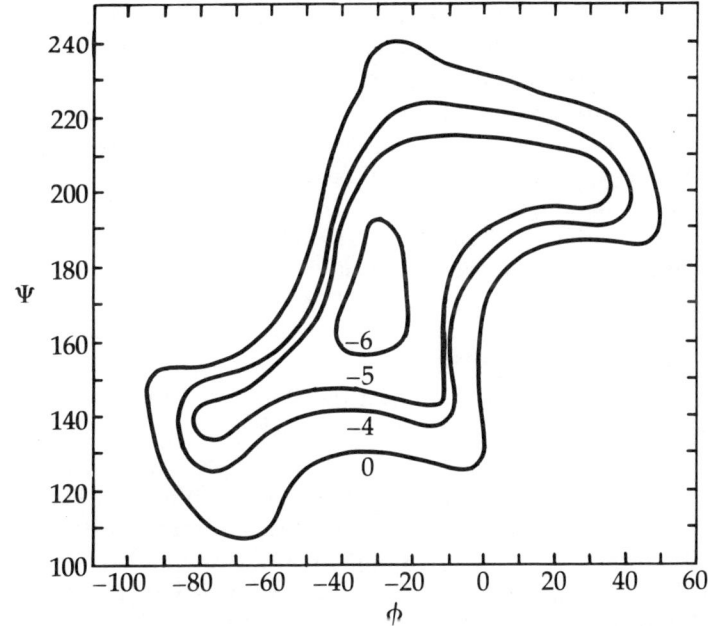

A two-dimensional conformational energy map (a Ramachandran map) may be used to diagram this process. Such a map is shown in Figure 7.1 for maltose (27). It shows an energy minimum at the torsion angles corresponding to those found in the amylose helix and in the cyclodextrine cyclohexaamylose. These facts indicate that the helix is an intrinsic property of the dimer unit, and does not result from a large number of long-range interactions in the polymer. Similar studies were reported for several other polysaccharides (45–51).

Academic Press

Figure 7.1. Energy contours in kcal/mol for maltose (27).

Rees and Smith also have extended their statistical method to di- and polysaccharides (52), adding a hydrogen-bonding potential to the force field, which is considered important in determining disaccharide conformations. A (Φ, Ψ)-potential energy map, again with the two C–O bonds of the glycosidic linkage as the variable parameters, shows good agreement between the predicted regions of minimum energy and the observed conformations of disaccharides and polysaccharides, and also with empirical studies on the allowed regions for amylose (27, 53–56). The number of hydroxyl conformations regarded as statistically significant could be diminished considerably by calculating the interactions between fixed atoms (the ring atoms and exocyclic oxygens) and floating atoms (hydroxyl protons) separately. If the energy of the fixed atoms was much higher than the energy calculated previously for other conformations, the conformation was omitted. Yet the computer time expended (12 hours of CPU time for the maltose calculation) seems prohibitive for a statistical treatment of sidechain rotational freedom in macromolecules.

Recent calculations employed a more detailed force field and full geometry optimization for the determination of the geometry of β-maltose (57), and β-gentiobiose (58) and this force field gave better agreement with experimental data (x-ray structure, NMR data) than did the earlier studies.

The experimental elucidation of the crystal and molecular structures of polysaccharides is hampered by the fact that these polymers crystallize in an imperfect manner, yielding overlapping diffraction patterns that give only low quality intensity data. The number of residues per helix turn in polysaccharide helices, and the height of its repeat unit along the helix axis are readily available from the x-ray data, but for a precise location of the atoms the intensity data are insufficient. Therefore, structure determination depends on a reasonable method of model building. In the early work, the problem was simplified by using monomers with a fixed geometry as found in crystal structures of the monomer molecules. Only the Φ and Ψ torsional angles of the glycosyl linkage and the arrangement of the helices in the crystal were varied. From energy contour maps of disaccharides, the most reasonable of these torsion angles are available.

Calculations for the generation of a starting geometry for optimization can be simplified considerably by the introduction of **virtual bonds** (59) [earlier called ring vectors (60)]. Under the assumption that the monomer unit is rigid (in bond lengths and angles), the O1–O4 distance is fixed, and the connecting vector is called the virtual bond. If the repeat height and the number of molecules per repeat unit are known for a polysaccharide, a model can be generated easily by constructing a sequence of virtual bonds, and only the rotational position of the monosaccharide units about this "bond" must be optimized. Restric-

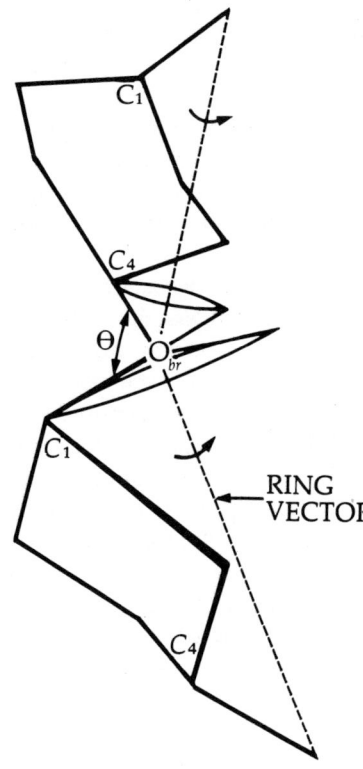

Figure 7.2. Dependence on $C_1O_{br}C_4$ bond angle, θ, on the rotation of the anhydroglucose residue about the ring vector (60). The bond angle is the angle between the surfaces of the two cones of rotation.

tions for the optimum geometry come not only from the (Φ, Ψ)-maps, but also from the COC bond angle at the oxygen atom connecting two virtual bonds. An example (60) is shown in Figure 7.2.

In this approach, the geometry is determined by molecular mechanics calculations, after x-ray crystallography defines the helix dimensions, and before it can be used for the determination of the chain packing. This approach, first used by Jones (61), was employed extensively by Marchessault and coworkers in the elucidation of the structure of several polysaccharides (59, 60, 62, 63).

More recently, an extension of these structure determination methods has been developed in which the geometry of the monomer unit is also optimized simultaneously on a molecular mechanics force field (to minimize strain) and on the experimental diffraction data (64–68).

Nucleosides, Nucleotides, and Nucleic Acids

Many calculations have been carried out to elucidate the conformations of nucleic acid fragments; these have employed potential functions for nonbonded interactions known from molecular mechanics. While the main aim has been to predict the conformations of nucleic acids, determination of the conformations of monomeric nucleosides of pharmacological interest also has been considered important. While early calculations were concerned mainly with the position of the aglycone with relation to the sugar ring (the conformation about the glycosyl bond), more recent calculations have focused on conformational energy maps of dinucleotides and their application to the determination of the helix geometries of polymeric nucleic acids.

This shift in the objectives of the calculations was accompanied by major improvements in the calculational techniques. The first calculations were of the hard-sphere type (69–71), but 6–12 van der Waals potential functions were introduced at a relatively early date. The parameters used for the potential functions were often those of contemporary force fields for peptides developed by Scheraga and coworkers (see pp. 274–278). In addition to van der Waals interactions, electrostatic (point charge) interactions and torsional potentials were used. Until recently (72) bond lengths and bond angles were not optimized. Analytical energy minimization has been utilized recently to optimize torsion angles (73). This seems mandatory in dinucleoside monophosphates where seven torsion angles had to be optimized simultaneously, which is impossible to do by hand. In more recent calculations on ribose (88, 153, 154), all internal degrees of freedom were optimized simultaneously.

Calculations of the rotational potential about the glycosyl bond reported by different groups (74–76) agree in general on the relationship between the sugar ring pucker and the allowed regions for the aglycone position in nucleosides. But, due to the rough approximations involved in the calculations (rigid rotation), the energies of the rotational barriers cannot be quantitatively correct, as was erroneously assumed in one study (77). Only rough information about the potential, restricted to the geometries near the energy minima, should be drawn from calculations at this stage. The calculations were carried out with the sugar ring in conformations now known to be close neighbors in the pseudorotational interconversion process (78). The C3'-endo and C2'-exo conformations are close to $^3T_2$, while C2'-endo and C3'-exo are close to $^2T_3$. (For definitions of nomenclature, see Ref. 79). Both conformations are probably in the same potential well, and the true energy minima are expected to lie somewhere between those proposed by these simple rigid sugar calculations.

For purine nucleosides the calculations indicate regions of low energy in the C3'-endo conformation at a glycosyl torsion angle [following the definition of Donohue and Trueblood (80) of about $\Phi_{CN} = -40°$ (anti) and $\Phi = +150°$ (syn)] (note that the different authors do not use identical definitions of the sign of Φ_{CN}!) (Figure 7.3). In the similar

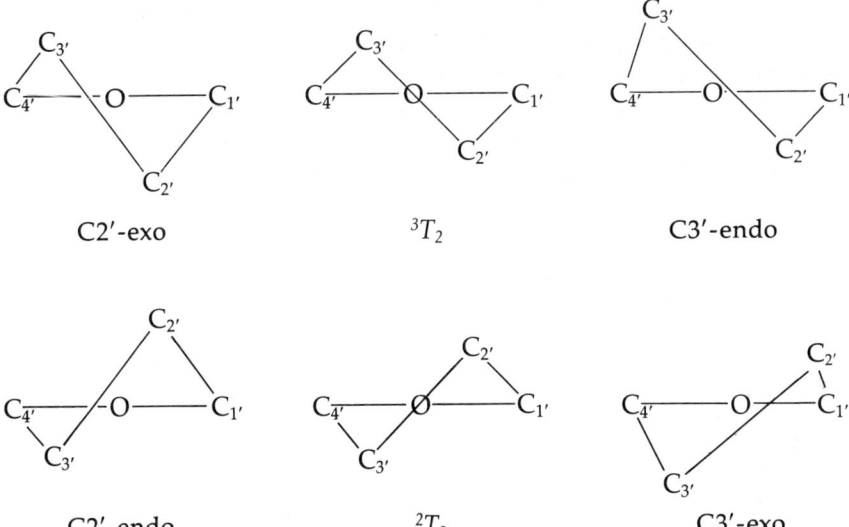

C2'-exo conformation, the energy comes to a minimum only at the *anti* conformation ($\Phi_{CN} \approx 0°$), with a slightly higher energy [~2 kJ (74)] than at the C3'-endo conformation. It was concluded that purine nucleosides would have both *syn* and *anti* conformations with similar probabilities, with a pucker in the sugar ring close to the $^3T_2$ form. On the other hand, it was calculated that for pyrimidine nucleosides with the C3'-endo sugar, the *anti* conformation would be about 10.5 kJ lower in energy than the *syn* minimum with the C2'-exo sugar, which is in agreement with the observed preference of these nucleosides for the *anti* conformation in the crystal.

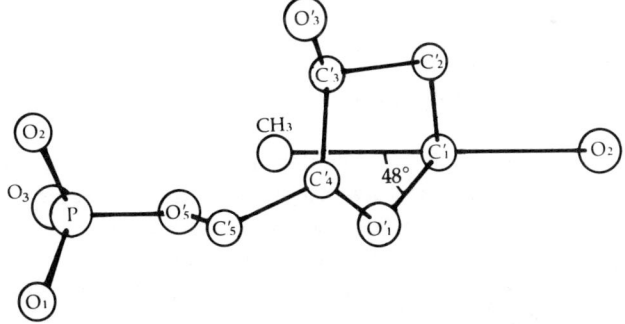

Figure 7.3. The conformation at the glycosyl bond of a nucleoside. Torsion angles follow the definition of Donohue and Trueblood (Φ here is $-48°$).

With the sugar in the C2'-endo conformation ($^2T_3$), the *syn* and *anti* conformations about the glycosyl bond were calculated to have comparable energies for both the purine and the pyrimidine nucleosides, while with the closely related C3'-exo conformation, the purines are again allowed in both conformations, but the pyrimidines were only *anti*. The energy difference of the minima in C2'-endo and C3'-exo is only about 2 kJ/mol (74).

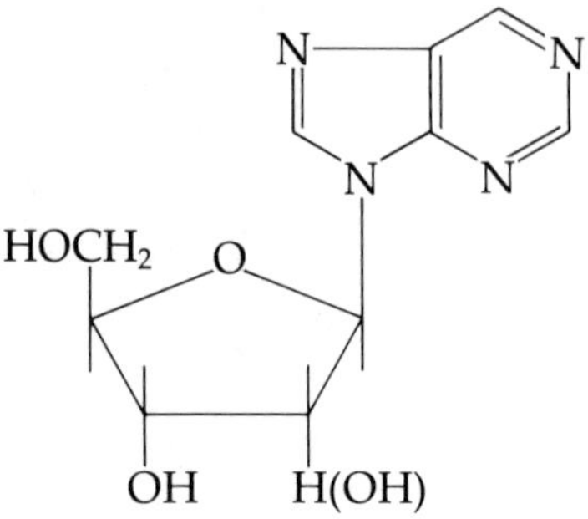

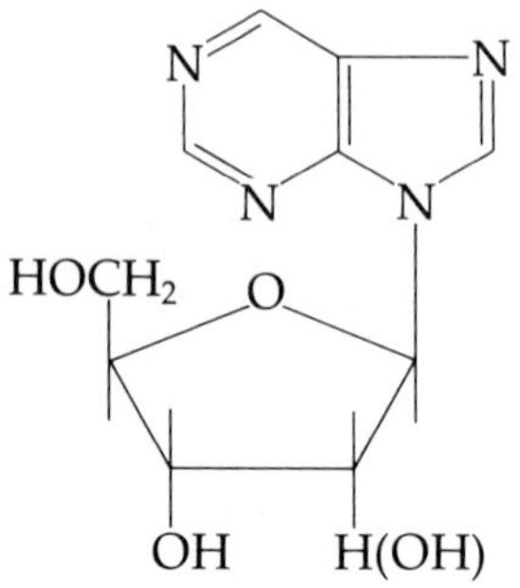

Purine nucleoside

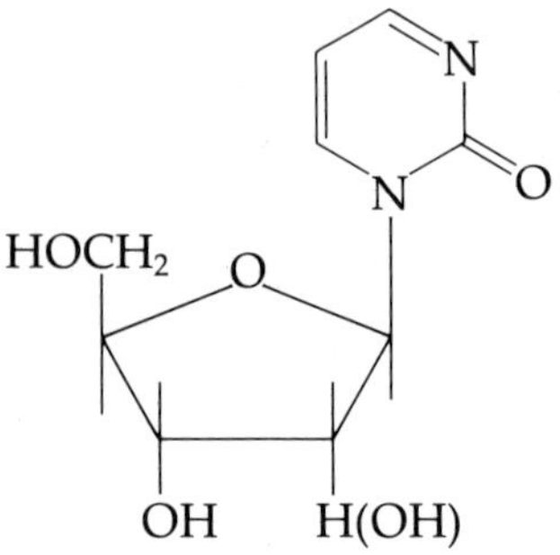

Pyrimidine nucleoside

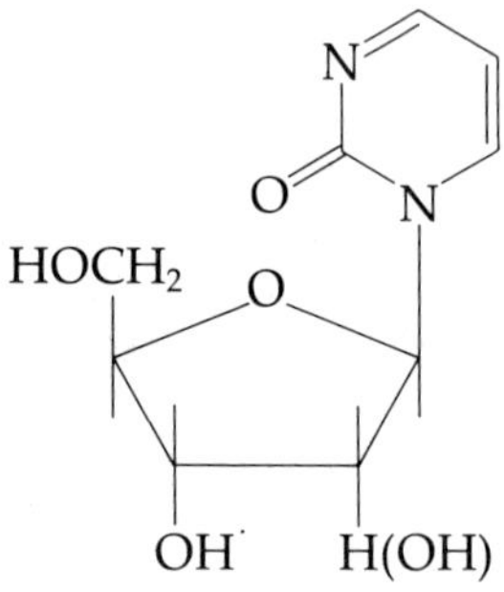

The conclusion drawn from these calculations (and from a large number of x-ray crystal structures of nucleosides) was that two regions of minimum energy exist for all nucleosides with regard to rotation about the glycosyl bond, *syn* and *anti*. Purine nucleosides have no preference for either the *syn* or *anti* conformation, regardless of the pucker in the sugar ring, but pyrimidine nucleosides have a preference for the *anti* form, at least when the sugar adopts a conformation approximating $^3T_2$. For pyrimidine nucleosides with $^2T_3$ sugars, one would conclude that the ring pseudorotates to a C2'-endo conformation if the pyrimidine is in the *syn* conformation. The geometry at the energy minimum depends mainly on van der Waals forces, and the good agreement between the results of these simple calculations and the trends observed in x-ray crystal structures of nucleosides confirms the model. As has often been observed in molecular mechanics calculations, electrostatic interactions are unimportant in determining the exact geometry of minimum energy, but may be quite important in determining conformational energies and rotational barriers.

The conformation about the C4'−C5' bond has also been studied as a function of the ring pucker. No preference for any one conformation was found with the sugar in C3'-endo and C2'-endo conformations, but a preference for the *gauche*$^+$ conformation, the one usually found in crystals of nucleosides, was calculated for the C3'-exo sugar pucker (75).

gauche$^+$

The influence of substituents at O2' on the conformation of nucleosides was studied using 2'-OCH$_3$ adenosine as a model (81). Some influence on the torsion angles about the glycosyl bond and the conformational equilibrium of C4'−C5' rotamers was found. However, the main effect of O2'-substituents in ribonucleic acids was proposed to be the restriction of the conformational freedom of the O3' phosphate group of the nucleotides due to van der Waals repulsions (81).

Later, the same computational methods were applied in conformational studies of mono- and dinucleotides, and generalizations were made to polynucleotides. Electrostatic interactions play a very important role because of the high charges in the phosphate group and the aglycone. A problem with calculations concerning nucleotides is that the torsional potentials about P−O bonds are not very well known. Usually a threefold potential was used for the torsional energy term, in addition to van der Waals and electrostatic interactions, although *d*-orbital participation might require terms of different periodicity (82).

Regardless of the pucker of the sugar ring (C2'-endo or C3'-endo), the conformation that has the optimum electrostatic interactions between the phosphate and the base has the *gauche*$^+$ conformation about the C4'−C5' bond with the base in the *anti* conformation. This conformation is much more favored in the polynucleotides than in the nucleosides. This result led Sundaralingam to propose that 5'-nucleotides exist in a rigid conformation (82, 83). Similar results had been found in earlier calculations, although with a smaller energy difference between conformations (*84*). A preference for the *anti* conformation was calculated for the phosphate group, as was also the case for 2'-O-methyl nucleotides (*81, 85*). At first, guanosine-5'-nucleotides seemed to be an exception to the rule, apparently favoring the *syn/gauche*$^+$ conformation over the *anti* (*86*). This result now seems to be caused by an overestimation of the attractions of the amino group in the earlier calculations, and it has been retracted (*87*).

In the studies discussed so far, the ribose unit was treated like a rigid ring. More recently studies have appeared dealing with the flexibility of the ribose ring, where full geometry optimization was taken into consideration (*88, 89, 153, 154*). Calculation of the pseudorotational potential of the ribose ring by Warshel and Levitt gave energy minima corresponding to $^3T_2$ and $^2T_3$ conformations, but the barrier obtained was lower than 2 kJ/mol, which speaks for extreme flexibility of the ribose unit in nucleosides (*88*). This, of course, sheds doubt on the results of calculations where the ribose ring was kept rigid. In another study (*89*), the conformational mobility around the C3'-O3' bond of nucleosides was studied as a function of the sugar ring pucker. Conformational energy maps derived in this way revealed several energy minima, with barriers no higher in nucleotides than in nucleosides. It was concluded, therefore, that the nucleotides are not as rigid as assumed earlier, but that the conformations found earlier (*83*) are simply at the (rather flat) energy minima of the present conformational energy maps (*89*). More recent calculations on ribose and deoxyribose have again indicated quite large barriers to pseudorotation (*153, 154*). A statistical analysis of these results and of the experimental data were interpreted in favor of a higher pseudorotational barrier than Warshel and Levitt suggested.

A study of the conformation of adenosine triphosphate (ATP) involved rigid rotations about the glycosyl bond, and all the bonds of the side chain, in 30° increments (*90*). The derived conformational energy maps were used to determine the global energy minimum of this multicoordinate problem. The authors report that the resulting geometry (Figure 7.4) is the same as that found both in the crystal and in a quantum mechanical study. It has the aglycone *anti*, the side chain CCO fragment *gauche/gauche*, and the P–O bonds mostly in the *gauche*$^-$ form.

The conformations about the glycosyl bond of uridine-2',3'-cyclo-oxyphosphorane, as well as of its complex with ribonuclease-S, have been calculated. The ribose ring is assumed to adopt the *O*-1'-endo conformation, and the *syn* and *anti* forms were found to have the same energy in the free nucleotide. However, the *anti* form has the lower energy in the complex (*91*).

Most natural di- and polynucleotides have a linkage between the 3' and 5' positions, although 5',5' bridging is also common, as in nicotinamide adenine dinucleotide (NAD$^+$). The preferred conforma-

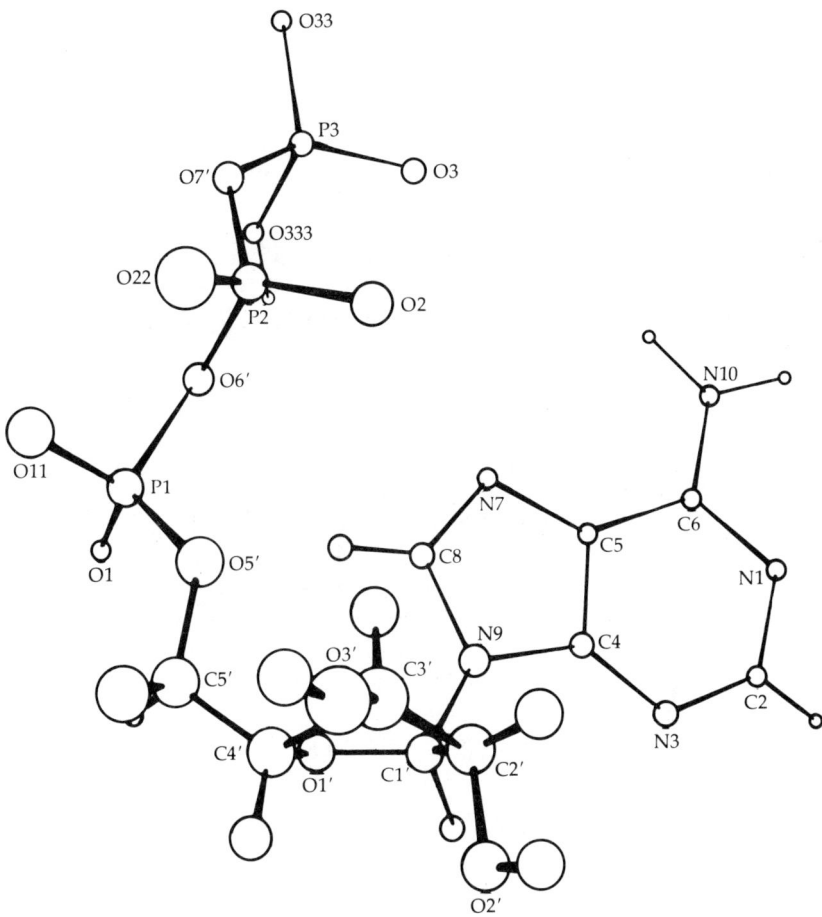

Figure 7.4. Calculated lowest-energy conformation of ATP (90).

tions have been calculated for 3'- and 5'-adenosine monophosphates, as well as for nicotinamidemononucleotide (NMN$^+$) (92), and the 5',5'-dinucleotides diadenosine pyrophosphate (93) and NAD$^+$ (94). The torsion angles were optimized by a simplex procedure, but bond lengths and angles were kept fixed. In the NMN$^+$ species, the strong electrostatic interactions between the aglycone and the phosphate residues determine the conformation. For the dimers, structures with stacked nucleobases were calculated to be the preferred ones, although other structures with relatively low energies were also found, especially in NAD$^+$, where the nicotinamide–phosphate interactions are strongly attractive. Solvation in polar solvents, neglected in these studies, is expected to favor unstacked structures, however. Thus for solutions, a distribution between folded and extended structures is predicted (94).

Nucleoside: $R_3 = R_5 = H$
3'-nucleotide: $R_3 = PO_3^{2-}$, $R_5 = H$
5'-nucleotide: $R_3 = H$, $R_5 = PO_3^{2-}$
3',5'-diphosphate: $R_3 = R_5 = PO_3^{2-}$

5',5'-dinucleoside pyrophosphate

3',5'-dinucleoside triphosphate

Conformational energy maps of dinucleoside phosphates have been very useful for the analysis of nucleic acid conformations. Such maps have been calculated by several research groups, mostly with a fixed sugar ring geometry (*73, 82, 95–102*), but more recently with optimized sugar ring pucker and bond angles (*72*). Most of these calculations concern dinucleoside monophosphates and deoxydinucleoside monophosphates. In one case the backbone of a triphosphate was calculated (with the bases omitted) (*82*). In the latter study the number of energy minima was smaller, and the potential wells were deeper, for triphosphates than for monophosphates. So the additional phosphates seem to restrict the mobility of the molecule. The authors therefore cautioned against generalization of the results of studies on monophosphates to nucleic acids (*82*).

Different single- and double-helix secondary structures have been discussed for deoxyribonucleic acids. The A form containing the ribose ring in the $^3T_2$ conformation (C3'-endo pucker) is also accessible to ribonucleic acids, while conformations with the sugar in the $^2T_3$ form (the B and Watson–Crick helices) were thought to be possible only for the deoxyribonucleic acids, because the O2' atom of the ribose ring suffers steric repulsions in the $^2T_3$ ring pucker. In most calculations on ribodinucleoside monophosphates (*73, 95, 100, 102–104*) and deoxyribodinucleoside monophosphates (*72, 89, 96, 101*), energy minima are found at the torsion angles necessary for these helix forms. For the ribo derivatives, a distinct preference for the A type helix was obtained (*100, 103*). In addition to these energy minima, however, several others show up on the energy maps, corresponding to extended conformations that are found experimentally. Furthermore, left-handed helices are calculated to be most favorable if the nucleobases are in the so-called high-*anti* conformation about the glycosyl bond (*98, 103*). Refined right- and left-handed double helix models have been derived from calculations, which also fit x-ray diffraction data well (*89*). All these calculations considered only helices that are continuous repetitions of identical mononucleotide conformations. Helices like the Z-DNA found recently (*105*), in which a dinucleotide with different conformations of the two consecutive mononucleotides is the repeating unit, have not been considered calculationally so far.

Another result of the calculations was that the preference for one or the other helix depends on the nucleoside sequence in the nucleic acid. Base stacking and base pairing in different helix types generated from the dinucleotides considered as energy minima were studied, and in most minimum energy conformations, base stacking with a large degree of overlap of the bases was found. Base pairing decreases the overlap because of the additional geometric restrictions introduced by the hydrogen bonds. Another consequence of these restrictions is the fact that single helices, with the better base stacking, pack more closely than can the double helices. Chains with the bases in the syn conformation do not form double helices at all, but left-handed helices can be formed with the bases in the high-*anti* conformation (*103*).

In a recent study, the mechanism of intercalation of the drugs ethidium, 9-aminoacridine, and proflavine into dinucleotides as models of nucleic acids has been examined using force field calculations (*152*).

Peptides and Proteins

The many energy minimization calculations on peptides based on molecular mechanics that have appeared over the years could fill a book of their own. This chapter can attempt only to outline some important aspects of the subject, while a number of review articles are recommended for the details and many examples of these calculations (*106–115*). The spontaneous refolding of a denaturized peptide to its native conformation has led to the postulate that the amino acid sequence contains sufficient information to determine the full conformation of the polypeptide, in regular (helix, sheet) and coil parts of peptide chains. Still, it is a matter of dispute if this native conformation corresponds to the global minimum of the Gibbs free energy, or if it also depends on the folding pathway, possibly under the influence of the ribosome, at least when it is in a partially synthesized state. Indeed, the time required for a reasonably large polypeptide to check all of the possible conformations and to choose the most stable one would be much longer than the time it actually needs to refold. It is usually assumed, however, that while the folding pathway is important, the resulting geometry probably corresponds to the global Gibbs energy minimum. This minimum is reached in separate stages by geometry optimization in limited regions, followed by adjustments to optimize long-range interactions.

Calculations on peptides are usually done with rigid bond lengths and angles, and in only a few studies (mainly on cyclic peptides) was the geometry fully optimized (*116–118*). In addition to those assumptions discussed in previous sections, an additional restriction is frequently introduced for peptides, namely, that the peptide bond is fixed in the planar *anti* conformation of Pauling and Corey (*119, 120*). With these assumptions, the only coordinates that can be varied (as for a polysaccharide) are two torsion angles per monomeric residue in the backbone and those coordinates that involve the conformation of the side chain. The justification for these assumptions is dubious because relaxations, which in such a rigid molecule are possible only at a high cost of torsional and van der Waals energies, may be possible with modest bond angle bending or torsion about the peptide bond. Unless adjusted manually, geometries with a *cis*-peptide bond cannot be reached at all. On the other hand, the optimization of all coordinates for large peptides would be extremely time-consuming with present computers and programs.

The potential functions needed for peptide calculations have been improved repeatedly—most recently by fitting the van der Waals parameters to obtain agreement with experimental crystal structure data on amino acids and oligopeptides (*121–123*). Of special importance in peptide calculations are the electrostatic interactions that result from the polar nature of the amino acids. These are usually treated in terms of point charge interactions and hydrogen bonds.

The question as to whether or not the electrostatic and ordinary van der Waals interactions are sufficient to reproduce experimental geometry and energy data of hydrogen-bonded systems has found opposing answers. Force fields from the Scheraga group contain additional van der Waals-type potential functions between hydroxyl and amino hydrogens and hydrogen-bond acceptors (*124*). In contrast, least-squares parameter optimization induced Hagler and coworkers to omit all van der Waals interactions of amide hydrogens (*122*), and, later, also of carboxyl hydrogens (*125*), for the calculation of the geometries of crystals of amides and carboxylic acids. (*See also* pp. 222–224 for a discussion of hydrogen bonding in alcohols.)

A disadvantage causing more uncertainty in peptide calculations than that resulting from the use of rigid amino acids is found in what has been termed the **multiple minima problem**. With current methods it is impossible to check each and every conformation for large peptides. Even if only two torsional angles are treated as variables, the number of minima involved is 3^{2n}, where n is the number of amino acid residues. If we adhere to the assumption that the native conformation is the global energy minimum, we must either find a method to start energy minimization initially in the correct potential well, or develop a method to determine the global energy minimum on an economical basis, passing over saddle points, from any specific starting geometry. Much work has been invested in the latter approach by Scheraga and coworkers. Although they investigated many different kinds of criteria for the movement on a multidimensional potential surface in an effort to locate the global minimum in a general way, they were not successful (*107, 126–130*). In the case of tetraglycine, for example, they found quite different global minima when starting with different geometries and applying different methods for approaching the global energy minimum (*127*). Scheraga also questioned whether, in case a method could be found that would pass over saddle points towards the global minimum, it would be too time-consuming to be applicable to peptides containing more than about ten amino acid residues (*131*).

The alternate approach, which consists in a well-founded guess at the approximately correct conformation, appears to be more promising. The known preferences of the building blocks of peptides, like di- or tripeptides, for certain conformations are used, assuming that the conformation is determined primarily by short-range interactions of the side chains on the amino acids and their own peptide backbone, but not by interactions between parallel backbones (108, 111). This approach is reminiscent of the methods used in calculations involving polysaccharides and nucleic acids. The potential energy maps necessary go back to hard-sphere calculations by Ramachandran and coworkers, who determined allowed and disallowed regions in the (Φ, Ψ)-conformational energy maps of amino acids (132). Such plots were later generated with soft-sphere van der Waals potentials, and also with different torsional energy functions, with increasing agreement between

the low-energy torsion angle combinations and the experimental structures of peptides [for example (119, 133, 134), see Figure 7.5 (119)]. Especially interesting, of course, are the energy minima at torsion angles corresponding to the helix and pleated sheet secondary structures. From the conformational energy maps of different combinations of amino acids, a classification of "helix-making" and "helix-breaking" amino acids is possible. In principle, this should allow for the prediction of the most probable conformations of the component fragments, which are then subjected to a minimization of the long-range interaction energies only.

A third approach to the multiple-minima problem avoids direct comparison of different conformations of the polypeptide, but seeks to simulate protein folding. This has been demonstrated by de Coen, assuming that folding leads to the native structure during protein synthesis on the ribosome (135). The conformation of the polypeptide chain here is determined mainly by the shape of a dipeptide conformational energy map and only secondarily by long-range interactions.

Protein folding has also been simulated by Levitt and Warshel (136–138) with a simple protein model, which is considerably simpler than the usual molecular mechanics potential. They added thermal

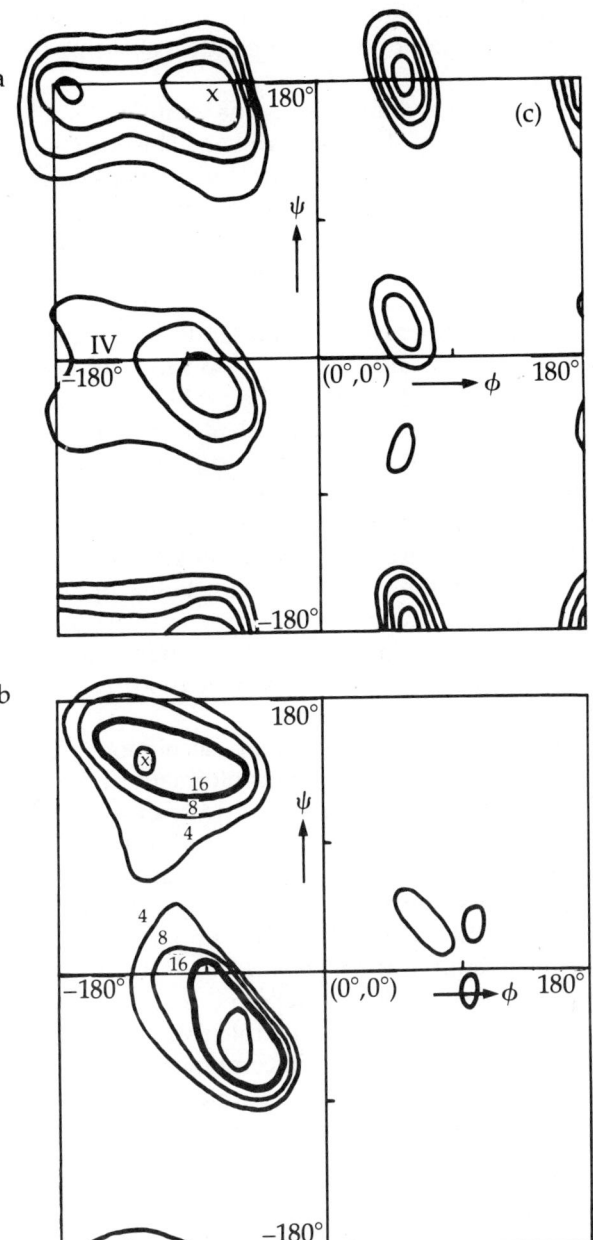

Figure 7.5. a, Energy map for internal rotation in an alanyl dipeptide; b, distribution of the conformations observed in a number of crystal structures of globular proteins (119).

energy to the peptide, and by solving the classical equations of motion they were able to map the trajectory of the folding process starting from an arbitrary trial geometry.

As stated above, the peptide chain adopts the conformation that minimizes the Gibbs free energy of the whole system. This means that not only the free energy of the peptide itself (including the effects of vibration and rotation about single bonds) but also that of the solvent must be considered. The generation of a cavity in water, and the hydration of the peptide, are both nontrivial in a peptide conformational energy calculation. This was realized early, and the structure of water and its relation to peptide conformations is presently under intense investigation (139). (For recent calculations on the solvation of nucleic acids, see Refs. 155 and 156.) However, neither the problem of multiple minima nor the shortcomings of solvent-effect calculations have hampered the application of the calculational method to the refinement of x-ray crystal structures of peptides, as obtained at low resolution. Studies of this type have been quite numerous (for a review, see Reference 110). These calculations were done using rigid bond lengths and angles. But because in a large molecule the torsional angles being optimized have very long lever arms, small changes in torsion angles can lead to significant improvements in energy. Geometry optimization in all coordinates (by the block-diagonal Newton–Raphson method) can effect major improvements in geometry refinement (140).

Another area in which conformational energy calculations can be applied without the above limitations is in the conformations of cyclic peptides. The number of possible conformations is reduced considerably by the condition that the ring must be closed. Methods have been developed for a complete, systematic mapping of this conformational space in large rings, where the number of conformations is very large (141, 143–146). Full energy minimization has been employed (116) for rings containing two to six amino acids (116, 142–147); full energy minimization for a cyclododecapeptide was reported recently (148). A more limited application of molecular mechanics calculations makes use of conformational energy maps of dipeptides in combined experimental (e.g., NMR) and theoretical studies of cyclic peptide conformations [e.g., (149–151)].

Literature Cited

1. Allinger, N. L.; Tribble, M. T.; Yuh, Y. *Steroids* **1975**, *26*, 398.
2. Schneider, H.-J.; Gschwendtner, W.; Weigand, E. F. *J. Am. Chem. Soc.* **1979**, *101*, 7195.
3. Altona, C.; Faber, D. H. *Top. Curr. Chem.* **1975**, *45*, 1.
4. Schwenzer, G. M. *J. Org. Chem.* **1978**, *43*, 1079.

5. Schmit, J. P.; Rousseau, G. G. *J. Steroid Biochem.* **1977**, *8*, 921.
6. Schneider, H.-J.; Gschwendtner, W.; Weigand, E. F. *J. Am. Chem. Soc.* **1979**, *101*, 7195.
7. Boeyens, J. C. A.; Bull, J. R.; Tuinman, A.; van Rooyen, P. H. *J. Chem. Soc., Perkin Trans. 2* **1979**, 1279.
8. Lane, G. A.; Allinger, N. L. *J. Am. Chem. Soc.* **1974**, *96*, 5825.
9. Allinger, N. L.; Burkert, U.; DeCamp, W. H. *Tetrahedron* **1977**, *33*, 1891.
10. Midgley, J. M.; Parkin, J. E.; Whalley, W. B. *J. Chem. Soc., Perkin Trans. 1* **1977**, 834.
11. Burkert, U.; Allinger, N. L. *Tetrahedron* **1978**, *34*, 807.
12. Patterson, D. G.; Djerassi, C.; Yuh, Y.; Allinger, N. L. *J. Org. Chem.* **1977**, *42*, 2365.
13. Dougherty, D. A.; Mislow, K.; Huffman, J. W.; Jacobus, J. *J. Org. Chem.* **1979**, *44*, 1585.
14. Allinger, N. L.; Chang, S. H.-M.; Glaser, D. H.; Hönig, H. *Isr. J. Chem.* **1980**, *20*, 51.
15. Whalley, W. B.; Ferguson, G.; Khan, M. A. *J. Chem. Soc., Perkin Trans. 2* **1980**, 1183.
16. Bucourt, R.; Cohen, N. *Bull. Soc. Chim. Fr.* **1970**, 2015.
17. Allinger, N. L.; Wu, F. *Tetrahedron* **1971**, *27*, 5093.
18. Van Horn, A. R.; Djerassi, C. *J. Am. Chem. Soc.*, **1967**, *89*, 651.
19. Eliel, E. L.; Allinger, N. L.; Angyal, S. J.; Morrison, G. A. "Conformational Analysis"; Wiley-Interscience: New York, 1965.
20. Altona, C.; Hirschmann, H. *Tetrahedron* **1970**, *26*, 2173.
21. Liang, C. D.; Baran, J. S.; Allinger, N. L.; Yuh, Y. *Tetrahedron* **1976**, *32*, 2067.
22. Flory, P. J. "Statistical Mechanics of Chain Molecules"; Interscience: New York, 1969.
23. Mark, J. E. *Acc. Chem. Res.* **1979**, *12*, 49.
24. Reeves, R. E. *Adv. Carbohydr. Chem.* **1951**, *6*, 107.
25. Angyal, S. J. *Aust. J. Chem.* **1968**, *21*, 2737.
26. Angyal, S. J. *Angew. Chem.* **1969**, *81*, 172; *Angew. Chem. Int. Ed. Engl.* **1969**, *8*, 157.
27. Rao, V. S. R.; Sundararajan, P. R.; Ramakrishnan, C.; Ramachandran, G. N. In "Conformation of Biopolymers"; Ramachandran, G. N., Ed.; Academic: London, 1967; p. 721.
28. Sundararajan, P. R.; Marchessault, R. H. *Biopolymers* **1972**, *11*, 829.
29. Sundararajan, P. R.; Rao, V. S. R. *Tetrahedron* **1968**, *24*, 289.
30. Rao, V. S. R.; Vijayalakshmi, K. S.; Sundararajan, P. R. *Carbohydr. Res.* **1971**, *17*, 341.
31. Vijayalakshmi, K. S.; Rao, V. S. R. *Carbohydr. Res.* **1972**, *22*, 413.
32. *Ibid.*, **1973**, *29*, 427.
33. Vijayalakshmi, K. S.; Yathindra, N.; Rao, V. S. R. *Carbohydr. Res.* **1973**, *31*, 173.
34. Rao, V. S. R.; Vijayalakshmi, K. S. *Carbohydr. Res.* **1974**, *33*, 363.
35. Vijayalakshmi, K. S.; Rao, V. S. R. *Proc. Indian Acad. Sci., Sect. A* **1973**, *77*, 83.
36. Rees, D. A.; Smith, P. J. C. *J. Chem. Soc., Perkin Trans. 2* **1975**, 830.

37. Dunfield, L. G.; Whittington, S. G. *J. Chem. Soc., Perkin Trans. 2* **1977**, 654.
38. Lugovskoi, A. A.; Dashevskii, V. G. *Zh. Strukt. Khim.* **1972**, *12*, 112.
39. Kildeby, K.; Melberg, S.; Rasmussen, K. *Acta Chem. Scand. Ser. A* **1977**, *31*, 1.
40. Jeffrey, G. A.; Park, Y. J. *Carbohydr. Res.* **1979**, *74*, 1.
41. Jeffrey, G. A.; Taylor, R. *J. Comput. Chem.* **1980**, *1*, 99.
42. Taga, T.; Osaki, K. *Bull. Chem. Soc. Jpn.* **1975**, *48*, 3250.
43. Burkert, U. *J. Comput. Chem.* **1980**, *1*, 192.
44. Burkert, U.; Gohl, A.; Schmidt, R. R. *Carbohydr. Res.* **1980**, *85*, 1.
45. Lemieux, R. U. *Chem. Soc. Rev.* **1978**, *7*, 423.
46. Potenzone, R., Jr.; Hopfinger, A. F. *Carbohydr. Res.* **1975**, *40*, 323.
47. Rao, V. S. R.; Yathindra, N.; Sundararajan, P. R. *Biopolymers* **1969**, *8*, 325.
48. Yathindra, N.; Rao, V. S. R. *Biopolymers* **1970**, *9*, 783.
49. Goebel, C. V.; Dimpfl, W. L.; Brant, D. A. *Macromolecules* **1970**, *3*, 644.
50. Sathyanarayana, B. K.; Rao, V. S. R. *Biopolymers* **1972**, *11*, 1379.
51. Tvaroska, I.; Perez, S.; Marchessault, R. H. *Carbohydr. Res.* **1978**, *61*, 97.
52. Rees, D. A.; Smith, P. J. C. *J. Chem. Soc., Perkin Trans. 2* **1975**, 836.
53. Rao, V. S. R.; Yathindra, N.; Sundararajan, P. R. *Biopolymers* **1969**, *8*, 325.
54. Blackwell, J.; Sarko, A.; Marchessault, R. H. *J. Mol. Biol.* **1969**, *42*, 379.
55. Giacomini, M.; Pullman, B.; Maigret, B. *Theor. Chim. Acta* **1970**, *19*, 347.
56. Sathyanarayana, B. K.; Rao, V. S. R. *Biopolymers* **1971**, *10*, 1605.
57. Melberg, S.; Rasmussen, K. *Carbohydr. Res.* **1979**, *69*, 27.
58. Melberg, S.; Rasmussen, K. *Carbohydr. Res.* **1980**, *78*, 215.
59. Sundararajan, P. R.; Marchessault, R. H. *Can. J. Chem.* **1975**, *53*, 3563.
60. Sarko, A.; Marchessault, R. H. *J. Am. Chem. Soc.* **1967**, *89*, 6454.
61. Jones, D. W. *J. Polym. Sci.* **1958**, *32*, 371.
62. Settineri, W. J.; Marchessault, R. H. *J. Polym. Sci., Part C* **1965**, *11*, 253.
63. Sundararajan, P. R.; Marchessault, R. H.; Quigley, G. J.; Sarko, A. *J. Am. Chem. Soc.* **1973**, *95*, 2001.
64. Zugenmaier, P.; Sarko, A. *Biopolymers* **1976**, *15*, 2121.
65. Ibid., **1973**, *12*, 435.
66. Zugenmaier, P. *Biopolymers* **1974**, *13*, 1127.
67. Sarko, A.; Mugglin, R. *Macromolecules* **1974**, *7*, 486.
68. Zugenmaier, P.; Sarko, A. *Acta Crystallogr., Sect. B* **1972**, *28*, 3158.
69. Haschemeyer, A. E. V.; Rich, A. *J. Mol. Biol.* **1967**, *27*, 369.
70. Sasisekharan, V.; Lakshminarayanan, A. V.; Ramachandran, G. N. In "Conformation of Biopolymers"; Ramachandran, N., Ed.; Academic: New York, 1967; Vol. 2, p. 641.
71. Lakshminarayanan, A. V.; Sasisekharan, V. *Biochim. Biophys. Acta* **1970**, *204*, 49.
72. Broyde, S.; Wartell, R. M.; Stellman, S. D.; Hingerty, B. *Biopolymers* **1978**, *17*, 1485.
73. Broyde, S. B.; Stellman, S. D.; Hingerty, B.; Langridge, R. *Biopolymers* **1974**, *13*, 1243.
74. Lakshminarayan, A. V.; Sasisekharan, V. *Biopolymers* **1969**, *8*, 475.
75. Wilson, H. R.; Rahman, A. *J. Mol. Biol.* **1971**, *56*, 129.
76. Kang, S. In "Conformation of Biological Molecules and Polymers"; (Proc. Jerusalem Symp. Quant. Chem. Biochem. 5, E. D. Bergmann and B. Pullman, Eds.) Jerusalem (1973) p. 271.

77. Jordan, F. J. Theor. Biol. **1973**, 41, 375.
78. Altona, C.; Sundaralingam, M. J. Am. Chem. Soc. **1972**, 94, 8205.
79. Schwartz, J. C. P. Chem. Commun. **1973**, 505.
80. Donohue, J.; Trueblood, K. N. J. Mol. Biol. **1960**, 2, 363.
81. Prusiner, R.; Yathindra, N.; Sundaralingam, M. Biochim. Biophys. Acta **1974**, 366, 115.
82. Yathindra, N.; Sundaralingam, M. Proc. Natl. Acad. Sci. USA **1974**, 71, 3325.
83. Yathindra, N.; Sundaralingam, M. Biopolymers **1973**, 12, 297.
84. Lakshminarayanan, A. V.; Sasisekharan, V. Biopolymers **1969**, 8, 489.
85. Stellman, S. D.; Hingerty, B.; Broyde, S.; Langridge, R. Biopolymers **1975**, 14, 2049.
86. Yathindra, N.; Sundaralingam, M. Biopolymers **1973**, 12, 2075.
87. Sundaralingam, M. In "Structure and Conformation of Nucleic Acids and Protein-Nucleic Acid Interactions"; Sundaralingam, M.; Rao, S. T., Eds.; University Park Press: Baltimore, 1975; p. 487
88. Levitt, M.; Warshel, A. J. Am. Chem. Soc. **1978**, 100, 2607.
89. Battabiraman, N.; Rao, S. N.; Sasisekharan, V. Nature **1980**, 284, 187.
90. Millner, O. E.; Andersen, J. A. Biopolymers **1975**, 14, 2159.
91. Lipkind, G. M.; Karpeiskii, M. Y. Mol. Biol. **1978**, 12, 282.
92. Thornton, J. M.; Bayley, P. M. Biochem. J. **1975**, 149, 585.
93. Thornton, J. M.; Bayley, P. M. Biopolymers **1976**, 15, 955.
94. Ibid., **1977**, 16, 1971.
95. Yathindra, N.; Sundaralingam, M. Biopolymers **1973**, 12, 2261.
96. Hingerty, B.; Broyde, S. Nucleic Acids Res. **1978**, 5, 127.
97. Kister, A. E.; Dashevskii, V. G. Biopolymers **1976**, 15, 1009.
98. Fujii, S.; Tomita, K. Nucleic Acids. Res. **1976**, 3, 1973.
99. Calascibetta, F. G.; Dentini, M.; de Santis, P.; Morosetti, S. Biopolymers **1975**, 14, 1667.
100. Broyde, S. B.; Wartell, R. W.; Stellman, S. D.; Hingerty, B.; Langridge, R. Biopolymers **1975**, 14, 1597.
101. Thiyagarajan, P.; Ponnuswami, P. K. Biopolymers **1978**, 17, 533.
102. Kister, A. E.; Dashevskii, V. G. Mol. Biol. **1975**, 9, 443.
103. Olson, W. K. Biopolymers **1978**, 17, 1015.
104. Stellman, S. D.; Hingerty, B.; Broyde, S. B.; Subramanian, E.; Sato, T.; Langridge, R. Biopolymers **1973**, 12, 2731.
105. Wang, A. H.-J.; Quigley, G. J.; Kolpak, F. J.; Crawford, J. L.; van Boom, J. H.; van der Marel, G.; Rich, A. Nature **1979**, 282, 680.
106. Scheraga, H. A. Adv. Phys. Org. Chem. **1968**, 6, 103.
107. Scheraga, H. A. Chem. Rev. **1971**, 71, 195.
108. Scheraga, H. A. In "Current Topics in Biochemistry 1973" Anfinsen, C. B.; Schechter, A. N., Eds.; Academic: New York, 1974; p. 1.
109. Anfinsen, C. B.; Scheraga, H. A. Adv. Protein Chem. **1975**, 29, 205.
110. Robson, B.; Osguthorpe, D. J. Amino Acids, Pept. Proteins **1978**, 9, 196.
111. Scheraga, H. A. Pure Appl. Chem. **1973**, 36, 1.
112. Maxfield, F. R.; Scheraga, H. A. Biochemistry **1976**, 15, 5138.
113. Hopfinger, A. J. "Conformational Properties of Macromolecules"; Academic: New York, 1973.
114. Nemethy, G.; Scheraga, H. A. Q. Rev. Biophys. **1977**, 10, 239.

115. Ingwall, R. T.; Goodman, M. In "Int. Rev. Sci. Org. Chem." Ser. Two, Vol. 6; Butterworths: London, 1976; p. 153.
116. White, D. N. J.; Morrow, C. *Tetrahedron Lett.* **1977**, 3385.
117. Levitt, M.; Lifson, S. *J. Mol. Biol.* **1969**, *46*, 269.
118. Warshel, A.; Levitt, M.; Lifson, S. *J. Mol. Spectrosc.* **1970**, *33*, 84.
119. Ramachandran, G. N. In "Peptides, Polypeptides and Proteins"; Blout, E. R.; Bovey, F. A.; Goodman, M.; Lotan, N., Eds.; Wiley-Interscience: New York, 1974; p. 14.
120. Zimmerman, S. S.; Scheraga, H. A. *Macromolecules* **1976**, *9*, 408.
121. Nemethy, G.; Scheraga, H. A. *J. Phys. Chem.* **1977**, *81*, 928.
122. Hagler, A. T.; Huler, E.; Lifson, S. *J. Am. Chem. Soc.* **1974**, *96*, 5319.
123. Hagler, A. T.; Leiserowitz, L.; Tuval, M. *J. Am. Chem. Soc.* **1976**, *98*, 4600.
124. McGuire, R. F.; Momany, F. A.; Scheraga, H. A. *J. Phys. Chem.* **1972**, *76*, 375.
125. Lifson, S.; Hagler, A. T.; Dauber, P. *J. Am. Chem. Soc.* **1979**, *101*, 5111.
126. Crippen, G. M.; Scheraga, H. A. *Arch. Biochem. Biophys.* **1971**, *144*, 453, 462.
127. Crippen, G. M.; Scheraga, H. A. *J. Comput. Phys.* **1971**, *12*, 491.
128. Gibson, K. D.; Scheraga, H. A. *Proc. Natl. Acad. Sci. USA* **1969**, *63*, 9.
129. Gibson, K. D.; Scheraga, H. A. *Comput. Biomed. Res.* **1970**, *3*, 375.
130. Crippen, G. M.; Scheraga, H. A. *Proc. Natl. Acad. Sci. USA* **1969**, *64*, 42.
131. Gibson, K. D.; Scheraga, H. A. *Proc. Natl. Acad. Sci. USA* **1969**, *63*, 9, 242.
132. Ramachandran, G. N.; Ramakrishnan, C.; Sasisekharan, V. *J. Mol. Biol.* **1963**, *7*, 95.
133. Zimmerman, S. S.; Pottle, M. S.; Nemethy, G.; Scheraga, H. A. *Macromolecules* **1977**, *10*, 1.
134. Zimmerman, S. S.; Shipman, L. L.; Scheraga, H. A. *J. Phys. Chem.* **1977**, *81*, 614.
135. de Coen, J. L. *J. Mol. Biol.* **1970**, *49*, 405.
136. Levitt, M.; Warshel, A. *Nature* **1975**, *253*, 694.
137. Warshel, A.; Levitt, M. *J. Mol. Biol.* **1976**, *106*, 421.
138. Levitt, M. *J. Mol. Biol.* **1976**, *104*, 59.
139. Scheraga, H. A. *Acc. Chem. Res.* **1979**, *12*, 7.
140. Wertz, D. H., private communication.
141. White, D. N. J.; Morrow, C. *Comput. Chem.* **1979**, *3*, 33.
142. Karplus, S.; Lifson, S. *Biopolymers* **1971**, *10*, 1973.
143. Gō, N.; Scheraga, H. A. *Macromolecules* **1970**, *3*, 178, 188.
144. Niu, G. C.-C.; Gō, N.; Scheraga, H. A. *Macromolecules* **1973**, *6*, 91.
145. Gō, N.; Scheraga, H. A. *Macromolecules* **1973**, *6*, 273.
146. Ibid., **1973**, *6*, 525.
147. Urry, D. W.; Khaled, M. A.; Renugopalakrishnan, V.; Rapaka, R. S. *J. Am. Chem. Soc.* **1978**, *100*, 696.
148. Madison, V. *Biopolymers* **1978**, *17*, 1601.
149. Deber, C. M.; Madison, V.; Blout, E. R. *Acc. Chem. Res.* **1976**, *9*, 106.
150. Tonelli, A. E.; Brewster, A. I. R. *Biopolymers* **1973**, *12*, 193.
151. Madison, V. *Biopolymers* **1973**, *12*, 1837.
152. Nuss, M. E.; Marsh, F. J.; Kollman, P. A. *J. Am. Chem. Soc.* **1979**, *101*, 825.
153. Olson, W. K.; Sussman, J. L. *J. Am. Chem. Soc.* **1982**, *104*, 270.

154. Olson, W. K. *J. Am. Chem. Soc.* **1982,** *104*, 278.
155. Clementi, E.; Corongiu, G. *Biopolymers* **1981,** *20*, 551.
156. Ibid., **1981,** *20*, 2427.
157. Wellman, K. M.; Djerassi, C. *J. Am. Chem. Sóc.* **1965,** *87*, 60.
158. Allinger, N. L.; Crabbe, P.; Perez, G. *Tetrahedron,* **1966,** *22*, 1615.
159. Schmit, J.-P.; Rousseau, G. G. *J. Steroid Biochem.* **1978,** *9*, 909.
160. Cohen, N. C. *Tetrahedron* **1971,** *27*, 789.
161. Duax, W. L.; Griffin, J. F.; Rohrer, D. C. *J. Am. Chem. Soc.* **1981,** *103*, 6705.
162. Profeta, S., Jr.; Kollman, P. A.; Wolff, M. E., unpublished data.

Stereochemistry and Rates of Chemical Reactions

THE APPLICATIONS OF MOLECULAR MECHANICS calculations to studies of conformational equilibria and the dynamics of conformational interconversions have been described in preceding sections of this book. Similar calculations may be used to study chemical equilibria between stereoisomers, which can be compared directly by their steric energies, and other equilibria, where the participants must be compared through their heats of formation or strain energies (pp. 173–184). Thus molecular mechanics may be used directly to predict the product ratios of chemical reactions under *thermodynamic control*. However, chemists are often interested in product ratios obtained under *kinetic control*, and also in reaction rates themselves. This chapter deals with applications of molecular mechanics to the problems of the dynamics of chemical reactions.

Calculation of reaction rates on a molecular basis, following Eyring's theory of the activated complex, requires the calculation of the geometry and energy of a model of the transition state. While there have been many molecular mechanics calculations of conformational transition states in which the energy is maximized in one degree of freedom by geometry optimization (pp. 72–76), no calculations of the transition state of a chemical reaction by partial energy maximization have been reported. In principle, such a calculation is straightforward in quantum mechanics, where there is no basic difference between conformational interconversions and chemical transformations (1). In most cases, however, too little is known about transition states of chemical reactions to allow a reliable parameterization of a complete force field. Most studies, therefore, have looked at reaction *intermediates* as close as possible to the transition states in energy, and have optimized their force fields to obtain the best fit with experimental reaction rates.

The S_N2 Reaction

One reaction for which clearly defined ideas about the geometry of the transition state exist is the bimolecular nucleophilic substitution (S_N2). One of the very first applications employing the concepts of molecular mechanics was a calculation of steric effects on the rate of the S_N2 reaction of alkyl (methyl, ethyl, n-propyl, isobutyl, neopentyl, isopropyl, and t-butyl) bromides with ethoxide. The transition state was defined in these calculations by fixing the bond lengths [and, in the early work, also the angle (2)] of the carbon–halogen bonds. These calculations, published by Ingold and coworkers in 1946 (2), were very crude compared with those utilizing present force fields. A severe limitation was that the calculations were done by hand with no energy minimization attempted. The model consisted of a trigonal bipyramid with the nucleophile and the leaving group at the apical positions, and in a linear arrangement with the carbon at the center. The lengths of the bonds that are half-broken in this transition state (half-bonds) were taken as the average of the covalent and ionic radii of the interacting atoms. Therefore, the only adjustable parameters were the van der Waals properties of the atoms and the shape of the nonbonded interaction function. The function that was employed was criticized later as being too soft (3), but more surprising is the fact that the potential was not taken to be spherical. Rather, [contrary to present belief (4)] repulsions were softer at angles of less than 90° with the bond than at larger angles. The errors in the different assumptions and approximations appear to have canceled, however, and reasonable agreement was found between the calculated energies of the transition states and the relative rates of the reactions. Later the model was refined (5) by allowing for relaxation of the angles at which the groups were entering and leaving in the trigonal bipyramid [the "plastic model" (5)]. Applied to Finkelstein halide/halide exchange, this model led to improved agreement with experiment. The additional parameters necessary for this improvement have, however, evoked some criticism (3). More recent calculations on the same basis, but with different parameters in which no polar effects of the alkyl groups were necessary, confirmed the validity of the approach (6).

An application of sophisticated force fields and energy minimization techniques to the problem of S_N2 reaction rates has been reported (7). The force field was based on Schleyer's hydrocarbon force field, and halide parameters were derived from those of Allinger and Meyer. Full relaxation, not only of the alkyl groups, but also of the half-bond distances and angles was taken into consideration. Unlike the calculations discussed above, this approach does not calculate the structure of the transition state (which is an energy maximum in one degree of freedom, and which in earlier work was reached by fixing the half-bond distances), but rather it minimizes the energy to find what is calculationally an intermediate with the geometric characteristics of the transition state. In these calculations an excellent correlation was obtained between the energy differences of the halides and transition states for a series of primary halides that differed in the size of their alkyl group (the β-series). But going from methyl to t-butyl halides (the α-series), the experimental activation energies for the S_N2 reaction do not increase as much as those calculated (see Table 8.1). Therefore, it was concluded that methyl groups have an additional accelerating effect on the reaction rate, which is different from the steric effect. Hybridization and polar (inductive) effects have been proposed as the most probable causes of this acceleration (7).

This approach to the calculation of characteristics of S_N2 transition states has been further extended to the activation enthalpies of the S_N2 cyclization of ω-haloamines (8). Reaction rates varying over a range of nearly 10^9 were reproduced to within a factor of two. According to these calculations, the role of entropy effects on the rates of ring closures has generally been overestimated. Enthalpies, which were calculated by molecular mechanics, were also found to account for the majority of the gem-dialkyl effect, the enhancement of cyclization observed in alkylated systems (8).

Table 8.1

Calculated and Experimental Reaction Rates of the Nucleophilic Substitution (S_N2 Mechanism) of Alkyl Halides (7)

Compound	Relative Activation Enthalpy (kJ/mol)	
	Experimental	Calculated (7)
α-Series		
MeX	−10.13	−25.94
EtX	0.0	0.0
i-PrX	10.96	19.29
t-BuX	16.78	48.95
β-Series		
MeCH$_2$X	0.0	0.0
EtCH$_2$X	1.17	0.17
i-PrCH$_2$X	8.33	7.82
t-BuCH$_2$X	28.66	29.41

Reactions Involving Carbenium Ions

For solvolysis reactions going through carbenium ion intermediates, several rate-influencing factors are known, one of which is the change in strain from a tetrahedral halide, or tosylate, to a trigonal, planar carbenium ion. In a series of papers (9–12, 14–18), Schleyer and coworkers have shown that the reaction rates for many compounds can be very well correlated with this energy difference. They compared the steric energy calculated for the parent hydrocarbon with that for a carbenium ion with slightly shortened C^+–C bonds and a preference for the planar trigonal geometry ($\Theta_0 = 120°$). The initial work aimed at an explanation of the reaction rates of bridgehead derivatives, which are known to solvolyze much more slowly than open-chain analogs (9, 12, 13). Without taking account of hyperconjugation, solvation, or other effects, a close relationship between the observed solvolysis rates and the calculated strain increase on ionization was found. This can be interpreted as indicating that the differences in hyperconjugation and other effects throughout the series are negligible compared with the differences in steric effects. While this is perhaps not surprising, the present authors find it puzzling that the effects of solvation also appear to cancel out. Solvation energies are enormous numbers, so that even relatively small (percentagewise) changes in them should show up very clearly. The rather different molecular shapes involved certainly suggest that they would have quite different solvation energies. It was also shown that participation by cyclopropane rings is not important for carbenium ion stabilization in cases where the carbenium ion site is over the face of the cyclopropane ring, since the observed solvolysis rates in such molecules agreed very well with those predicted by calculations based only on strain (14, 15).

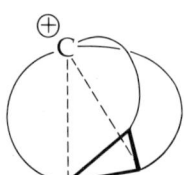

On the other hand, a case in which the electronic effects must be different from those in most other carbenium ions studied was found in 1, which has a solvolysis rate that is 10^9 times slower than predicted from molecular mechanics (10, 16). It was claimed that solvation and entropy effects could be excluded, so hyperconjugation, which depends on the torsion angles around the carbenium ion, remained as the sole explanation. In fact, the torsional situation around the reaction center is different from that in most other cases that were studied, as the bonds

1

are partially eclipsed. Later, compounds showing an increased solvolysis rate because of a strain *relief* when going from the tetrahedral starting material to the carbenium ion were added to the study (*19*). With very few such instructive exceptions, the calculated and experimental solvolysis rates differed by not more than a factor of two over a range of 25 orders of magnitude for a large number of examples (*18–20*).

In more recent work (*21*) the correlation obtained between relative solvolysis rates of alcohols in sulfuric acid and the strain energy difference between the parent hydrocarbon and carbenium ion was not as good, except for 2-adamantanols. The much increased solvolysis rate of p-nitrobenzoates as compared with the alcohols could be mimicked by replacing the leaving group (in the above studies, hydrogen) with a larger one (methyl) (*21*).

Solvolysis rates of cycloalkyl tosylates (5–12-membered rings) in trifluoroethanol have been correlated with the strain energy difference between a tosylate model (methylcycloalkane) and a trigonal intermediate model (cycloalkanone) (*22*).

The interconversion of hydrocarbons, via carbenium ions under equilibrating conditions that tend to yield the most stable isomer (stabilomer), is a problem in both thermodynamics and kinetics. In principle, the stabilomer can be found by molecular mechanics by calculating the heats of formation of all possible isomers. However, often there is a large number of possible structures. Group theoretical methods for the determination of the structures of all isomers of a C_nH_m hydrocarbon, first developed by Whitlock and Siefken (*23*), were used in a computerized form in the studies to be discussed (*24*). It is expedient to presort the structures on the basis of obvious arguments. Thus, for the compounds $C_{11}H_{18}$ there are 2889 possible isomers if compounds containing a methyl group are allowed. If all carbon atoms are to be members of a ring, only 434 isomers remain, and if all compounds containing a three- or four-membered ring are deleted as being too highly strained, only 43 isomers remain. Some of these can exist in several conformations, which in this case required the calculation of 69 different structures to find the stabilomer (*25*). But even after the application of such a selection procedure, often too many isomeric geometries result. A valuable approach then is to simulate the rearrangement process and construct the graph of intermediate carbenium ions, which often leads to the stabilomer in a small number of steps. For an intermediate carbenium ion, all possible reaction products that can be formed by a 1,2-alkyl shift are determined, and it is assumed that the carbenium ion with the lowest energy is the next intermediate, which serves as the starting point for the next step, and so on.

The first reaction to which this method was applied was the rearrangement of perhydrodicyclopentadiene to adamantane, the $C_{10}H_{16}$ stabilomer (26). The lack of experimental detection of intermediates in the interconversion was explained on the basis that the first rearrangement is the slowest one, which was consistent with the calculated heats of formation of possible intermediates. Therefore, all intermediates exist only in trace amounts.

This concept was also applied to the formation of the stabilomer of $C_{14}H_{20}$, diamantane, from tetrahydro-binor-S (24, 27), of ethanonoradamantanes $C_{11}H_{16}$ (28, 29), of tricycloundecanes $C_{11}H_{18}$ (25, 30), of ethanoadamantanes $C_{12}H_{18}$ (31), and of bisethanodiamantane $C_{16}H_{20}$ (32), and to the thermolysis of homocubanes (33).

Calculations were also carried out with Schleyer's force field on more flexible carbenium ions like the 1-methylcyclohexyl cation. The conformations obtained agreed with those derived from low-temperature ^{13}C NMR data (34).

Free Radical Reactions

The free radical is another type of reaction intermediate that usually prefers a more-or-less trigonal planar geometry. Molecular mechanics calculations have been used to determine the strain in free radicals in the same way as was done for carbenium ions. The decomposition rates of azo and peroxy compounds via radical intermediates were correlated with the energy differences between tetrahedral starting material and trigonal intermediate (20).

$$R_2-\underset{\underset{R_3}{|}}{\overset{\overset{R_1}{|}}{C}}-\underset{\underset{R_3}{|}}{\overset{\overset{R_1}{|}}{C}}-R_2 \longrightarrow 2\,R_1-\overset{\cdot}{C}\underset{R_3}{\overset{R_2}{\diagup}}$$

The range of thermolysis rates that has been observed for ethane derivatives corresponds to activation energies of 125–300 kJ/mol. In a series of papers by Rüchardt and coworkers (35–41) an excellent corelation was found between the total strain present in the ethanes and their thermolysis rates. It was not necessary to evaluate the strain remaining in the two radicals that form on thermolysis; they were taken to be strainless, which apparently is not a bad approximation in the cases examined. In connection with this study, the crystal structure of (n-Bu$_2$CPh—)$_2$ was determined (42). The central bond has a length of 1.638 Å, which is the longest so far known for a carbon–carbon single bond. (The MM2 value is 1.625 Å.)

Actually, most radicals are not completely strainless. The strain energies, heats of formation, and geometries of alkyl radicals have been studied by Profeta and Rahman (43). Conventional methods for studying structure (x-ray, electron diffraction, microwave) cannot be conveniently applied to studies on radicals, because radicals are too reactive. Nor can the usual thermochemical methods (heats of combination and hydrogenation) be used for radicals, for the same reason. Heats of formation of radicals can be determined by more roundabout methods (81), although the accuracy is much less than that for hydrocarbons. The structures of small radicals have been well-determined by ab initio methods (82, 83, 86). Thus sufficient information was available to permit formulation of a force field for alkyl radicals (43). Molecular mechanics studies will likely be especially fruitful in cases such as these, where the experimental procedures are applicable only with very great difficulty.

Addition-Elimination Reactions

Another type of reaction to which molecular mechanics calculations have been successfully applied is the carbonyl addition reaction, which involves interconversion of a trigonal carbonyl compound and a tetrahedral derivative. Thus it was possible to correlate the rates of ester formation (especially lactonization of carboxylic acids) and ester hydrolysis with the strain energy differences between the carboxylic acid and the intermediate ortho ester derivative. Studies by DeTar (44–47) showed that a considerable amount of information about relative reaction rates in molecules containing heteroatoms can be obtained from calculations on the related hydrocarbons. Lactonization is accelerated in some molecules by steric compression in the hydroxy acid, and a correlation was found between this steric acceleration and the decrease of strain in a similar hydrocarbon, where the hydroxyl group was replaced by a methyl, and the carboxyl by an isopropyl group (46). Similarly, the isopropyl group served as a model for the carboethoxy group,

and the t-butyl group for the ortho ester intermediate, in a study of ester hydrolysis rates. The calculated quantities related to the reaction rates were the differences in the heats of formation of the isopropyl- and t-butyl derivatives of the hydrocarbon, and a correlation was found, although with considerable scatter, probably caused in part by the very approximate model (44, 47). However, the agreement could be in-

$$R-C{\overset{O}{\underset{OEt}{\Big<}}} \longrightarrow R-C{\overset{OH}{\underset{OEt}{\underset{|}{\Big<}}}}OH$$

$$R-CH{\overset{CH_3}{\underset{CH_3}{\Big<}}} \longrightarrow R-C{\overset{CH_3}{\underset{CH_3}{\underset{|}{\Big<}}}}CH_3$$

creased dramatically by using an approximate force field for the ester and the adduct with parameters for oxygen. A very good correlation was obtained between the calculated differences of the steric energies of the esters and intermediates and the relative rates of reaction (47).

With a similar calculation, Winans and Wilcox (48) compared the influence of stereopopulation control and conventional steric effects on the lactonization of hydrocoumarinic acids. The large difference in the rates of lactonization of tetramethyl-o-hydroxyhydrocinnamic acid 2, and the parent unmethylated compound 3, had been ascribed earlier to an exclusive population of the conformation necessary for lactonization in 2 (49). In a detailed study including the calculation of the contribution of vibrations to the enthalpy (pp. 169–174), and the calculation of the entropy and Gibbs free energy of activation, Winans and Wilcox found complete agreement of their calculated Gibbs activation energies (with

2   3

the orthoacyl derivative as the transition state) with the experimental ones. It was pointed out that entropy effects account for nearly one-third of the rate difference. However, the observed acceleration is not due to stereopopulation control, because the latter results from the entropy of mixing of rotational isomers, while the calculated entropy difference is due to vibrations, possibly the torsional vibrations about the C–C bond between the benzene ring and the sidechain (48).

The rate of chromic acid oxidation of secondary alcohols to ketones has been the subject of many experimental studies. The formation of the chromic acid ester is fast compared with the oxidation step, which involves the removal of the hydrogen atom from the carbinol carbon. A question still unresolved is the geometry of the transition state of the reaction. If it is more similar to the ketone (a late transition state), the total strain difference between the alcohol and ketone determines the reaction rate. If it resembles the alcohol, and the carbon is still sp^3 hybridized (early transition state), the strain felt by the *iso*-hydrogen to be removed in the oxidation is more important (*50*). The difference in strain calculated for the ketone and (as a model for the alcohol) the methyl hydrocarbon correlated well with the observed relative reaction

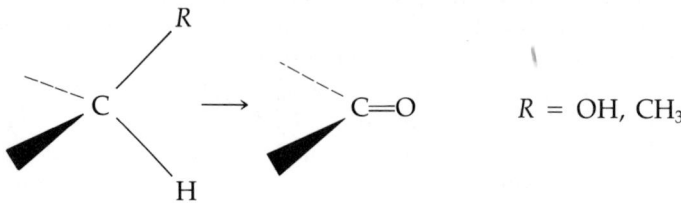

$R = OH, CH_3$

rate for a large number of compounds, which would be in good agreement with a late transition state (*51, 52*). However, the absolute rates of oxidation span a range of only about four orders of magnitude, which corresponds to relative activation enthalpy differences of only 24 kJ/mol, while the calculated strain differences between alcohol and ketone vary over 63 kJ/mol (*51, 52*). Thus, although the oxidation rates have a close relationship to the change of total strain, and the strain differences can be used to predict oxidation rates, the transition state is not close to the ketone. Hence only limited conclusions can be drawn about the geometry of the transition state from this correlation (*51, 52*). Recent studies have presented experimental values for alcohol ⇌ ketone rate and equilibrium constants (*77, 78*), and have compared such data with molecular mechanics (MM1) values (*78*). While fair correlations were found, there is clearly a need for improvement here. The MM2 force field, which is generally superior to MM1, and which has been paramaterized for alcohols and ketones, might give much better results.

The stereochemistry of nucleophilic additions to carbonyl groups has been a favorite in discussions of steric effects in chemical reactions ever since the Cram and Prelog rules were first proposed (*53*). For the reaction of a large nucleophile with any carbonyl compound, and for the reaction between a small nucleophile and a sterically congested carbonyl compound, the product ratio appears to be determined by steric approach control with an early transition state. The product ratio, which depends on the relative rates of reaction on the diastereotopic sides of the carbonyl group, can be correlated for large nucleophiles with the steric accessibility of the two sides (54). For small nucleophiles no such correlation exists, and product development control with a late transition state was postulated for most complex hydride reductions of carbonyl groups.

A weak point in the original Cram and Prelog rules was their foundation on the conformational equilibrium of the carbonyl compound, which violates the Curtin–Hammett principle (53). Instead, a model must compare the energies of the diastereomeric transition states that lead to the different products (55, 56). Such a late transition state is assumed in a molecular mechanics model by Perlberger and Müller (57, 58) for ketone reductions with borohydride. As in their work on alcohol oxidation (51, 52), their calculations employed hydrocarbon analogs because of the lack of oxygen parameters. The transition state was approximated by a pyramidal carbonyl center and a nucleophile attacking on a line perpendicular to the original carbonyl plane. The strain energy difference between the starting ketone and the transition state can be correlated with the reaction rates over eight orders of magnitude, and

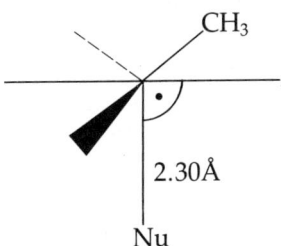

can also predict stereoisomer ratios for diastereotopic reductions. The geometry of the transition state was chosen to obtain the best agreement between the calculated and experimental reaction rates. (For a qualitative discussion of hydride reductions in these terms see also Reference 50.)

A different concept for the prediction of the product ratio, one which tries to cover the whole range of diastereotopic additions for different kinds of nucleophiles, was developed by Wipke and Gund (59, 60). Instead of assuming a product-like transition state for the reaction of a carbonyl group with small nucleophiles, they employed the model of Chérest and Felkin (61) in which steric approach control is included as previously, but the torsional strain of the incoming group is added as a second factor. To deal with reactions under steric approach control, a congestion function was also added. This function differs for the two sides of the carbonyl plane and is defined by a cone of accessibility around the axis perpendicular to the carbonyl plane. The size of this cone is determined by its avoidance of all atoms on the same side of the carbonyl plane. The cone angle, Θ, for an atom, i, just touches the van

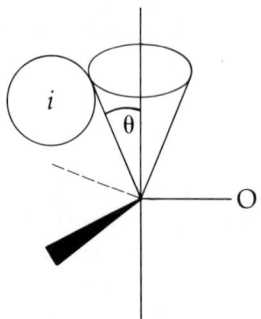

der Waals (hard) sphere of the atom. Accessibility due to atom i is then defined as $A_i = 2\pi r(1 - \cos \Theta)$, where r is a unit radius around the carbonyl carbon atom. Congestion of a side is the sum of the reciprocal accessibilities due to all atoms:

$$C = \sum_i \frac{1}{A_i}$$

The authors could correlate this fully empirical function with the product ratio observed for additions of small nucleophiles (complex hydrides) to sterically demanding carbonyl groups, but additions to unhindered carbonyl groups require the additional torsional correction function, $C_{tor} = 65 \cos(2.572\omega)$, be added to steric congestion, where ω is the torsion angle between the axis perpendicular to the carbonyl plane (the line of attack of the nucleophile) and the bonds to the α carbon atoms.

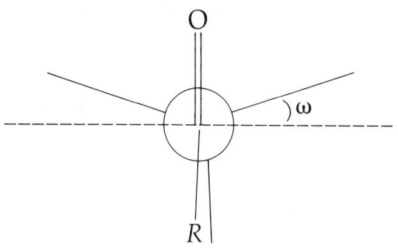

The model gives an excellent correlation with the observed stereoselectivities but does not give a good correlation of the relative reaction rates of different substrates. The authors have proposed, therefore, that their congestion function is related more to the activation entropy than to the activation enthalpy. To calculate the latter, we would have to know the position of the transition state along the reaction coordinate. This depends on both the substrate and the nucleophile, and is therefore different for every reaction, but not for the addition from the diastereotopic sides of the carbonyl plane (59, 60).

The importance of torsional energy contributions for the stereoselectivity of ketone reductions was further substantiated in a study in which the molecular geometries of cyclic ketones were calculated by molecular mechanics. Because of different heteroatoms in the ring, and different ring sizes, the equatorial hydrogen atoms α to the carbonyl group were either above or below the carbonyl plane. These geometries led to larger torsion angles (preferred torsional strain) between the incoming nucleophile and either the ring carbon or the axial hydrogen, and addition on the side with the larger torsion angle was always preferred (62, 80).

Treatment of 3-keto terpenoids and 3-keto steroids with benzaldehyde and base yields benzylidene derivatives; the rate of reaction depends on the substitution of the B and C rings, varying within two orders of magnitude, and this variation is said to result from conformational transmission (84, 85). This reaction was studied by molecular mechanics (63), and agreement of the relative rates found experimentally with the calculated energy difference between the 3-keto steroids and a properly chosen model for the transition state was found to be within a factor of three. The transition state was assumed to be close to the enolate formed in the rate-determining step, but the available force field did not allow calculations on carbanions to be carried out. The main structural feature of the enolate was thought to be the partial double bond between the carbonyl and the carbanion carbons, so a reasonable model for the transition state was found in the Δ^2-steroids. The correlation found is fair, with the observed scatter perhaps caused by neglect of differential solvation and vibrational effects.

Still and Galynker have recently published (79) a description of extensive studies aimed at obtaining highly stereoselective formations of new asymmetric centers. They examined kinetic enolate alkylations, dimethylcuprate additions (conjugate additions), and catalytic hydrogenations applied to a variety of monosubstituted 8- to 12-membered macrocyclic ketones and lactones. The MM2 force field was used, and semiquantitative prediction of the product distributions were provided in every case. The least motion reactions connecting starting material conformations with closely related product conformations were chosen in each case, and the product distributions were based upon either the starting material energies (early transition state model) or the product energies (late transition state model).

The final reaction we wish to consider in the group involving a change from trigonal to tetrahedral geometry is double-bond hydrogenation. Changes in bond-angle strain and in torsional strain during hydrogenation were calculated with a very simple force field in an investigation of the relative rates of diimide reductions of olefins (64). The reaction was simulated by going from a planar olefin through a range of intermediate structures to the cis-hydrogenated, saturated hydrocarbon by changing the standard bond angles at the double bond continuously from 120° to the tetrahedral value. Correlation between the observed reaction rates and the calculated angle-bending energy was best when the transition state was assumed to be where the CCC bond angle compression was 34% complete. The rate data covered a range of three orders of magnitude, and the calculated values showed a deviation of less than a factor of two, giving very satisfactory agreement with experiment, although the ad hoc optimization of the position of the transition state on the reaction coordinate may be open to criticism.

The reactions discussed in the preceding pages involve carbenium ions, radicals, carbonyl compounds, and olefins, and are characterized by transition states in between tetrahedral and trigonal geometries at the reaction centers. If we assume that the different substituents are more or less independent of each other, we can develop transferable substituent constants similar to Taft's parameters (65) to evaluate the steric effects of a substituent for accessibility and for carbenium ion or radical formation. Such parameters were developed by Beckhaus, who calculated the difference in the heats of formation of the compounds formed between a substituent residue and methyl, and between the same residue and t-butyl (66). These parameters could be successfully related to the relative rates of radical formation by halogen abstraction from CCl_4 and CCl_3Br (67). In these two papers, the steric requirement of a substituent group was treated as though it were independent of conformational effects. In the first paper (66) it was demonstrated that deviations, which may depend on such effects, exist between experimental and predicted reaction rates (of the classical reaction studied by Taft, acid-catalyzed esterification). A detailed topological study by Dubois et al. showed that indeed, this is the case (68). Linear correlations of steric effects (derived from other than molecular mechanics sources) and reaction rates are possible only within a series of carboxylic acids having the same (eclipsed) conformation. With increasing steric demand of the alkyl groups bound to the carbon of the carboxylic acid, staggered conformations become more favored. This was also found with model calculations on analogous methyl alkyl ketones where the resulting

eclipsed

staggered

correlations were nonlinear. The absolute numbers obtained for steric energies and torsion angles around the carbonyl group in this study were not discussed in any detail because of the approximate nature of the force field, but it was shown that conformational effects play a decisive role in the interpretation of Taft parameters.

Electrocyclic Reactions

The complete potential surfaces for the Cope rearrangements of some simple dienes have been simulated by molecular mechanics (69). In addition to the potential terms usually used in a molecular mechanics calculation, the expression for the energy included a Hückel treatment of the pi system and an energy term for the breaking and forming of the sigma bonds as a function of the distance between the bond-forming atoms. The absolute height of the activation barrier cannot be expected to be calculated accurately by such a simple model, but qualitative agreement was obtained for the relative rates of rearrangement of cyclohexa-1,5-diene, *cis*-1,2-divinylcyclopropane, and *cis*-1,2-divinylcyclobutane.

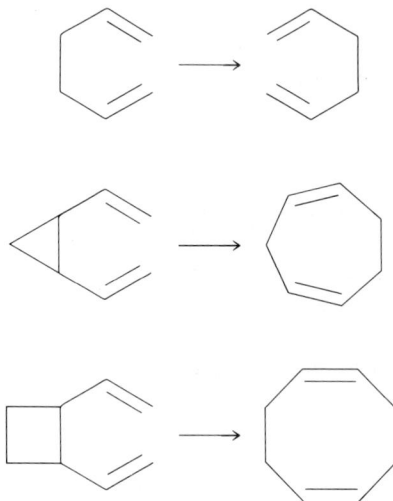

In the first case, the chair-like transition state was calculated to be 22.2 kJ more favored than the boat-like transition state. Another calculation concerning a Cope rearrangement [the conversion of germacrene (1) into elemene (2)] employed a force field for the calculation of the transition state geometry in which the forming bond and the breaking bond have assumed force constants and standard bond lengths. The calculations give information about the conformations from which the rearrangement begins, the transition state, and the immediate product (all are double-chairs) (70).

[Structures 1 and 2 with CH₃ substituents]

1   2

For some kinetically controlled reactions, the product ratio could be correlated well with the relative stabilities of the possible reaction products, assuming a late transition state. Among these was the hydrogenolysis of cage hydrocarbons like basketane (71), the selective dimerization of adamantene (72), and the ring bridging on addition of electrophiles to molecules having two juxtaposed, isolated double bonds (73). The latter reaction was studied mostly with bridged 1,5-cyclooctadienes, and a 1,4-cyclohexadiene derivative, norbornadiene. In most cases, the cross-bridged product was formed preferentially. This

[Cage hydrocarbon structures with N and E labels]

had been explained earlier on the basis of orbital interactions (74), but no explanation was offered for the varying amounts of product with parallel bridging. While orbital-interaction studies usually neglect strain effects completely, and argue on the basis of an early transition state, the molecular mechanics study (73) invokes strain as the only cause for the course of the reaction, assuming a late transition state. When the cross-bridged product is more stable than the parallel product by at least 42 kJ/mol, it is formed exclusively. This is not the case for norbornadiene, which contrary to orbital perturbation arguments, forms the parallel product exclusively. Exceptions to the strain-controlled reaction path were found in reactions that are initiated by a ring opening of an oxirane ring. The stereochemistry of the orbital overlap, which also forms the basis of Baldwin's rules for ring closure (75), is very unfavorable for a cross-bridging in the intermediate protonated epoxide (73). In such cases, the transition state is early on the reaction coordinate, and the stabilities of the two products are of minor importance.

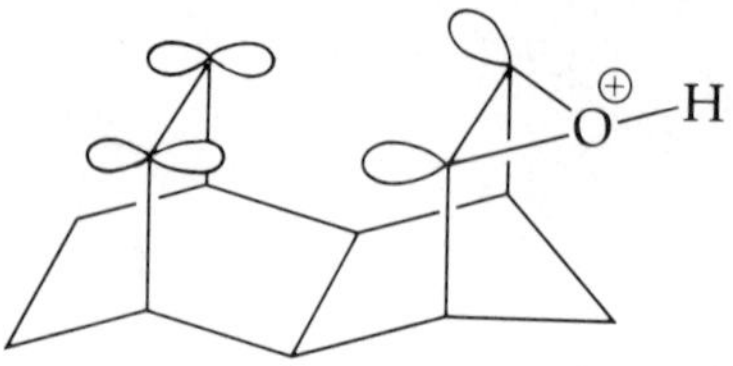

For the thermal isomerization of preisolamendiol (3), which is 86% stereospecific, molecular mechanics calculations of the ground state and an assumed transition state geometry (4) were carried out. The ground state is a mixture of several conformations, and the stereospecificity, therefore, cannot result from an early transition state. One transition state was calculated to be populated with about 80% of the molecules, and this was claimed to explain the high stereospecificity (76).

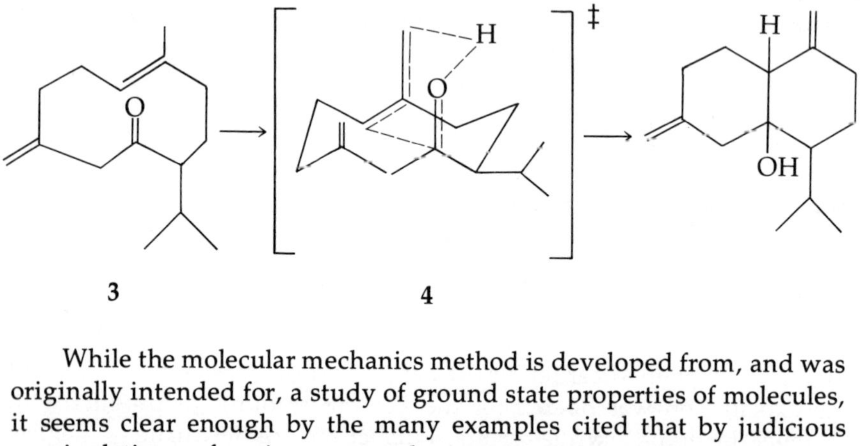

3 4

While the molecular mechanics method is developed from, and was originally intended for, a study of ground state properties of molecules, it seems clear enough by the many examples cited that by judicious manipulations of various sorts, the transition state of a reaction can often be mimicked sufficiently well that one can study kinetic as well as thermodynamic phenomena. It was certainly not obvious a priori that this would be the case, but since it clearly was found to be so, there is no doubt that this area will be subject to substantial exploitation in the future.

Literature Cited

1. McIver, J. W. *Acc. Chem. Res.* **1974**, 7, 72.
2. Dostrovski, J.; Hughes, E. D.; Ingold, C. K. *J. Chem. Soc.* **1946**, 173.

3. Westheimer, F. H. In "Steric Effects in Organic Chemistry"; Newman, M. S., Ed.; John Wiley & Sons: New York, 1956; Chap. 12.
4. Kitaigorodsky, A. I. *Chem. Soc. Rev.* **1978,** *7,* 133.
5. de la Mare, P. B. D.; Fowden, L.; Hughes, E. D.; Ingold, C. K.; Mackie, J. D. H. *J. Chem. Soc.* **1955,** 3200.
6. Abraham, M. H.; Grellier, P. L.; Hogarth, M. J. *J. Chem. Soc., Perkin Trans. 2* **1975,** 1365.
7. DeTar, D. F.; McMullen, D. F.; Luthra, N. P. *J. Am. Chem. Soc.* **1978,** *100,* 2484.
8. DeTar, D. F.; Luthra, N. P. *J. Am. Chem. Soc.* **1980,** *102,* 4505.
9. Gleicher, G. J.; Schleyer, P. v. R. *J. Am. Chem. Soc.* **1967,** *89,* 582.
10. Bingham, R. C.; Schleyer, P. v. R. *J. Am. Chem. Soc.* **1971,** *93,* 3189.
11. Fry, J. C.; Engler, E. M.; Schleyer, P. v. R. *J. Am. Chem. Soc.* **1972,** *94,* 4628.
12. Schleyer, P. v. R.; Isele, P. R.; Bingham, R. C. *J. Org. Chem.* **1968,** *33,* 1239.
13. Dauben, W. G.; Poulter, C. D. *J. Org. Chem.* **1968,** *33,* 1237.
14. Bingham, R. C.; Sliwinski, W. F.; Schleyer, P. v. R. *J. Am. Chem. Soc.* **1970,** *92,* 3471.
15. Sherrod, S. A.; Bergman, R. G.; Gleicher, G. J.; Morris, D. G. *J. Am. Chem. Soc.* **1972,** *94,* 4615.
16. Bingham, R. C.; Schleyer, P. v. R. *Tetrahedron Lett.* **1971,** 23.
17. Karim, A.; McKervey, M. A.; Engler, E. M.; Schleyer, P. v. R. *Tetrahedron Lett.* **1971,** 3987.
18. Lenoir, D.; Hall, R. E.; Schleyer, P. v. R. *J. Am. Chem. Soc.* **1974,** *96,* 2138.
19. Farcasiu, D. *J. Org. Chem.* **1978,** *43,* 3878.
20. Parker, W.; Tranter, R. L.; Watt, C. I. F.; Chang, L. W. K.; Schleyer, P. v. R. *J. Am. Chem. Soc.* **1974,** *96,* 7121.
21. Lomas, J. S.; Luong, P. K.; Dubois, J.-E. *J. Org. Chem.* **1979,** *44,* 1647.
22. Schneider, H.-J.; Thomas, F. *J. Am. Chem. Soc.* **1980,** *102,* 1424.
23. Whitlock, H. W., Jr.; Siefken, M. W. *J. Am. Chem. Soc.* **1968,** *90,* 4929.
24. Gund, T. M.; Schleyer, P. v. R.; Gund, P. H.; Wipke, W. T. *J. Am. Chem. Soc.* **1975,** *97,* 743.
25. Osawa, E.; Aigami, K.; Takaishi, N.; Inamoto, Y.; Fujikura, Y.; Majerski, J.; Schleyer, P. v. R.; Engler, E. M.; Farcasiu, M. *J. Am. Chem. Soc.* **1977,** *99,* 5361.
26. Engler, E. M.; Farcasiu, M.; Sevin, A.; Cense, J. M.; Schleyer, P. v. R. *J. Am. Chem. Soc.* **1973,** *95,* 5769.
27. Gund, T. M.; Schleyer, P. v. R. *Tetrahedron Lett.* **1973,** 1959.
28. Godleski, S. A.; Schleyer, P. v. R.; Osawa, E.; Inamoto, Y.; Fujikura, Y. *J. Org. Chem.* **1976,** *41,* 2596.
29. Godleski, S. A.; Schleyer, P. v. R.; Osawa, E. *Chem. Commun.* **1976,** 38.
30. Takaishi, N.; Inamoto, Y.; Aigami, K.; Fujikura, Y.; Osawa, E.; Kawanisi, M.; Katsushima, T. *J. Org. Chem.* **1977,** *42,* 2041.
31. Farcasiu, D.; Wiskott, E.; Osawa, E.; Thielecke, W.; Engler, E. M.; Slutsky, J.; Schleyer, P. v. R.; Kent, G. J. *J. Am. Chem. Soc.* **1974,** *96,* 4669.
32. Osawa, E.; Furusaki, A.; Matsumoto, T.; Schleyer, P. v. R.; Wiskott, E. *Tetrahedron Lett.* **1976,** 2463.
33. Osawa, E.; Aigami, K.; Inamoto, Y. *J. Chem. Soc., Perkin Trans. 2* **1979,** 181.
34. Harris, J. M.; Shafer, S. G.; Smith, M. R.; McManus, S. P. *Tetrahedron Lett.* **1979,** 2089.

35. Rüchardt, C.; Beckhaus, H.-D.; Hellmann, G.; Weiner, S.; Winiker, R. *Angew. Chem.* **1977**, *89*, 913.
36. Beckhaus, H.-D.; Rüchardt, C. *Chem. Ber.* **1977**, *110*, 878.
37. Beckhaus, H.-D.; Hellmann, G.; Rüchardt, C. *Chem. Ber.* **1978**, *111*, 72.
38. Eichin, K.-H.; McCullough, K. J.; Beckhaus, H.-D.; Rüchardt, C. *Angew. Chem.* **1978**, *90*, 987.
39. Beckhaus, H.-D.; Hellmann, G.; Rüchardt, C.; Kitschke, B.; Lindner, H. J.; Fritz, H. *Chem. Ber.* **1977**, *111*, 3764.
40. Beckhaus, H.-D.; Hellmann, G.; Rüchardt, C.; Kitschke, B.; Lindner, H. J.; *Chem. Ber.* **1978**, *111*, 3780.
41. Rüchardt, C.; Beckhaus, H.-D. *Angew. Chem.* **1980**, *92*, 417.
42. Littke, W.; Drück, U. *Angew. Chem.* **1979**, *91*, 434.
43. Profeta, S., Jr.; Rahman, M., unpublished data.
44. DeTar, D. F. *J. Am. Chem. Soc.* **1974**, *96*, 1254.
45. Ibid., 1255.
46. DeTar, D. F.; Tenpas, C. J. *J. Am. Chem. Soc.* **1976**, *98*, 4567.
47. Ibid., 7903.
48. Winans, R. E.; Wilcox, C. F., Jr. *J. Am. Chem. Soc.* **1976**, *98*, 4281.
49. Milstien, S.; Cohen, L. A. *J. Am. Chem. Soc.* **1972**, *94*, 9158.
50. Wertz, D. H.; Allinger, N. L. *Tetrahedron* **1974**, *30*, 1579.
51. Müller, P.; Perlberger, C. *J. Am. Chem. Soc.* **1975**, *97*, 6862.
52. Ibid., **1976**, *98*, 8407.
53. Eliel, E. L.; Allinger, N. L.; Angyal, S. J.; Morrison, G. A. "Conformational Analysis"; Wiley-Interscience: New York, 1965.
54. Ashby, E. C.; Yu, S. H.; Roling, P. V. *J. Org. Chem.* **1972**, *37*, 1918.
55. Karabatsos, G. J. *J. Am. Chem. Soc.* **1967**, *89*, 1367.
56. Zioudrou, C.; Moustakali-Mavridis, I.; Chrysochou, P.; Karabatsos, G. J. *Tetrahedron* **1978**, *34*, 3181.
57. Perlberger, J.-C.; Müller, P. *J. Am. Chem. Soc.* **1977**, *99*, 6316.
58. Müller, P.; Perlberger, J.-C. *Helv. Chim. Acta* **1976**, *59*, 1880.
59. Wipke, W. T.; Gund, P. *J. Am. Chem. Soc.* **1976**, *98*, 8107.
60. Ibid., **1974**, *96*, 299.
61. Chérest, M.; Felkin, H. *Tetrahedron Lett.* **1968**, 2205.
62. Kobayashi, Y. M.; Lambrecht, J.; Jochims, J. C.; Burkert, U. *Chem. Ber.* **1978**, *111*, 3447.
63. Allinger, N. L.; Lane, G. A. *J. Am. Chem. Soc.* **1974**, *96*, 2937.
64. Garbisch, E. W., Jr.; Schildcrout, S. M.; Patterson, D. B.; Sprecher, C. M. *J. Am. Chem. Soc.* **1965**, *87*, 2932.
65. Taft, R. W., Jr. In "Steric Effects in Organic Chemistry"; Newman, M. S., Ed.; John Wiley & Sons: New York, 1956; p. 556.
66. Beckhaus, H.-D. *Angew. Chem.* **1978**, *90*, 633; *Angew. Chem. Int. Ed.* **1978**, *17*, 593.
67. Giese, B.; Beckhaus, H.-D. *Angew. Chem.* **1978**, *90*, 635; *Angew. Chem. Int. Ed.* **1978**, *17*, 595.
68. Dubois, J.-E.; MacPhee, J. A.; Panaye, A. *Tetrahedron* **1980**, *36*, 919.
69. Simonetta, M.; Favini, G.; Mariani, C.; Gramaccioni, P. *J. Am. Chem. Soc.* **1968**, *90*, 1280.
70. Terada, Y.; Yamamura, S. *Tetrahedron Lett.* **1979**, 3303.

71. Osawa, E.; Schleyer, P. v. R.; Chang, L. W. K.; Kane, V. V. *Tetrahedron Lett.* **1974**, 4189.
72. Slutsky, J.; Engler, E. M.; Schleyer, P. v. R. *Chem. Commun.* **1973**, 685.
73. Osawa, E.; Aigami, K.; Inamoto, Y. *Tetrahedron* **1978**, *34*, 509.
74. Inagaki, S.; Fujimoto, H.; Fukui, K. *J. Am. Chem. Soc.* **1976**, *98*, 4054.
75. Baldwin, J. E. *J. Chem. Soc., Chem. Commun.* **1976**, 734.
76. Terada, Y.; Yamamura, S. *Tetrahedron Lett.* **1979**, 1623.
77. Müller, P.; Blanc, J. *Helv. Chim. Acta* **1980**, *63*, 1759.
78. Müller, P.; Blanc, J. *Tetrahedron Lett.* **1981**, *22*, 715.
79. Still, W. C.; Galynker, I. *Tetrahedron* **1981**, *37*, 3981.
80. Burkert, U. *Habilitationsschrift, Konstanz* **1981**.
81. Houle, F. A.; Beauchamp, J. L. *J. Am. Chem. Soc.* **1979**, *101*, 4067.
82. Yoshimine, M.; Pacansky, J. *J. Chem. Phys.* **1981**, *74*, 5168.
83. Paddon-Row, M. N.; Houk, K. N. *J. Am. Chem. Soc.* **1981**, *103*, 5046.
84. Barton, D. H. R.; Head, A. J.; May, P. J. *J. Chem. Soc.* **1957**, 935.
85. Barton, D. H. R.; McCapra, F.; May, P. J.; Thudium, F. *J. Chem. Soc.* **1960**, 1297.
86. Pacansky, J.; Dupuis, M. *J. Chem. Phys.* **1980**, *73*, 1867.

Applications to the Solid State

MOLECULAR MECHANICS was originally intended as a method for the calculation of the geometries and energies of isolated molecules, that is, in the gas phase. Because of the importance of nonbonded interactions in the method, it is not surprising that the application of force field calculations to the determination of crystal packing, and to the calculation of the structures of solids, was proposed at an early date. It was assumed that the same potential functions and parameters could be applied to intramolecular and intermolecular nonbonded interactions (1, 2). The field of crystal structure calculations is very broad and is still in an active growth phase (3). It ranges from the investigation of the influence of crystal packing on molecular structure, past the a priori and molecular mechanics calculations of packing patterns and the solution of x-ray structural problems (especially of large molecules like polysaccharides and polypeptides), to the determination of thermodynamic and dynamic properties of crystals (4, 5). For molecular mechanics, crystal packing calculations have a special importance in the determination of nonbonded interaction parameters.

Geometry optimization in a crystal can be done in three stages. The first is to keep the unit cell parameters fixed, for example, by locating the molecules at positions found by experiment, and to optimize only intramolecularly. Such a calculation elucidates the influence of crystal forces on the geometries of the molecules. Related studies are discussed on page 308. The second method keeps the intramolecular coordinates fixed, and varies the six unit-cell parameters (three lengths and three angles). However, such crystal packing calculations contain considerable uncertainty because of their assumption of a rigid molecule. The third and most powerful approach is the simultaneous optimization of intramolecular and intermolecular coordinates. These types of calculations are discussed in pp. 309–311. Applications of molecular mechanics for the solution of x-ray crystallographic problems in general are discussed in pp. 311–314. Calculations of dynamic and thermodynamic crystal properties will not be discussed here.

Influence of Crystal Packing on Molecular Structure

The structures of molecules obtained by gas phase measurements, such as microwave or electron diffraction studies, generally agree very well with those found in crystals. Deviations here, or in comparisons of molecular structures in crystals with the results of quantum mechanical or molecular mechanics calculations, are most often explained by crystal packing effects. A direct experimental check on the effect of crystal packing is sometimes possible when more than one independent molecule is present in the crystallographic unit cell, or when the molecule forms two different kinds of crystals (polymorphism). The differences between the structures of these molecules must then be ascribed to packing forces. A theoretical approach starts from the argument that a molecule in a crystal assumes its geometry of minimum energy in the force field of the surrounding neighbors. To study the effect of crystal packing on the geometry of an individual molecule, we locate the molecules with approximately correct geometry in the positions assumed in the crystal, and minimize the energy on the intramolecular coordinates. This procedure has been applied by Simonetta et al. (6) to a classical case, that of biphenyl. Biphenyl has a twisted conformation in the gas phase, but is planar in the crystal. Molecular mechanics calculations of an isolated molecule gave a conformation with a central bond length and torsion angle close to the gas phase value (6, 7, 64), but energy minimization starting with a large array of biphenyl molecules held at the positions they have in the crystal, but allowing the central bond length and torsion angle to vary, converged to a planar geometry. This structure shows the stretching of the central bond found experimentally, which is caused by the repulsion between the ortho hydrogens (6, 7). The calculation of the crystal packing of p,p'-bitolyl (8) and of bi-p-nitrophenyl (9) also gave the experimental torsion angles within 1° (both molecules are twisted in the crystal with a central torsion angle of about 35°). The contrast between these molecules and biphenyl is noteworthy, and is accurately reproduced by the calculations.

Related compounds studied by this method include several 1,1-diarylethylenes. Calculations for an isolated molecule of 1,1-di-p-tolylethylene showed torsion angles different from those found in the crystal, but calculation of a larger assemblage of molecules gave exactly the experimental angles (10). In di-p-nitrophenylethylene, the calculated positions of the phenyl rings for an isolated molecule agreed with the experimental value in the crystal (11, 12), but the position of the nitro group came into full agreement only when the crystal packing calculation was performed as described above.

Crystal Packing Calculations

The unit cell dimensions of a crystal are determined by the intermolecular nonbonded interactions. Molecular mechanics potential functions, therefore, must be suitable for the calculation of unit cell dimensions. Hence, crystal properties are important experimental criteria for the determination of these potential functions. Early calculations of the packing in the n-hexane crystal (unit cell parameters and heat of sublimation) done on a trial and error basis served for the determination of van der Waals parameters for carbon and hydrogen (13). Least squares optimization of nonbonded interaction parameters (including electrostatic interactions) for hydrocarbons was described by Williams (14–20). Instead of minimizing the difference between the calculated and experimental unit cell parameters, which requires geometry optimization by the methods outlined in Chapter 3—a time-consuming, expensive procedure even for moderately large molecules—Williams [and later others (21, 22)] determined the parameters that minimized the sum of the forces calculated at the experimental equilibrium geometry. However, this approach was criticized by Lifson and coworkers (23, 24), who found parameters determined by this procedure were not well suited for obtaining good agreement between calculated and experimental unit cell dimensions. Instead, they minimized the difference between the calculated and experimental structures, not only on the six unit-cell parameters, but also with respect to the intramolecular Cartesian coordinates. Thus, both the crystal packing and the influence of the crystal forces on the geometry were taken into account in their calculations. This approach was later extended to the calculation of the geometry (and the influence of crystal packing on the geometry) of molecules with delocalized pi electrons (25). For larger molecules, a more economical procedure was developed by Hagler and Lifson (24). Their procedure applied a Newton–Raphson type of approximation to the quantity to be minimized, which was the difference between calculated and experimental crystal structure (24). The main disadvantage of the earlier calculations (23), the lengthy computation time, arose in part because the derivatives of the energy had to be recalculated many times during energy minimization. In the later approach (24) these must be calculated only once, at the experimental equilibrium geometry. This procedure was employed to develop parameters for the calculation of crystal structures of peptides. A similar procedure was employed by Momany and coworkers (26–28) to determine the repulsive part of the van der Waals potential of a force field for hydrocarbons, carboxylic acids, amines, and amides.

The van der Waals interactions between two molecules in a crystal are relatively small, compared with many of the intramolecular van der Waals interactions met with in molecular mechanics calculations. As the molecules become further apart, these small interactions fall off proportional to $1/r^6$. This might lead us to think that we could study a crystal as an assemblage of just a few nearest neighbor molecules. This is certainly not the case when the absolute value of the lattice energy is to be determined, because the number of interactions increases as $4\pi r^3/3$. The energy required to sublime a molecule of n-hexane from a crystal of the substance was calculated by using cubical blocks containing different numbers of chains. A total of 3375 (15 × 15 × 15) such chains was needed to give a calculated energy estimated to be within 0.4 kJ/mol of that which would be obtained from an infinite block (29).

Warshel and Lifson have also pointed out that a considerable error in cell dimensions arises when the calculation of van der Waals interactions is cut off at too small a distance. Using the force field of Williams, who neglected all interactions farther apart than 6 Å, they obtained (with a 10-Å cutoff) unit cell parameters that were up to 0.24 Å different from those obtained by Williams (23). Therefore, long-range van der Waals interactions seem to be more important for crystal packing than is often assumed. The sensitivity of these crystal parameters to the size of the assemblage under study can be understood by realizing that the packing forces result from a very large number of quite small interactions. These combine to give a function with a rather flat minimum, so the exact position of the minimum is influenced by relatively small forces. Considering the molecules packed around a given center, the size of the individual forces falls off rapidly as we move outward from the center; the number of molecules packed in the expanding sphere increases, tending to compensate, to a significant extent, the rather rapid falloff of the individual forces. On the other hand, long-range Coulomb interactions, although falling off only with r^{-1}, cancel in the calculation of the crystal packing (unless the crystal consists of ions), because the attractive and repulsive electrostatic interactions from different polar parts of the same molecule cancel at relatively short intermolecular distances (23).

Even when all of the above is taken into consideration, the calculated unit cell parameters sometimes differ from the experimental values by up to 5%, much more than the experimental error. The main cause of error here stems from the effect of molecular vibrations, which cause the crystal to expand thermally. Warshel and Lifson have given a value of 10% of the calculated crystal spacing for the thermal expansion at room temperature (23), which seems to the present authors to be somewhat high. In any case, this effect must be considered when calculating a crystal structure at 0 K, and transferring the results to room temperature.

Crystal packings of a large number of compounds have been studied with force fields derived by the procedures described above. In most cases, the geometries of the individual molecules were kept rigid. Examples of such calculations include work on hydrocarbons (*14–20, 23*), especially adamantane (*30*), hexamethylbenzene (*31*), cycloheptatriene derivatives (*32, 33*), hexa(bromomethyl)benzene (*34*), cyclooctasulfur (*35*), N-methylacetamide (*36, 37*), monomeric and polymeric saccharides (*38*) (for a discussion of polysaccharides see pp. 263–265), and also more polar compounds such as amino acids (*27–28, 39*), peptides (pp. 274–278), and others (*40–44, 65*). In an especially noteworthy study, the preference for different conformations of the same compound in different crystallographic environments (conformational polymorphism) was explained (*45, 66*).

In a recent application crystal packing calculations were combined with electron microscopy to determine the crystal structure of a metastable phase of anthracene that could not be studied by x-ray crystallography (*48*). An important aspect of this study was the check on the crystal vibrational frequencies, as imaginary frequencies would point to incomplete energy minimization or stationary points other than energy minima (*48*).

The results of these calculations are very encouraging; they indicate that crystal packing calculations are a reliable tool for the prediction of crystal structures. However, it is necessary that the force field employed has been carefully parameterized. Although sets of parameters have been proposed for practically all elements of the periodic table, and even for strongly electrostatically interacting species (ions) (*46*), such universal procedures are still quite preliminary.

In many calculations the crystal energy from the van der Waals interactions is taken to be real, not only assuming that the crystal adopts the packing with the lowest calculated crystal energy, but also relating the crystal energy to the heat of sublimation. Although Coiro and coworkers have criticized the quantitative interpretation of the energy as being unreliable because of the approximate nature of the van der Waals potential (*47*), we believe that a quantitative utilization of the calculated energy is justified, as successful applications in lattice-energy calculations indicate (*23*). This requires, however, that the van der Waals potentials be derived as carefully as in the hydrocarbon force fields, as was discussed in Chapters 4 and 5. At present, this does not often seem to be the case with the potentials employed for crystal packing calculations. Furthermore, the crystal/vibrational energy (mostly neglected in the past) must be included.

Molecular Mechanics as a Tool for the Solution of X-ray Crystallographic Problems

Molecular mechanics has useful applications in assisting in the solution of x-ray structures in two fields: the correction of the structure optimized by least squares in cases where an apparently disordered or otherwise unreasonable structure results; and, the phase problem. An example of the first application is the structure determination of 1,1,5,5-tetramethylcyclodecane-8-carboxylic acid. Several crystal structures had been determined on cyclodecane derivatives containing few and small substituent groups. In each case the same (boat–chair–boat) conformation was found. When the crystal structure of the derivative mentioned above was examined, it seemed at first to give a hitherto unknown geometry, but with several unusual bond lengths and angles, and very large temperature factors, which was interpreted as indicative of a disordered structure (49). Molecular mechanics energy minimizations starting with roughly this structure resulted, surprisingly, in two different, new conformations, which were only 8–13 kJ above the global minimum for the parent cyclodecane (50). It was shown that the existance of a 4:1 mixture of these two conformations in the crystal could account for the observed diffraction data for the tetramethyl derivative (49).

A similar case was found in a diethoxy Meisenheimer salt, which in the initial crystal structure determination seemed to indicate disordered ethoxy groups (51). However, packing calculations showed that the energy minimum obtained from the usual least-squares fitting of the diffraction data was not the only one for these groups. In one of the two independent groups in the unit cell, one of the ethoxy groups had the other conformation found by the packing calculation, and reasonable bond lengths and angles, as well as temperature factors, were obtained (52).

$$O_2N \underset{\underset{NO_2}{\diagup}}{\overset{\overset{EtO \quad OEt}{\diagdown \diagup}}{\diagdown}} NO_2 \quad K^+$$

An even more satisfying result was obtained when the geometry of 2-bromo-1,1-diphenylprop-1-ene was calculated and compared with the geometry determined by x-ray crystallography. The experiment was hampered because of the heavy bromine atom, and resulted in a structure with unreasonably short C=C and long C–C bond lengths. When the experimental structure was subjected to a crystal packing calculation, again not allowing any translational movement of molecules as a whole, a slightly different geometry was obtained. Further least-squares optimization of the diffraction data led directly to an alternative structure with reasonable bond lengths. Obviously, there were two minima for the least-squares process, but due to the poor starting structure, the procedure had initially converged to the wrong one (53).

In the calculations just discussed, molecular mechanics merely improved the geometry of the molecule in the crystal after it had already been determined by one of the classical methods of x-ray crystallography. It is, however, also possible to use a molecular mechanics crystal structure as a starting point for the least-squares optimization of the x-ray structure. This structure can be used to determine the phases of the reflections, and the refinement is then carried out in the usual way. For the solutions of the structures of macromolecules when the available crystallographic data are insufficient to determine the structure completely, this approach of building a model of the unit cell and comparing it with the experimental structure factors is the only way to proceed (see pp. 263–265 for related work with polysaccharides). This approach has also been successfully applied to simpler molecules, and it offers a potentially valuable alternative. Pioneering work in this direction was reported by Rabinovich and Schmidt (54), who reported that they solved eight crystal structures by calculation of crystal packings; however, not with an atom–atom interaction approach, but by treating the whole molecule as one entity of ellipsoidal geometry. The method was further refined by Liquori and coworkers, who were able to solve the structure of androstan-3, 17-dione, for example, by packing calculations (55, 56). Several examples of this approach have been given by Simonetta and coworkers, who solved crystal structures of bridged annulenes (51, 57, 58), and by other groups working with nucleotides (59, 60). Other work in this field has been reviewed by Kitaigorodsky (4, 5).

In connection with this approach to the prediction of crystal packing are studies involving clusters containing relatively small numbers of molecules. The structures of microclusters of spherical atoms have been studied in some detail (61). For nonspherical molecules, the situation is more complex. The packing of benzene molecules has been studied in clusters containing various numbers (2–15) of molecules (62). The favored packing arrangement differs from one cluster to the next larger. With an increasing number of molecules in the cluster, the minimum energy structure approaches that found in the benzene crystal. In this study, additional benzene rings were added to the smaller clusters stepwise, reoptimizing the geometry after each new addition. Usually only one structure was considered for each cluster size. It would seem that, in the general case, the different possible ways of packing molecules into a crystal would be quite large, and to find the preferred packing arrangement without additional information would involve solution of the familiar multiple minimum problem encountered with peptide conformations (pp. 274–278). It now appears that the magnitude of the problem here is much smaller than in the protein case; perhaps here it is manageable.

Finally, the calculation of crystal vibrations by molecular mechanics has been employed for the solution of x-ray problems. Often it is not clear if large anisotropic temperature factors, in most cases caused by vibrations of the atoms in the crystal, really originate from wide-amplitude vibrational motions, or if they are a sign of a disordered structure. Therefore, the intramolecular vibrations in the crystal have been studied by force field calculations to indicate what the temperature factors ought to look like. This information can be used to judge if the observed temperature factors are reasonable (63).

Literature Cited

1. Kitaigorodsky, A. I. "Chemical Organic Crystallography"; Plenum: New York, 1961.
2. Dauber, P.; Hagler, A. T. *Acc. Chem. Res.* **1980,** *13,* 105.
3. Ramdas, S.; Thomas, J. M. *Surf. Defect Prop. Solids* **1978,** *7,* 31.
4. Kitaigorodsky, A. I. *Chem. Soc. Rev.* **1978,** *7,* 133.
5. Kitaigorodsky, A. I. "Molecular Crystals and Molecules"; Academic: New York, 1973.
6. Casalone, G. L.; Mariani, C.; Mugnoli, A.; Simonetta, M. *Mol. Phys.* **1968,** *15,* 339.
7. Allinger, N. L.; Sprague, J. T. *J. Am. Chem. Soc.* **1973,** *95,* 3893.
8. Casalone, G. L.; Mariani, C.; Mugnoli, A.; Simonetta, M. *Acta Crystallogr., Sect. B* **1969,** *25,* 1741.
9. Gavezzotti, A.; Simonetta, M. *J. Chem. Soc., Perkin Trans. 2* **1973,** 342.
10. Casalone, G. L.; Mariani, C.; Mugnoli, A.; Simonetta, M. *Acta Crystallogr.* **1967,** *22,* 228.
11. Casalone, G.; Simonetta, M. *J. Chem. Soc. B* **1971,** 1180.
12. Gramaccioli, C. M.; Destro, R.; Simonetta, M. *Acta Crystallogr., Sect. B* **1968,** *24,* 129.
13. Allinger, N. L.; Miller, M. A.; Van-Catledge, F. A.; Hirsch, J. A. *J. Am. Chem. Soc.* **1967,** *89,* 4345.
14. Williams, D. E. *J. Chem. Phys.* **1965,** *43,* 4424.
15. Williams, D. E. *Science* **1965,** *147,* 605.
16. Williams, D. E. *J. Chem. Phys.* **1966,** *45,* 3770.
17. Ibid., **1967,** *47,* 4680.
18. Williams, D. E. *Acta Crystallogr., Sect. A* **1972,** *28,* 84.
19. Ibid., **1972,** *28,* 629.
20. Williams, D. E.; Starr, T. L. *Comput. Chem.* **1977,** *1,* 173.
21. Momany, F. A.; Vanderkooi, G.; Scheraga, H. A. *Proc. Natl. Acad. Sci. USA* **1968,** *61,* 429.
22. Minicozzi, W. P.; Stroot, M. T. *J. Comput. Phys.* **1970,** *6,* 95.
23. Warshel, A.; Lifson, S. *J. Chem. Phys.* **1970,** *53,* 582.
24. Hagler, A. T.; Lifson, S. *Acta Crystallogr., Sect. B* **1974,** *30,* 1336.

25. Warshel, A.; Huler, E.; Rabinovich, D.; Shakked, Z. *J. Mol. Struct.* **1974**, *23*, 175.
26. Momany, F. A.; Carruthers, L. M.; McGuire, R. F.; Scheraga, H. A. *J. Phys. Chem.* **1974**, *78*, 1595.
27. Momany, F. A.; Carruthers, L. M.; Scheraga, H. A. *J. Phys. Chem.* **1974**, *78*, 1621.
28. McGuire, R. F.; Vanderkooi, G.; Momany, F. A.; Ingwall, R. T.; Crippen, G. M.; Lotan, N.; Tuttle, R. W.; Kashuba, K. L.; Scheraga, H. A. *Macromolecules* **1971**, *4*, 112.
29. Allinger, N. L.; Miller, M. A.; Van-Catledge, F. A.; Hirsch, J. A. *J. Am. Chem. Soc.* **1967**, *89*, 4345.
30. Liquori, A. M.; Giglio, E.; Mazzarella, L. *Nuovo Cimento B* **1968**, *55*, 476.
31. Giglio, E.; Liquori, A. M. *Acta Crystallogr.* **1967**, *22*, 437.
32. Stegemann, J.; Lindner, H. J. *Acta Crystallogr., Sect. B* **1979**, *35*, 2161.
33. Stegemann, J., Ph.D. Thesis, Darmstadt, 1979.
34. Giglio, E. *Z. Kristallogr., Kristallgeom., Kristallphys., Kristallchem.* **1970**, *131*, 385.
35. Giglio, E.; Liquori, A. M.; Mazzarella, L. *Z. Nuovo Cimento B* **1968**, *56*, 57.
36. Ramachandran, G. N.; Sarathy, K. P.; Kolaskar, A. S. *Z. Naturforsch., Teil A* **1973**, *28*, 643.
37. Deutini, M.; DeSantis, P.; Morosetti, S.; Piantanida, P. *Z. Kristallogr., Kristallgeom., Kristallphys., Kristallchem.* **1972**, *136*, 305.
38. Zugenmaier, P.; Sarko, A. *Acta Crystallogr., Sect. B* **1972**, *28*, 3158.
39. Hagler, A. T.; Lifson, S. *J. Am. Chem. Soc.* **1974**, *96*, 5327.
40. Ahmed, N. A.; Kitaigorodsky, A. I.; Mirskaja, K. V. *Acta Crystallogr., Sect. B* **1971**, *27*, 867.
41. Ahmed, N. A.; Kitaigorodsky, A. I. *Acta Crystallogr., Sect. B* **1972**, *28*, 739.
42. Caillet, J.; Claverie, P. *Acta Crystallogr., Sect. A* **1975**, *31*, 448.
43. Caillet, J.; Claverie, P.; Pullman, B. *Acta Crystallogr., Sect. A* **1977**, *33*, 885.
44. Bates, J. B.; Busing, W. R. *J. Chem. Phys.* **1974**, *60*, 2414.
45. Bernstein, J.; Hagler, A. T. *J. Am. Chem. Soc.* **1978**, *100*, 673.
46. Skorczyk, R. *Acta Crystallogr., Sect. A* **1976**, *32*, 447.
47. Coiro, V. M.; Giglio, E.; Quagliata, C. *Acta Crystallogr., Sect. B* **1972**, *28*, 3601.
48. Ramdas, S.; Parkinson, G. M.; Thomas, J. M.; Gramaccioli, C. M.; Filippini, G.; Simonetta, M.; Goringe, M. J. *Nature* **1980**, *284*, 153.
49. Dunitz, J. D.; Eser, H. *Helv. Chim. Acta* **1967**, *50*, 1565.
50. Dunitz, J. D.; Eser, H.; Bixon, M.; Lifson, S. *Helv. Chim. Acta* **1967**, *50*, 1572.
51. Simonetta, M. *Acc. Chem. Res.* **1974**, *7*, 345.
52. Destro, R.; Gramacciolo, G. M.; Simonetta, M. *Acta Crystallogr., Sect. B* **1968**, *24*, 1369.
53. Casalone, G. L.; Mariani, C.; Mugnoli, A.; Simonetta, M. *Theor. Chim. Acta* **1967**, *8*, 228.
54. Rabinovich, J.; Schmidt, G. M. J. *Nature* **1966**, *211*, 1391.
55. Damiani, A.; Giglio, E.; Liquori, A. M.; Mazzarella, L. *Nature* **1967**, *215*, 1161.
56. Coiro, V. M.; Giglio, E.; Lucano, A.; Puliti, R. *Acta Crystallogr., Sect. B* **1973**, *29*, 1404.

57. Gavezzotti, A.; Mugnoli, A.; Raimondi, M.; Simonetta, M. *J. Chem. Soc., Perkin Trans. 2*, **1972**, 425.
58. Gramaccioli, C. M.; Mugnoli, A.; Pilati, T.; Raimondi, M.; Simonetta, M. *Acta Crystallogr., Sect. B* **1972**, *28*, 2365.
59. Giglio, E. *Nature* **1969**, *222*, 339.
60. Stellman, S. D.; Hingerty, B.; Broyde, S. B.; Subramanian, E.; Sato, T.; Langridge, R. *Biopolymers* **1973**, *12*, 2731.
61. Hoare, M. R. *Adv. Chem. Phys.* **1979**, *40*, 49.
62. Williams, D. E. *Acta Crystallogr., Sect. A* **1980**, *36*, 715.
63. Shmueli, U.; Goldberg, I. *Acta Crystallogr., Sect. B* **1973**, *29*, 2466.
64. Gustav, K.; Sühnel, J.; Wild, U. P. *Helv. Chim. Acta* **1978**, *61*, 2100.
65. Visser, R. J. J.; Vos, A.; Engberts, J. B. F. N. *J. Chem. Soc., Perkin Trans. 2* **1978**, 634.
66. Hagler, A. T.; Bernstein, J. *J. Am. Chem. Soc.* **1978**, *100*, 6349.

Appendix

Computer Programs Available from the Quantum Chemistry Program Exchange

Large computer programs are necessary for molecular mechanics calculations, as is the case for every current method of calculating molecular structures and energies. A number of such programs have been developed over the years, and several of them have been made generally available through the Quantum Chemistry Program Exchange (QCPE), Indiana University, Chemistry Building 204, Bloomington, Indiana 47401. The programs available from QCPE as of March 1981 that are of immediate utility for molecular mechanics work are listed below. These include programs for molecular mechanics calculations, all of which include geometry optimization, as well as several programs that are useful to generate input coordinates or to analyze the results of a calculation. Different force fields contain not only different parameters, but also slightly different potential functions. The programs listed are, for the most part, tailored to one force field, with those particular parameters stored in the program.

Programs for Molecular Mechanics Calculations

QCPE Number	Name	Authors, Description
247	QCFF/PI	A. Warshel and M. Levitt, a program for calculations with the Lifson/Warshel force field (Consistent Force Field) (1), and also for delocalized pi systems.
286	ECEPP	H. A. Scheraga et al., a program for calculations of peptide structures.

Programs for Molecular Mechanics Calculations (Continued)

QCPE Number	Name	Authors, Description
318	MMI/MMPI	N. L. Allinger and Y. Yuh, a program for general molecular mechanics calculations with the 1973 force field (2). (IBM version. For a version adapted to VAX 11/780 computers see QCPE No. 400, for a UNIVAC version see QCPE No. 404.) MMPI is the version of the program for calculations on molecules containing delocalized pi systems (see pages 150–156).
325	MCA	E. Huler, R. Sharon and A. Warshel, a program for the calculation of crystal packing. An extension of QCPE 247.
348	BIGSTRN	K. Mislow et al., a program that can carry out calculations with the older force fields of Allinger [1971 (3) and 1973 (2)] or of Engler, Andose, and Schleyer (4).
361	UNICEPP	Updated version of QCPE 286 with the new force field parameters.
373	PCK5/PCK6	D. E. Williams, a program for determining crystal structure from structural parameters.
395	MM2	Same as MMI, but for the more recent MM2 force field (5).[†]
410	BIGSTRN-2	K. Mislow et al., in addition to the force fields of QCPE 348, the MM2 force field of Allinger (5) and the MUB-2 force field of Bartell (6) can be used.

[†] The MMP2 program, together with MM2(82) (an updated and expanded version of MM2), is available from Molecular Design, Ltd., 1122 B Street, Hayward, CA 94541.

Auxiliary Programs

QCPE Number	Name	Authors, Description
300	UDRAW	W. E. Brugger and P. C. Jurs, routine for the generation of input data from a CRT display terminal.
370	NAMOD	Y. Beppu, a computer graphics program that produces perspective diagrams of molecules (for use in conjunction with molecular mechanics programs).
419	COORD	K. Müller, the most recent and advanced of a series of programs for the generation of coordinate input data.

Literature Cited

1. Lifson, S.; Warshel, A. *J. Chem. Phys.* **1968,** *49* 5116.
2. Wertz, D. H.; Allinger, N. L. *Tetrahedron* **1974,** *30,* 1579.
3. Allinger, N.L.; Tribble, M. T.; Miller, M. A.; Wertz, D. H. *J. Amer. Chem. Soc.* **1971,** *93,* 1637.
4. Engler, E. M.; Andose, J. D.; von R. Schleyer, P. *J. Amer. Chem. Soc.* **1973,** *95,* 8005.
5. Allinger, N. L. *J. Amer. Chem. Soc.* **1977,** *99,* 8127.
6. Fitzwater, S.; Bartell, L. S.; *J. Amer. Chem. Soc.* **1976,** *98,* 5107.

INDEX

A

Ab initio calculations
  atom slope 31
  butane 52
  with extended basis sets 11
  geometries, optimized 81
  hydrogen molecule 43
  modified 12
Acetals
  electrostatic interactions 217–224
  gauche effect 219
Activation energy(ies)
  congestion function 297
  S_N2 reaction, alkyl halides 287
Activation enthalpies, S_N2 cyclization of ω-haloamines 287
Acyclic pentavalent phosphorus compounds, van der Waals interactions 243
Acyclic saturated compounds, geometry and relative energy.79–88
Adamantane, heats of formation.. 178
Adamantane conformation, geometry and relative energy ...115–116
Addition–elimination reactions, calculations 292–300
Additivity of conformational energies, cyclohexane system 92
Additivity rule for conformational analysis of alkanes 87
3′- and 5′-Adenosine monophosphates, conformation 271
Adenosine triphosphate conformation 270
  calculated lowest energy 271f
Alanyl dipeptide, internal rotation energy map 277f
Alcohols
  conformation 222
  electrostatic interactions 211–224
  heats of formation 182t

Aldopyranose pentaacetates and tetraacetates, conformation .. 259
Alkane(s)
  additivity rule for conformational analysis 87
  heats of formation 176t
  heats of solution 180
Alkenes
  conformation geometry and relative energy 121–143
  heats of formation 179
  heats of hydrogenation 179
  heats of solution 176t
Alkyl halides
  conformational energies 206
  electrostatic interactions 205
  nucleophilic substitution reaction rates 287t
Alkyl radical
  force field 292
  strain energies, heats of formation, and geometries 291
Alkylated cyclopentanes, pseudorotation 91
Alkylcyclohexane(s)
  conformations 93–98
  geometry and relative energy ..93–98
  ring flattening 95
2,3- and 3,5-O-Alkylidene derivatives of pentofuranoses 262
Alkynes, conformation geometry and relative energy 143–144
Allinger
  buttressing 95
  force field (MM2) 51
  force field, heats of formation calculations 175
  hydrocarbon force field; thioethers 235
  hydrocarbon force field, thiols .. 235
Allinger and Sprague approach, conjugated pi-electronic systems.. 151

A

Allinger and Wuesthoff scheme,
  dipole moments 197
Amino acid conformation of polypeptide 274
Androstan-3,17-dione packing calculations 313
Androstanes, stereoisomers, relative energies 255
Androsterone, geometry 253
Anharmonic stretching and bending 23f
Anharmonicity
  effects 18
  vibrational frequencies 171
Anhydroglucose residue, rotation about ring vector 265f
1,6-Anhydropyranoses, structure . 261
Anthracene, crystal packing calculations 311
Aromatic, definitions 189
Aromaticity
  definition 190
  thermochemical definition 189
Aryl group, rotameric behavior ... 145
Asymmetric centers, stereoselective formation 298
Atomic charges treatment models.. 195
Azabicyclo[4.4.0]decanes, conformation 233, 234
9-Azabicyclo[3.3.1]nonane, conformation 233
Azabicyclo[3.3.3]undecane, conformation 233
Azetidine, conformation 231

B

Bartell MUB-1 force field
  butane 48
  gauche butane energy 49
Basketange, product ratio 301
Bending force constants 38
Benzene
  heat of formation 189
  molecules, packing 313
  parameterization, force fields ... 53
  rings, clusters 313
Benzocycloheptene, conformation geometry and relative energy. 129
Berry pseudorotation mechanism.. 243
Bicyclo[4.2.1]- and -[5.1.1]nonenes, stabilities 142
Bicyclo[5.3.0]decane, conformation geometry and relative energy. 110
Bicyclo[3.3.2]decane, conformation geometry and relative energy. 113

Bicyclo[5.5.0]dodecahexaene,
  geometry and stability 152
Bicyclo[2.2.1]heptane, conformation geometry and relative energy 111–112
Bicyclo[3.2.0]heptene, conformation and energy 141
Bicyclo[3.1.0]hexane, conformation geometry and relative energy. 118
Bicyclo[2.2.0]hexane, conformation geometry and relative energy. 120
Bicyclo[4.3.0]nonane, conformation geometry and relative energy. 109
Bicyclo[3.3.1]nonane, conformation geometry and relative energy 112–113
Bicyclo[2.2.2]octane, conformation geometry and relative energy. 112
Bicyclo[3.3.0]octane, conformation geometry and relative energy. 109
Bicyclo[4.2.0]octane, conformation geometry and relative energy 120, 121
Bicyclo[5.3.1]undecapentaene
  calculated bond lengths and torsional angles 154f
  geometry and bond lengths 154
Bicyclo[4.4.1]undecapentaene
  geometry 153f
  structure and bond lengths 153
Bimolecular nucleophilic substitution 286
Biphenyl molecule, molecular mechanics calculations 308
1,2-Bis(2,6-dimethylphenyl)-1,2-di-*t*-butylethane, Newman projection 88
Block diagonal
  matrix 70f
  method, computer time 69
  Newton–Raphson method, minimization 71
Boltzmann distribution equation.. 174
Bond angle
  axial ligands 243
  bonding, molecular mechanics model 22
  N-butane 80
  CCC, experimental and calculated 83f
  deformation, cyclobutanone 27
  deformation parameters in current force fields 40t
  HCC, experimental and calculated 84f
  interconversion of Cartesian and internal coordinates 63
  measurement 6

Bond energy increments
  Franklin scheme 174
  heat of formation calculation .. 174
Bond length
  C–C, experimental and calculated 81f
  C–H, experimental and calculated 82f
  deformation parameters in current force fields 39t
  of diffraction radial distribution function 8
  measurement 6
  of quantum mechanical calculations 8
  and torsional angles in bicyclo-[5.3.1]undecapentaene 154f
Bond stretching
  and angle bending parameters, force fields 38–39
  molecular mechanics model 22
Bond-order–bond-length relationship 151, 153
Born–Oppenheimer
  approximation 2
  surface 1–6, 10
Bredt olefins, geometry and relative energy 141, 142
Bridged 1,5-cyclooctadines, product ratio 301
Bridged olefins, geometry and relative energy 141, 142
Bridgehead derivatives, reaction rates 288
2-Bromo-1,1-diphenylprop-1-ene, molecular mechanics calculation 312
Buckingham curve 31
Buckingham potential function ... 31
  equation 30
Butane
  ab initio calculations 52
  Bartell MUB-1 force field 48
  comparison of force fields 48, 49
  gauche energy, MM1 force field 49
  gauche energy, Schleyer force field 49
  hydrogen–carbon interaction ... 48
  methyl–hydrogen interaction ... 48
  MM1 force field 48
  molecular mechanics 51
n-Butane, bond angles 80
2-Butanone, conformation 211
Butene *gauche* energy, Bartell MUB–1 force field 49
1-Butene, conformation geometry and relative energy 121–122

Butylamine, low periodicity torsional energy 230
N-*t*-Butylpseudopelletierine, conformation 234
δ-Butyrolactone, conformation 225

C

Carbenium ion
  electronic effects 288
  hyperconjugation 288
  interconversion of hydrocarbons 289
  intermediates, solvolysis reactions 288
  reactions 288–290
Carbocyclic 4-membered rings, conformation geometry and relative energy 120
Carbohydrates, conformational energies and anomeric equilibria 257–265
Carbonyl addition reaction 292
Carbonyl compound
  conformational equilibrium 295
  heats of formation 183
  product ratio 294
Carbonyl oxygen, 1,4-nonbonded interactions 209
Carboxylic acids and esters, hydrocarbon force field 224–228
Cartesian coordinates energy minimization 64
Cartesian and internal coordinates, interconversion, vector formulas 63
Caryophyllene, conformation and energy 141
Catechol derivatives, geometry .242, 243
C–C bond length, experimental and calculated 81f
CCC bond angle, experimental and calculated 83f
Central force field, definition and use 20
CFF-3, vibrational contribution to enthalpy 50
C–H bond length, experimental and calculated 82f
Charges, peptide, semiempirical method 196
Chelate complex, conformational and configurational equilibria. 244
Chemical reactions, rates, stereochemistry 285–305
Chirality inversion, transition states 147f
Chlorocyclohexane, conformational energy 207

4-Chlorocyclohexanone, conformation ... 208
Chromic acid ester, formation ... 294
Chromic acid oxidation, rate ... 294
Complete energy minimization program ... 64
Complete neglect of differential overlap (CNDO) calculation ... 13
Computer input, molecular geometry ... 63
Computer programs, molecular mechanics
   calculations ... 317
   flow diagram ... 62f
   and quantum mechanics calculations ... 61
Computer programs, quantum chemistry program exchange ... 317–319
Computer time, block diagonal method ... 69
Conformation(s)
   of ATP, calculated lowest-energy ... 271f
   definition ... 3
   of cyclohexane, energy surface cross section ... 93f
   definition ... 3
   geometries, molecular mechanics ... 84
   at nucleoside glycosyl ... 267f
   minimum energy ... 60
   silabutanes ... 203
   symmetrical, of cyclooctane ... 101f
Conformational analysis of
   alkanes, additivity rule ... 87
   carbohydrates ... 258
   cis,cis-1,5-cyclooctadiene ... 134f
   1-silabutane ... 202
Conformational and configurational equilibria
   chelate complexes ... 244
   pentofuranose derivatives ... 262
Conformational effects, Taft parameters ... 300
Conformational energy(ies)
   1,4; 1,3; and 1,3,5 ... 95
   alkyl halides ... 206
   calculation, cyclic peptides ... 278
   chlorocyclohexane ... 207
   1,3-dioxanes ... 221
   map, maltose ... 263
   maps, dinucleoside phosphates ... 273
   maps, dinucleotides ... 266
   methyl groups, silylcyclohexane ring ... 204
   2-methylbutane ... 87
   nonadditivity ... 87

Conformational energy(ies)— *Continued*
   nucleosides ... 269
   prediction ... 86
   silycyclohexane and trimethylsilylcyclohexane ... 203
Conformational equilibrium, carbonyl compound ... 295
Conformational interconversion
   Hilderbrandt ... 74
   mapping ... 75
   pathways ... 72–76
Conformational potential function for 1,2,4,5-tetrathiane ... 238f
Conformational properties, vicinal dichlorides ... 207
Conformational stabilities and rotational barriers ... 206
Conformer, definition ... 3
Congestion function and activation entropy ... 297
Conjugated pi-electronic systems(s)
   Allinger and Sprague approach ... 151
   general treatment ... 150–156
   heats of formation calculations ... 192t
   resonance energies, heats of formation ... 189–192
Conjugated system, planar, structure calculations ... 151
Conjugation energy of force fields ... 54
Consistent force field (CFF–3) ... 38, 50, 224
   linear least squares parameter optimization ... 36, 37
Constitutional equilibrium, pentofuranose derivatives ... 262
Contour map, cyclohexane energy surface ... 94f
Cope rearrangements of dienes, potential surfaces ... 300
Cost estimates, molecular geometry calculations ... 11
Coulomb interactions, crystal packing ... 310
Coulomb potential, equation ... 29
Cross-term, torsional energy ... 35
Crystal, geometry optimization ... 307
Crystal and molecular structures, polysaccharides ... 264
Crystal packing
   calculations ... 309–311
   crystal structure prediction ... 311
   ellipsoidal geometry ... 313
   metastable phase of anthracene ... 311
   Coulomb interactions ... 310
   effects ... 308

Coulomb interactions—*Continued*
  force field calculations 307
  forces 308
  and molecular structure 308
  van der Waals interactions 310
Crystal structure calculations,
  applications 307
Crystal structures of
  globular proteins 277f
  peptides 309
Crystal vibrational energy 311, 314
Cyclic acetals, geometries and
  energies 262
Cyclic ethers, molecular mechanics 218
Cyclic ketones, conformation ...211, 297
Cyclic peptide, conformational
  energy calculation 278
Cyclic peptide, geometry
  optimization 257
Cyclitols, isomer, conformation .. 260
Cycloalkanes, strain energies 187
Cycloalkene ring systems, con-
  formation geometry and
  relative energy 125–126
Cycloalkenes, conformation geom-
  etry and relative energy ...121–143
Cycloalkynes, energy 144
Cyclobutane(s)
  bond angle reduction 27
  conformational geometry and
    relative energy 119–121
  ring, van der Waals interaction.. 24
  Urey-Bradley interactions 24
  valence force field 24
Cyclobutanone, out-of-plane
  deformation 26
Cyclodecadienes, conformation
  geometry and relative
  energy 138–139
Cyclodecane
  conformations 105–106
  geometry and relative energy.105–106
  ring stain 226
Cyclodecane-1,6-dione, conformation 216
Cyclodecanone, conformation 214
Cyclodecasulfur, conformation ... 240
Cyclodecatetraene, conformation
  geometry and relative
  energy 139–140
Cyclodecene, conformation geom-
  erty and relative energy ...137–138
Cyclododecane conformations
  geometry and energy 107–108

Cyclododecasulfur, conformation . 240
Cyclodecatetraene, conformation
  geometry and relative
  energy 139–140
Cyclododecane
  conformations 107–108
  geometry and relative energy.107–108
Cyclododecapeptide, full energy
  minimization 278
Cyclododeca-1,5,9-triene, conformation geometry and relative
  energy 140
cis-Cyclododecene, conformation
  geometry and relative energy . 140
1,3-Cycloheptadiene, conformation
  geometry and relative energy . 130
Cycloheptane
  conformation, relative energies 99f
  conformations and pseudorota-
    tation 98
  derivative, structure 98
  geometry and relative energy ..98–100
Cycloheptanone, conformation ... 214
Cycloheptasulfur, conformation ... 239
1,3,5-Cycloheptatriene, non-
  neighbor resonance interaction 154
Cycloheptene
  conformation geometry and
    relative energy 129–130
  inversion 129f
Cycloheptyne, energy 144
1,9-Cyclohexadecanedione,
  square conformation 217
Cyclohexadiene
  conformation geometry and
    relative energy 127–129
  rearrangement rates 300
Cyclohexane
  conformation(s) 91
    energy surface cross section . 93f
    interconversion pathways 92
  energy surface contour map ... 94f
  geometry, equatorial methyl
    group 95
  geometry and relative energy ..91–93
  molecular mechanics calculations 91
  pseudorotation 92
    free, transition state 93
  geometry, equatorial *t*-butyl
    group 96
  geometry, torsion angles 97
  phenyl group, conformation .. 146
  torsion angle 92

Cyclohexane-1,4-dione,
  conformation 214
Cyclohexanone, conformation ..212, 214
Cyclohexan-1,3,5-trione,
  conformation 216
Cyclohexasilane, conformation ... 204
Cyclohexasulfur, conformation ... 239
Cyclohexan-1,3,5-trione,
  phloroglucin, conformation .. 216
Cyclohexene, conformation
  geometry and relative energy . 127
Cyclononadienes, conformation
  geometry and relative
  energy 135–137
Cyclononane
  conformations 104, 240
  relative energies 105*t*
  geometry and relative energy 104–105
  interconversion map 105
Cyclononasulfur, conformation ... 240
Cyclononatrienes, conformation
  geometry and relative
  energy 136, 137
Cyclononyne, conformations
  force fields calculation 144
Cyclooctadecane-1,9-dione,
  conformation 216
Cyclooctadiene(s)
  conformational analysis 134*f*
  conformation geometry and
  relative energy 132–134
  transition state energies 134*t*
Cyclooctane use of driving routine 100
Cyclooctane
  conformations 100
  relative energies 103*t*
  geometry and relative energy 100–104
  interconversions 102, 104
  pseudorotation and ring
  inversion 100, 102*f*
  symmetrical conformation 101*f*
  use of driving routine 100
Cyclooctanone, conformation 214
Cyclooctasulfur, conformation ... 239
Cyclooctatriene, conformation
  geometry and relative energy . 135
Cyclooctene
  conformation geometry and
  relative energy 130
  deformations 131
Cyclooctyne, geometry and
  relative energy 144
Cyclopentane
  conformations 89, 90
  geometry and relative energy ..89–91
  molecular mechanics calculations 90

Cyclopentane—*Continued*
  ring
    conformation 211
    pseudorotation pathway 89*f*
    torsion angles 90
Cyclopentanone, conformation 211
Cyclopentasilane, stability 204
Cyclopentasulfur, conformation .. 239
Cyclopentene, conformation
  geometry and relative
  energy 126–127
Cyclopropanes, conformation
  geometry and relative
  energy 116–118
Cyclotetradecane, conformation ... 108
Cyclotetrasulfur, conformation ... 239
Cyclotridecane, conformation 108
Cycloundecane, conformation
  geometry and relative energy .106–107

D

Decalin, conformation geometry
  and relative energy 109–110
Decalones
  conformational equilibria 214
  structure 212
Decomposition rates of azo and
  peroxy compounds 291
Deformation(s)
  bond angle, force field
    parameters 40*t*
  bond length, force field
    parameters 39*t*
  in bond lengths and angles,
    ethane derivatives 86
  trans-cyclooctene 131
  methylcyclohexane 95
2-Deoxy-2-aminohexo-pyranoses,
  conformational and anomeric
  energies 261
Deoxydinucleoside mono-
  phosphates, conformation 273
Deoxyribodinucleoside monophosphates, conformation 273
Deoxyribonucleic acids, structures 273
Deoxyribose, pseudorotation 270
Deuterium, molecular mechanics
  calculations 157
Deuterium in hydrocarbons,
  geometry and relative energy . 156
Dewar benzene, structure 126, 127
Diadenosine pyrophosphate,
  conformation 271
Diagonal valence force fields 21
Diamantane, heats of formation .. 179

INDEX 327

Diamantoid compounds,
conformation geometry and
relative energy 115–116
1,3-Diaminopropane complexes
with cobalt 244
Diasteromeric transition state,
energies 295
Dibenzo-1,5-cyclooctadiene,
conformation interconversion . 133
5H-8H-Dibenzo[d,f](1,2)dithiocin,
conformation 237
Di-t-butylcyclohexanes,
conformations 97
1,2-Di-t-butylethylene,
conformation geometry and
relative energy 123–125
trans-1,4-Dichlorocyclohexane,
conformation 207
1,2-Dichloroethane, dipole/dipole
interaction 200
Dielectric constant
bulk 199
effective 199
electrostatic interactions 200
solvent effect in molecular
mechanics 199
Diethoxy Meisenheimer salt,
structure 312
Diffraction radial distribution
function, bond length 8
Dihalides, solvent effects 209
9,10-Dihydroanthracene,
conformation geometry and
relative energy 128
Dihydrogen trisulfide, electrostatic
interactions 239
Dimerization of adamantane,
product ratio 301
5,5-Dimethyl-1,3-dithiane-1-oxide,
conformation 242
4,4-Dimethylandrostan-3-ones and
-androst-5-ene-3-ones, conformational equilibrium 254
2,3-Dimethylbutane, isomer
stability 87
Dimethylcyclohexanes 95
ring flattening 96
3,3-Dimethylpentane-2,4-dione,
reaction field model 201
4,4-Dimethyl-2-pentene,
conformation geometry and
relative energy 122–123
3,3-Dimethylthian-1-oxide,
conformation 241
Dimethyltrisulfide, electrostatic
interactions 239

Di-p-nitrophenylethylene 308
Dinucleoside monophosphates
conformation 273
torsion angles 266
Dinucleoside phosphates, conformational energy maps 273
Dinucleoside pyrophosphate, conformation 272f
Dinucleoside triphosphate, conformation 272f
Dinucleotides, conformational
energy maps 266
1,3-Dioxanes, conformational
energies 221
1,2-Diphenylethane, isomer
stability 148
Dipole/dipole interaction
1,2-dichloroethane 200
energy, equation 29
scheme, halides 206
Dipole method, procedure 196
Dipole model, atomic charges
treatment 195
Dipole moments
Allinger and Wuesthoff scheme . 197
induced, and molecular
mechanics geometries 198
induced, point charge
determination 197
induction and bond polarizability 197
Disaccharides
conformational analysis 263
potential energy map 264
Disulfides, conformational
analysis 237, 239
1,2-Dithiane, conformation 237
1,3-Dithiane-1-oxide, conformation 241
1,1-Di-p-tolylethylene, calculations 308
cis-1,2-Divinylcyclobutane,
rearrangement rates 300
cis-1,2-Divinylcyclopropane,
rearrangement rates 300
Dodecahedrane, conformation
geometry and relative
energy 114–115
Double-bond hydrogenation, bond-
angle and torsional strain ... 298

E

Eckart constraints 75
energy minimization 68
Electrocyclic reactions 300–302
Electron diffraction
structure determination 6, 80
terminology 8
and x-ray crystallography 6

Electronegativity and ligand
  conformation 242
Electronic effect, carbenium ions .. 288
Electrostatic interaction(s) 196
  acetals 217–224
  alcohols 217–224
  alkehydes 211–217
  alkyl halides 205
  conformational analysis,
    sulfur 235–242
  dielectric constant 200
  dihydrogen trisulfide 239
  dimethyltrisulfide 239
  ethers 217–224
  halogens 205–209
  heteroatoms 195–202
  and hydrogen bonding 222
  induction and polarizabilities .. 197
  ketones 211–217
  lone pair 210
  nitrogen 228–234
  nucleosides 269
  oxa derivatives 209
  oxygen 209–228
  peptide calculations 275
  phosphorus 242
  piperidones 234
  polyethers 219
  polyhalides 206
  polynucleotides 269
  reaction field model 200
  solvent effects 198
  sulfoxides 241
Energy calculation,
  hydrocarbons 79–167
Energy contours for maltose 263f
Energy decrease, reaction field
  model, equation 200
Energy diagram, qualitative
  cycloheptene inversion 129f
Energy difference of cyclohexane
  ring 97
Energy differences of azo and
  peroxy compounds 291
Energy map for alanyl dipeptide
  internal rotation 277f
Energy minimization
  block diagonal and full matrix
    procedures 71
  calculation, peptides 274
  Cartesian coordinates 64
  Eckart constraints 68
  first derivative techniques 66–67
  geomery change threshold 61
  internal coordinates 64
  iterative geometry optimization . 59
  molecular geometry 64–72

Energy minimization—Continued
  monosaccharide conformations . 261
  Newton–Raphson method 68
  procedures 60, 261
  reaction coordinate 75
  S_N2 reaction rates 286
  saddle point 72
  single geometry approach 60
  steepest descent method 66, 68
  technique 59
  transition vector 74
Energy, molecular 1–15
Energy, relative
  acyclic saturated compounds ... 79
  adamantane 115
  alkenes 121
  alkylcyclohexanes 93–98
  alkynes 143–144
  bicyclodecanes 110, 113
  bicycloheptane 111–112
  bicyclohexanes 118, 120
  bicyclononanes 109, 112–113
  bicyclooctanes 109, 112, 120, 121
  1-butene 121–122
  cyclic conjugated pi-electronic
    systems 144–156
  cycloalkenes 121, 125–126
  cyclobutanes 119
  cyclodecadienes 138–139
  cyclodecane 105–106
  cyclodecatetraenes 139–140
  cyclodecene 137–138
  cyclododecane 107–108
  cyclododeca-1,5,9-triene 140
  cis-cyclododecene 140
  1,3-cycloheptadiene 130
  cycloheptanes 98, 99f
  cycloheptenes 129–130
  cyclohexadienes 127–129
  cyclohexane 91
  cyclohexene 127
  cyclononadienes 135–137
  cyclononane 104, 105t
  cyclononatrienes 136, 137
  cyclononenes 135
  cyclooctadienes 132–134
  cyclooctane 100, 103t
  cyclooctatrienes 135
  cyclooctenes 130–132
  cyclopentane 89
  cyclopentene 126
  cyclopropanes 116
  cycloundecane 106–107
  decalin 109–110
  diamantoid compounds 115
  1,2-di-t-butylethylene 123–125
  4,4-dimethyl-2-pentene 122–134

INDEX

Energy relative—*Continued*
  dodecahedrane 114–115
  fused ring and other bicyclic
    hydrocarbons 108
  hydrocarbons containing
    deuterium 156f
  large ring hydrocarbons 108
  methylenecyclohexane 127
  perhydroanthracene 110
  perhydroazulene 110
  perhydrophenanthrene 111
  perhydroquinacene 113–114
  polycyclic olefins 141–143
  small ring hydrocarbons 116–121
Energy–distance relationship
  for bonds 8f
Enthalpy of mixture, equation 173
Entropy effect, ring closures
  rates 287
Equatorial *t*-butyl group, cyclohexane ring geometry 96
Equilibrium geometries 82
  hexamethylethane 25
Ermer–Lifson force field, structures
  and energies 195
Ester(s)
  force field 225
  formation rates and strain
    energy 292
  hydrolysis rates 292
  steric energy and reaction rates. 293
Ethane
  strain and thermolysis rates ... 291
  torsional potential, calculation . 33
Ethane derivatives
  deformations in bond lengths
    and angles 86
  thermolysis rates 291
Ether oxygen
  nonbonded interaction 209
  van der Waals potential 32
Ethers
  cyclic, molecular mechanics 218
  electrostatic interactions 217–224
  heats of formation 182t
  parameterization 220
Ethylamine, conformational equilibrium 230
Ethylene double bond, torsional
  potential 34
Ethylene torsion angles 35
Extended Hückel calculations 82
Extended valence force field 21

F

Factorization of bending potential. 26

First derivative techniques,
  energy minimization 66–67
Flow diagram of molecular mechanics
  computer program 62f
Folding pathway, polypeptide 274
Force constants 17
  of internal coordinates 20
Force field(s)
  alkyl radicals 291
  benzene parameterization 52–55
  bond angle deformation, param-
    eters 40t
  bond length deformation, param-
    eters 39t
  bond stretching and angle bend-
    ing parameters 38–39
  components 4
  conjugation energy 54
  hydrocarbons 51, 217
  molecular mechanics, torsional
    terms 33
  molecular mechanics, potential
    function 22–36
  naphthalene 52–55
  nonbonded and torsional energy
    parameters 39–52
  parameterization 36–52
  S_N2 reaction rates 286
  selection 50
  torsional energy 32
  vibrational spectroscopy and
    molecular mechanics .. 17–22, 26
  van der Waals function 44
  van der Waals potential 42
Force field calculation(s) 2, 4
  crystal packing 307
  mechanical model 49
  strained molecules 151
  sulfoxides 241
Force field with Hooks law harmonic function, equation 20
Franklin scheme, bond energy
  increments 174
Free radical reactions 291
Full F and block diagonal matrix .. 70f
Full-matrix method, transition
  state calculations 73
trans-Fused 3,5-*O*-alkylideneribo-
  furanoses, conformation 262

G

gauche effect 219
Gear effect
  hexamethylbenzene 86
  methyl-*N,N*-diisopropyldithio-
    carbamate 86

Gear effect—*Continued*
  partial rotation about single
    bonds 85
β-Gentiobiose, geometry 264
Geometry(ies)
  ab initio calculations 81
  acyclic saturated compounds ...79–88
  adamantane 115
  alkenes 121
  alkyl radicals 291
  alkylcyclohexanes 93–98
  alkynes 143–144
  bicyclodecanes 110, 113
  bicyclo[2.2.1]heptane 111–112
  bicyclohexanes 118, 120
  bicyclononanes 109, 112–113
  bicyclooctanes 109, 112, 120, 121
  bicycloundecapentaene 153f
  1-butene 121–122
  change threshold, energy
    minimization 61
  cyclic conjugated pi-electronic
    systems 144–156
  cyclic ketones 297
  cycloalkenes 121, 125–126
  cyclobutanes 119
  cyclodecadienes 138–139
  cyclodecane 105–106
  cyclodecatetraenes 139–140
  cyclodecene 137–138
  cyclododecane 107–108
  cyclododeca-1,5,9-triene 140
  cis-cyclododecene
  1,3-cycloheptadiene 130
  cycloheptanes 98
  cycloheptenes 129
  cyclohexadienes 127–129
  cyclohexane 91
  cyclohexene 127
  cyclononadienes 135–137
  cyclononane 104
  cyclononatriene 136, 137
  cyclononenes 135
  cyclooctadienes 132–134
  cyclooctane 100
  cyclooctatrienes 135
  cyclooctenes 130–132
  cyclopentane 89
  cyclopentene 126
  cyclopropanes 116–118
  cycloundecane 106–107
  diamantoid compounds 115
  1,2-di-*t*-butylethylene 123–125
  4,4-dimethyl-2-pentene 122–123
  dodecahedrane 114
  effect of crystal packing 308
  fused rings and other bicyclic
    hydrocarbons 108

Geometry(ies)—*Continued*
  helix, of polymeric nucleic acid . 266
  hydrindane 109
  hydrocarbons, containing
    deuterium 156f
  hydrocarbons, precision and
    reliability 83f
  large ring hydrocarbons 108
  methylenecyclohexane 127
  molecular 1–15
  perhydroanthracene 110
  perhydroazulene 110
  perhydrophenanthrene 111
  perhydroquinacene 113
  polycyclic olefins 141–143
  small ring hydrocarbons 116–121
  1,5,9,13-tetraazacyclohexadecane. 231
  tetraphenylmethane 148
Geometry optimization
  crystal 307
  peptides 278
  steroids 253
Germacrene to elemene, transition
  state geometry 300
Gibbs energies, pyranose ring 259
Global energy minimum 60, 275
Globular proteins, crystal structures, conformations 277f
D-Glucopyranoside, structures and
  energies 261
Glycol ethers 219
Glycosyl bond
  conformations 267f, 268
  rotational potential 266
Group theoretical method 290
Guanosine-5′-nucleosides,
  conformation 270

H

Halides
  dipole/dipole interaction 206
  point charge interaction model .. 206
  S_N2 reaction 1
ω-Haloamines, activation enthal-
  pies, S_N2 cyclization 287
Halogens, electrostatic inter-
  actions 205–209
Haloketone, solvent effects 209
Halophosphazenes, force field 244
Harmonic approximation 8f, 17
Harmonic force field equation 17
Harmonic potential 18
  Hooke's law function 22
  Urey–Bradley force fields 27
Harmonic stretching and bending . 23f

Harmonic stretching potential 25
Harmonic vibrations 18
Hartree–Fock approximation 11
HCC bond angle, experimental
  and calculated 84f
Heat of combustion 173
Heats of formation 169–194
  adamantane 178
  alcohols and ethers 182t
  alkanes 176t
  alkenes 179
  alkyl radicals 291
  benzene 189
  calculation(s) 174
    Allinger force field 175
    bond energy increments 174
    conjugated hydrocarbons 192t
    MMP1 force field 191
  carbonyl compounds 183t
  conjugated hydrocarbons 189–192
  description 173
  diamantane 179
  evaluation 174
  full statistical mechanical
    treatment 176
  hetero atom molecules 180
  low periodicity torsional
    energy terms 176
  2-methyladamantane 178
  molecular mechanics 179
  norbornane 180, 181f
  physical description 175
  potential function improvements 176
  radicals 291
  relative stability 190
  resonance energies 189–192
  statistical mechanics terms 176
  sulfur homologues 241
  Wertz force field 50
Heats of hydrogenation
  alkenes 179
  norbornadiene 180
  norbornene 180, 180f
Heats of solution, alkenes
  and alkanes 180
Helix geometries of polymeric
  nucleic acid 266
Heteroatoms
  heats of formation 180
  electrostatic interactions and
    solvation 195–202
Heterotrypticenes, chemical shifts . 244
Heat of vaporization 173
Hexahelicene, helix pitch 152
Hexamethylbenzene, gear effect ... 86
Hexamethylethane
  equilibrium geometry 25
  molecular relaxation 85

Hexaphenylethane, propeller
  conformations 150
Hexopyranoses, conformational
  and configurational equilibria . 258
Hexopyranoses geometries, conformational energies 260
Hilderbrandt conformational
  interconversion 74
Hybridization term, definition 196
Hydrindane, bicyclo[4.3.0]nonane,
  conformation geometry and
  relative energy 109
Hydrindanones, structure 212
Hydrocarbon(s)
  conformation geometry and
    relative energy 108–116
  conjugated, resonance energies,
    heats of formation 189–192
  force field 217
  geometries and relative
    energies 79–167
  interconversion via carbenium
    ions 289
  large ring, conformation geom-
    etry and relative energy ... 108
  optimization of interaction
    parameters 309
  strain energies 187, 188t
  structural data 80
Hydrocoumarinic acids,
  lactonization 293
Hydrogen bonding
  electrostatic and van der
    Waals interactions 222
  role in conformation analysis .. 223
  stability 223
Hydrogen–carbon interaction,
  butane 48
Hydrogen molecule ab initio
  calculation 43
Hydrogenation, bond-angle and
  torsional strain 298
Hydrogenolysis of cage hydrocarbons, product ratio 301
Hyperconjugation, carbenium ions 288
Hyperconjugative interactions 34

I

Indolizidine, conformation 232
Inherent strain, definition 186
Interaction(s)
  nonbonded 27–31
  potential, equation 30
  steric 1
  terms, Urey–Bradley force fields 22
1,4-Interactions of lone pairs,
  torsion angles 34

Interconversion of hydrocarbons
  via carbenium ions 289
Interconversion map of
  cyclononane 105
Interconversion pathway(s)
  cyclohexane conformations 92
  molecular geometry 72
  ring conformations 76
Intermolecular interaction, potential function 28
Internal coordinates, energy
  minimization 64
Internal coordinates, molecular
  geometry 19
Internal energy, definition 170
Internal rotation in ananyl
  dipetide, energy map 277f
Interrelations between the cyclo-
  heptanone conformations 99f
Intramolecular interaction, CFF ... 38
Inverse F matrix, equation 68
Inversion
  cycloheptene, qualitative
    energy diagram 129f
  transition states for chirality ... 147f
Isobutyric acid, conformation 225
Isopropylmesitylene, rotational
  barrier 145

J

Jeans' formula, equation 29
JTB force field 47

K

3-Keto steroids and terpenoids
  with benzaldehyde, reaction
  rates 297
Ketone
  electrostatic interaction 211–217
  reductions, stereochemistry,
    torsional energy 297
  strain differences, correlated to
    relative reaction rates 294

L

Lactone(s)
  conformation 225
  force field, stability 227
  8-membered ring, conformation
    and strain 226
  12-membered ring, strain 227
  MM2 force field 226

Lactonization
  of hydrocoumarinic acids 293
  steric compression 292
  of tetramethyl-o-hydroxy-
    hydrocinnamic acid 293
Late transition state, ketone
  reduction model 295
Least squares optimization of
  hydrocarbon interaction
  parameters 309
Leonard–Jones, potential function . 30
Lifson–Warshel force field, structures and energies 195
Ligand electronegativity and strain,
  conformation 242
Linear least squares parameter
  optimization, CFF 36, 37
London dispersion forces 28
Lone pair(s)
  electrostatic interaction 210
  on nitrogen 229
  role in oxygen calculations 210
  van der Waals properties 229
Low-order torsional potentials 34
Low periodicity torsional energy
  butylamine 230
  heats of formation 176
  n-propylamine 230
Low periodicity torsional potential,
  glycol ethers 219
Low periodicity torsional terms
  (V1 and V2) 34

M

Macrocyclic polyethers, geometries 34
Macroscopic effect, solvation 199
Maltose
  conformational energy map 263
  energy contours 263f
  geometry 264
  Ramachandran map 263
Manxine, conformation 233
Meisenheimer salt, diethoxy,
  structure 312
Metal chelate complexes,
  conformation 244
1,5-Methano-10-annulene 153, 154
Methyl ethyl ether, conformational
  energies 221
Methyl–hydrogen interaction,
  butane 48
2-Methyladamantane, heats of
  formation 178
Methly-D-aldopyranosides,
  conformation 259

Methylbicyclo[3.3.1]nonasilane,
  conformation 205
2-Methylbutane, conformational
  energy 87
Methyl-*t*-butylethylene, conformational geometry and relative
  energy 123–124
Methylcyclobutene and methylenecyclobutane, equilibrium 125
Methylcyclopentane, conformation 91
Methylenecyclohexane, conformation geometry and relative
  energy 127
Methyl-*N*,*N*-diisopropyldithio-
  carbamate, gear effect 86
Methylsilane, rotational barrier ... 202
3-Methyl-2-thiabutane,
  conformation 235
Microwave spectroscopy 6, 8
MINDO/3 program 13
Minimization characteristics, block-
  diagonal Newton–Raphson
  method 71
Minimization of function, Newton–
  Raphson procedure 65*f*
Minimum energy conformation ... 60
Mixed force field 21
MM1 force field
  axial methylcyclohexane energy . 47
  butane 48
  cyclodecane 47
  gauche butane energy 49
  modified, torsional term 51
MM2 force field 23, 51
MMP1
  approach 54
  force field, heats of forma-
    tion calculations 191
  naphthalene 53
MMP2 approach 54
Modified neglect of diatomic
  overlap (MNDO) 13
Molecular energy 1–15
Molecular force field calculation,
  constants 19
Molecular geometry 1–15
  calculations, cost estimates 11
  computation methods 59–73
  computer input 63
  conformational interconversion
    pathways 72–76
  definitions 169
  energy minimization 64–72
  internal coordinates 19
Molecular physics and molecular
  mechanics 5

Molecular properties, quantum
  mechanics 10
Molecular relaxation, hexamethylethane 85
Molecular structure
  computations 59–64
  determination 6
  experimental methods 6–10
  influence of crystal packing 308
Molecular vibrations, calculations . 310
Monosaccharide conformation,
  energy minimization 261
Morse function 8*f*, 18, 22, 23
MUB-1 force field 47
MUB-2 with low-order torsional
  terms 51

N

Naphthalene
  MMP1 53
  Pariser–Parr–Pople 53
  self-consistent field calculation . 53
Neutron diffraction measurement . 7
Newton–Raphson method
  block diagonal 70*f*
  full matrix 69
Newton-Raphson minimization
  disadvantages 68
  of function 65*f*
  second derivative techniques ... 67–72
  technique 65
Nicotinamide adenine dinucleotide, conformation 270
Nicotinamidemononucleotide,
  conformation 271
Nitrogen
  electrostatic interactions 228–234
  lone pair 229
Nonbonded interaction(s) 195
  carbonyl oxygen 209
  ether oxygen 209
  intramolecular and inter-
    molecular 27
  pair potentials 31
  parameters 39
  silabutanes 203
  silicon 202
1,3-Nonbonded interaction(s)
  Urey–Bradley force field 27
  valence force field 79
Nonbonded parameters, force
  fields 39–52
Non-neighbor resonance interaction of 1,3,5,-cyclohepta-
  triene 154

Nonpotential energy effect 173
Norborndiene
　heats of hydrogenation 180
　product ratio 301
Norbornane, heat of formation
　and hydrogenation 180, 181f
Normal coordinates 19
Nortropane, conformation 233
Nucleic acid
　conformations 265–274
　helix types 273
　polymeric, helix geometries 266
Nucleophilic addition to carbonyl
　groups, stereochemistry 294
Nucleophilic substitution of alkyl
　halides, reaction rates 287t
Nucleosides, conformations 265–274
Nucleotides, conformations 265–274

O

Octahedral metal complexes,
　force fields 244
Octalin octahydronaphthalene,
　geometry and relative
　energy 141
Olefin parameterization force
　field 37
Olefins, substituted, geometry
　and relative energy 123, 124
Orbital interactions 301
Out-of-plane deformation,
　cyclobutanone 26
Oxacycloheptane, conformation .. 219
Oxacyclooctane, conformation ... 219
Oxa derivatives, electrostatic
　interaction 209
Oxepane, conformation 219
11-Oxo-9α-estradiol ring
　system stability 256
Oxetane, conformation 218
Oxygen, electrostatic
　interactions 209–228

P

Pair potential, nonbonded
　interactions 31
2,2-Paracyclophane 155
Parameter optimization 36
Parameterization
　benzene, force fields 53
　electron diffraction measure-
　　ments 9
　ethers 220
　force fields 36–52
　x-ray measurements 9

Pariser–Parr–Pople (PPP),
　naphthalene 53
Partial rotation about single bonds 85
Pattern search, modified steepest
　descent technique 66
Pentaphenylethane, propeller
　conformations 150
Pentapyranoses, conformational
　and configurational
　equilibria 258
Pentavalent phosphorus
　compounds, geometry 242
Pentofuranose derivatives,
　equilibria 262
Peptide
　calculation 274, 275, 278
　chain conformation 278
　charges, semiempirical method . 196
　conformation 274–278
　crystal structures 309
　geometry optimization 278
　multiple minima 278
Perhalophosphazene cyclic
　oligomers, bond properties .. 244
Perhydroanthracene 110, 111t
Perhydroazulene, conformation
　geometry and relative energy . 110
Perhydrodicyclopentadiene, rear-
　rangement to adamantane 290
Perhydrophenanthrene, confor-
　mation geometry and
　relative energy 111
Perhydroquinacene, conformation
　geometry and relative
　energy 113–114
Perhydroquinolines, conformation . 233
Phenanthrene, experimental
　structures 153
Phenyl group, conformation
　geometry and relative
　energy 145–150
Phloroglucin, conformation 216
Phosphoranes, stretching and
　bending parameters 243
Phosphorous
　acyclic pentavalent interactions 243
　electrostatic interactions 242
Pi-electronic systems, conjugated
　Allinger and Sprague approach . 151
　conformation geometry and
　relative energy 144–156
Pi-sigma interaction effects,
　2,2-paracyclophane 155
Piperidine
　sauche/anti equilibrium 229
　N-H problem 231
Piperidone 234

Planar conjugated system,
   structure calculations 151
Point charge
   determination, induced dipole
      moments 197
   equilization of electronegativities 195
   interactions 195, 206
   model 195, 196
Polycyclic olefins, conformation
   geometry and relative
      energy 141–143
Polyethers, electrostatic
   interactions 219
Polyhalides, electrostatic
   interactions 206
Polynucleotides, electrostatic
   interactions 269
Polypeptide
   chain conformation 276
   folding pathway 274
   simulated protein folding 276
Polysaccharides
   conformational analysis 263
   crystal and molecular structures 264
   potential energy map,
      conformations 264
   ring vector 265
Polysilanes, electrostatic
   interactions 204
Polysulfides, conformational
   analysis 237
Potential energy
   equation 170
   map, disaccharides and
      polysaccharides 264
   surface 2
Potential function(s)
   Buckingham, equation 30
   conformational, for 1,2,4,5-
      tetrathiane 238f
   crystal properties 309
   with cubic term 23f
   of force fields 22–36
   intermolecular interaction 28
   Lennard–Jones 30
   molecular mechanics 26
   peptide calculations 275
Potential surfaces, Cope
   rearrangements of simple
      dienes 300
PRDDO program 13
Preisolamendiol, thermal
   isomerization 302
5α-Presnan-3β, 20-diol 256
Product ratios 285, 294–296
Progesterone, side chain
   conformation 255

Propeller conformation
   hexaphenylethane 150
   pentaphenylethane 150
n-Propyl chloride, enthalpy 207
n-Propylamine, low periodicity
   torsional energy 230
Protein folding, simulated 276
Proteins
   conformation 274–278
   globular, crystal structures 277f
Pseudorotation
   cyclohexane 92
   cyclopentane 89f
   deoxyribose 270
   ribose 270
   ring inversion in cyclooctane .. 102
   potential, ribose ring 270
Purine nucleosides,
   conformation 266–268
Pyranose ring enthalpies and
   Gibbs free energies 259
Pyranoses, conformational and
   configurational equilibria 258
Pyranosides, structure 261
Pyrimidine nucleosides,
   conformation 267, 268
Pyrrolidine, conformation and
   energy 231
Pyrrolizidine, conformation 232

Q

Quantitative theories, strain
   energy 184
Quantum chemistry program
   exchange 317–319
Quantum mechanics
   approach, simplifications 11
   basis of molecular mechanics
      calculation 10
   calculations 8, 10–13
   molecular properties 10
Quinolizidine, conformation 232
Quinuclidine, conformation 232

R

Racemization rate, halosubsti-
   tuted biphenyls 2
Radical, heats of formation 291
Ramachandran map, maltose 263
Rao's entropy scheme 262
Reaction coordinate, energy
   minimization 75
Reaction field model
   3,3-dimethylpentane-
      2,4-dione 201

Reaction field model—*Continued*
electrostatic interaction 200
Reaction rates
bridgehead derivatives 288
nucleophilic substitution of
alkyl halides 287t
Resonance energies, conjugated
hydrocarbons 189–192
Ribodinucleoside monophosphates, conformation 273
Ribose
internal degrees of freedom 266
pseudorotation 270
Ring
bridging, product ratio 301
flattening 95, 96
inversion and pseudorotation,
cyclooctane 102f
torsion angles, cyclopentane ... 90
vector 264, 265
Rotameric behavior of aryl
groups 145
Rotation of anhydroglucose
residue ring vector 265f
Rotational barriers
nucleosides 269
toluene isopropylmesitylene . 145
Rotational isomerism, side chain .. 256
Rotational potential, glycosyl bond 266

S

S_4 symmetry 85
Saddle point 3
energy minimization 72, 73
Scheraga's EPEN method, use of
electrostatic nonbonded
interactions 32
Schleyer force field 23
cyclooctane conformations 101
sauche butane energy 49
Schrodinger equation 2
gauche/anti butane 49
Second derivative technique,
Newton–Raphson minimi-
zation 67–72
Second-order perturbation
theory 28
Second virial coefficient,
nonbonded interactions 41
Secular determinant 18
Self-consistent field (SCF)
approximation 10
calculation, naphthalene 53
Shrinkage, definition 7

Silaalkanes, geometric features ... 202
Silabutanes 202, 203
Silacyclohexane ring, conformational energies of methyl
groups 204
Silacyclopentane, conformation .. 203
Silicon
electrostatic interactions ... 202–205
parametrizations 202
Silylcyclohexane, conformational
energies 203
Slater orbitals 11
S_N2
cyclization of ω-haloamines,
activation enthalpies 287
reactions 1, 286–287
Solid state, application of
molecular mechanics 307–316
Solid structure, force field
calculations 307
Solvation
effect, recent theories 199
energies 288
heteroatoms 195–202
macroscopic effects 199
Solvent effect
dielectric constant 199
dihalides and haloketones 209
equation 201
evaluation, electrostatic
interactions 198
Solvolysis rates and strain energy . 289
Solvolysis reactions, carbenium
ion intermediates 288
Spectroscopic quantities,
definitions 8
Stability
hydrogen bonding 223
relative 190
terpenoids, lactone force field .. 227
Statistical mechanics term, heats
of formation 176
Statistical thermodynamics and
vibrational frequencies 169–173
Steepest descent method 66, 67
Stereochemistry of nucleophilic
additions to carbonyl groups . 294
Stereopopulation control 293
Steric compression, lactonization .. 292
Steric effect(s)
in central force field 21
S_N2 reaction of alkyl bromides
with ethoxide 286
of substituents 299
Steric energies of esters,
correlated with relative
reaction rates 293

INDEX 337

Steric energy(ies) 169–194
  absolute value significance 185
  definition 4, 172
  and reaction rate of esters 293
Steroid-receptor interactions,
  geometries 253
Steroids
  B/C *cis* ring fusions, energies .. 255
  geometry optimization 253
  molecular mechanics calculations 253
$\Delta^{5,10}$ Steroids, A-ring
  conformers 255
STO-3G calculation 11, 13
Strain 169–194
  free radicals 291
  lactone ring 226
Strain energy(ies)
  alkyl radicals 291
  calculation 185
  cycloalkanes 187
  definitions 185
  for hydrocarbons 187, 188*t*
  quantitative theories 184
  reference points 185, 186
  and stability 189, 190
  sulfur homologues 241
  total 186
Strain energy differences and
  alcohol solvolysis rates 289
  ester formation rates 292
  cycloalkyl tosylates solvolysis
    rates 289
Stretching and bending 23*f*
Stretching force constants 25*f*, 38
Structural data, hydrocarbons 80
Structural feature increments,
  heat of formation
  calculations 185
Structure calculations for planar
  conjugated systems 151
Structure determination
  cycloheptane derivative 98
  electron diffraction 6
  microwave spectroscopy 6
  vibrational effects 7
  x-ray diffraction 6
Structure and energy, molecular
  mechanics calculations 5
Structures and energies of
  polysulfur cycles 239
Succinic acid, conformation 225
Sulfoxides 241
Sulfur, electrostatic interaction,
  conformational analysis ...235–242
Sulfur homologues 241
Symmetry in minimization
  techniques 60

T

T symmetry structures 85
Taft parameters, conformational
  effects 300
Terpenoids, lactone force field
  stability 227
1,5,9,13-Tetraazacyclohexadecane,
  geometry 231
Tetra-*t*-butylethane
  conformations, Newman
    projections 88
  transition state 75, 76
Tetra-*t*-buylphosphonium
  tetrafluoroborate, T conformation 85
Tetraglycine, global energy
  minimum 275
Tetrahydropyran, conformation .. 218
1,1,2,2,-Tetrakis(2,6-xylyl)ethane,
  conformation 149
Tetramethylbicyclo[4.2.2]deca-
  3,7,10-triene, conformation
  geometry and energy 142
1,1,5,5-Tetramethylcyclodecane-8-
  carboxylic acid, structure 312
Tetramethylethane, conformations
  and energies 149
Tetramethyl-*o*-hydroxyhydro-
  cinnamic acid, lactonization .. 293
1,1,2,2-Tetraphenylethane, conformation 149
Tetraphenylmethane 148, 149
Tetrasilane isomers, stability 204
1,2,4,5-Tetrathiane 237, 238
Tetroxocane, conformations 221
Thermal isomerization of
  preisolamendiol 302
Thermochemical definition of
  aromaticity 189
Thermochemical stability 190
Thermodynamics, statistical,
  calculations of
  vibrational frequencies 169–173
2-Thiabutane, conformation 235
Thiacyclobutane, structure 236
Thiacyclohexane, energy 236
Thiacyclopentane, conformation .. 236
Thioethers, Allinger hydrocarbon
  force field 235
Thiols, Allinger hydrocarbon
  force field 235
Toluene, rotational barrier 145
Torsion angle
  bicyclo [5.3.1] undecapentaene .. 154*f*
  cyclohexane ring 92
  ethylene 35

Torsion angle—*Continued*
  1,4-interactions of lone pairs .. 34
  interconversion of Cartesian and
    internal coordinates 63
  5-member puckered ring 90f
Torsion-bend interaction term,
  equation 35
Torsion about single bond,
  equation correction function . 33
Torsional deformations,
  hindered molecules 85
Torsional energy
  calculation 33–36
  contributions, stereoselectivity
    of ketone reductions 297
  cross-terms 35
  force fields 32
  parameters 39–52
  van der Waals interactions 33
Torsional potential
  cosine function with twofold
    symmetry 34
  double bond in ethylene 34
  for ethane, calculation 33
  function for rotation about a
    bond 33
  low-order 34
Torsional term
  dipole/dipole interaction hyper-
    conjugation, steric 33
  low periodicity (V1 and V2) ... 34
  molecular mechanics force field . 33
Transferable
  force fields 21
  substituent constants 299
Transition state
  chirality inversion 147
  determination 74
  energies, *cis,cis*-1,5-cycloocta-
    diene 134t
  full matrix calculation 73
  geometry 73–75, 287, 300
  structure, calculation 74
  tetra-*t*-butylethane 75, 76
  tetrahedral to trigonal
    geometries 299
Transition vector, energy
  minimization 74
Tri-*t*-butylmethane, geometry 230
Triethylenetetramine,
  conformation 245
1,3,5-Trimethylcyclohexane 95
Trimethylsilylcyclohexane,
  conformational energies 203
Tris-1,3-diaminopropane
  complex, conformation 244
1,2,3-Trithiane, conformation 238

U

Unit cell parameters,
  calculations 310
Urey–Bradley
  force field 21, 22, 24
  interaction, cyclobutane 24
  terms 4
Uridine-2′,3′-cyclooxyphosphorane,
  conformations 270

V

Valence force field 20, 24, 79
Valence shell electron pair
  repulsion (VSEPR) method,
  force fields 79
δ-Valerolactone, conformation 225
Variable electronegativity self
  consistent field (VESCF)
  calculation for pi system 55
van der Waals curve, calculations . 42
van der Waals energy plots 42f, 46f
van der Waals function,
  force fields 44
van der Waals interactions 1, 31
  acyclic pentavalent phosphorus
    compounds 243
  crystal packing 309, 310
  cyclobutane ring 24
  EAS force field 47
  and hydrogen bonding 222
  torsional energy 33
van der Waals potential
  aspherical 32
  of ether oxygen 32
  force fields 42
  functions 1
  nonbonded interaction 41
van der Waals properties of
  lone pairs 229
van der Waaals radius 41
Vector formulas, Cartesian and
  internal coordinates 63
Vibrational amplitudes and
  shrinkage, electron
  diffraction 172
Vibrational analysis 17, 19
Vibrational contribution to
  enthalpy 50
Vibrational effect, structure
  determinations 7
Vibrational energy 172
  conformational equilibria 172
  equation 169
  explicit evaluation 178
  heats of formation 172
  molecular thermodynamics 171

Vibrational force field, nonbonded
   interactions 22
Vibrational frequencies 171
   anharmonicity 171
   calculated, uses 171
   calculation 18
   equation 36, 170
   statistical thermodynamics,
      calculations 169–173
Vibrational motion, Morse curve .. 7
Vibrational spectroscopy, force
   fields 17–22, 26
Vicinal dichlorides, conformational
   properties 207
Virtual bonds, definition 264

W

Warshel–Karplus approach, conjugated pi-electronic systems .. 150

Wertz force field 50
Westheimer method, molecular
   mechanics 2

X

X-ray crystallography and
   electron diffraction 6
   molecular mechanics 311–314
X-ray diffraction
   measurement, atom slope 31
   measurement, parameterization . 9
   structure determination 6
X-ray structures, molecular
   mechanics solution 311
D-Xylose, conformation 260

Z

Z isomer of DNA 273

Copy and Production Editor: Robin Giroux
Indexer: L. Luan Corrigan
Jacket Artist: Kathleen Schaner

Typesetting: Bi-Comp, Inc., York, PA,
and Service Composition Co., Baltimore, MD
Printing: Maple Press Company, York, PA